Chromatography

The Author

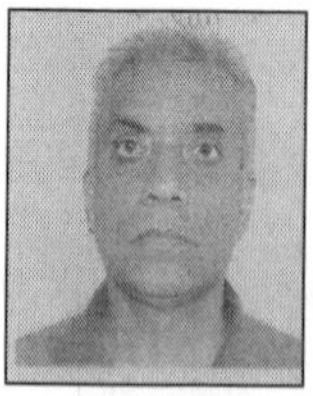

Dr. Javid Ahmad Sofi is a analytical chemist with over 25 years of experience in separation science. He holds a PhD in Chemistry from the Indian Institute of Agriculture(IARI) New Delhi, specializing in chromatographic techniques. Dr. Sofi has authored more than 30 peer-reviewed papers in international and National journals, and he has served as a consultant for establishing laboratories at state level.. Currently, he is a professor at the University of SKUAST-K, where he leads research on advanced HPLC, LCMSMS, GCMSMS, ICPMSMS, GC, AAS, Graphite Furnace methods for pesticide and heavy metal analysis. His passion for making complex principles accessible drives this book, aimed at students, researchers, and professionals entering the field.

Professor Haroon Naik

Director Research & Director Project and Planning

SKUAST-K

Foreword

In the vast landscape of analytical chemistry, few techniques have revolutionized scientific inquiry as profoundly as chromatography. From its humble beginnings in the early 20th century, when Mikhail Tswett first separated plant pigments using a column of chalk and alcohol, chromatography has evolved into an indispensable tool across disciplines-be it purifying biomolecules in biotechnology, detecting pollutants in environmental monitoring, or ensuring the safety of pharmaceuticals.

Dr. Javid Sofi's *Chromatography* and its principles arrives at a pivotal moment. As analytical demands grow more sophisticated with the rise of miniaturization, automation, and green chemistry, there is a pressing need for a resource that demystifies the fundamentals while bridging to cutting-edge applications. This book does precisely that. With clarity and precision, Dr. Sofi unpacks the theoretical underpinnings-partition, adsorption, ion-exchange, and beyond-while grounding them in practical examples that resonate with real-world challenges.

What sets this work apart is its pedagogical elegance. Diagrams, case studies, and step-by-step derivations make abstract concepts tangible, empowering readers from undergraduates to seasoned chromatographers. As someone who has mentored countless students in separation sciences, I can attest to the frustration of opaque texts; Dr. Sofi's approach alleviates that, fostering not just understanding but innovation.

I recommend this book to anyone seeking to master chromatography's artistry and science. It is more than a textbook-it's a gateway to discovery.

Preface

Very often the students, Personnel's in R&D sectors in food industries, other experts from different organization (including NGO's) ask for a book on analytical techniques that is suitable for them to establish a Food testing laboratory and /or test samples and recommend independently. The books covering practical aspect of Chromatography are very few in number in the market do not fully meet their need. Hence I have tried my best to prepare this book that is concise but precise and suitable for them.

Chromatography and Its Principles is born from that spark, refined over decades of teaching, research, and troubleshooting in labs from across the country. This book is not merely a compendium of techniques but an invitation to appreciate the principles that govern them. In these pages, we explore the thermodynamic heart of chromatography: the equilibrium constants, selectivity factors, and resolution equations that dictate success or failure in a separation.

First chapters introduces the principle and fundamentals of Chromatography with numerical along with solution for better understanding of the subject. Chapter 2-6 deals with different chromatographic techniques with proper experimental design layout so that students can perform these techniques in their laboratories in easy and very lucid manner. Chapter 7-9 discusses sophisticated techniques of chromatography like Gas Chromatography, High Performance liquid Chromatography. I've included worked examples, troubleshooting guides, and even a glossary to bridge theory and practice. Whether you are a student grappling with your first chromatogram or a researcher scaling up for industrial applications, my aim is to equip you with the intuition to innovate.

Chromatography teaches us patience and precision, much like life itself. I hope this book serves as your reliable companion on that path. Feedback is welcome; science thrives on collaboration.

Javid Ahmad Sofi

Contents

Chapter 1

Introduction to Chromatography

1.1 Introduction

In Separation Science there are several methods for separating an analyte from potential interferents. For example, in a liquid–liquid extraction the analyte and interferent initially are present in a single liquid phase. We add a second, immiscible liquid phase and thoroughly mix them by shaking. During this process the analyte and interferents partition between the two phases to different extents, effecting their separation. After allowing the phases to separate, we draw off the phase enriched in analyte. Despite the power of liquid–liquid extractions, there are significant limitations. There are two limitation of liquid liquid separation as given under

Suppose we have a sample that contains an analyte in a matrix that is incompatible with our analytical method. To determine the analyte's concentration we first separate it from the matrix using a simple liquid–liquid extraction. If we have several analytes, we may need to complete a separate extraction for each analyte. For a complex mixture of analytes this quickly becomes a tedious process. This is one limitation to a liquid–liquid extraction. A more significant limitation is that the extent of a separation depends on the distribution ratio of each species in the sample. If the analyte's distribution ratio is similar to that of another species, then their separation becomes impossible. For example, let's assume that an analyte, *A*, and an interferent, *I*, have distribution ratios of, respectively, 5 and 0.5. If we use a liquid–liquid extraction with equal volumes of sample and extractant, then it is easy to show that a single extraction removes approximately 83 per cent of the analyte and 33 per cent of the interferent. Although we can remove 99 per cent of the analyte with three extractions, we also remove 70 per cent of the interferent. In fact, there is no practical combination of number of extractions or volumes of sample and extractant that produce an acceptable separation.

Based on our experience with extraction we can define the distribution ratio, *D*, for a solute, *S*, is

$$D = \left[\frac{S_{ext}}{S_{samp}}\right] ----------- 1.1$$

where $[S]_{ext}$ is its equilibrium concentration in the extracting phase and $[S]_{samp}$ is its equilibrium concentration in the sample. We can use the distribution ratio to calculate the fraction of *S* that remains in the sample, q_{samp}, after an extraction

$$q_{samp} = \frac{V_{samp}}{DV_{ext+V_{samp}}} ----------- 1.2$$

where V_{samp} is the volume of sample and V_{ext} is the volume of the extracting phase. For example, if $D = 10$, $V_{samp} = 20$, and $V_{ext} = 5$, the fraction of *S* remaining in the sample after the extraction is

$$q_{samp} = \frac{20}{10\ x\ 5 + 20} = 0.29 ------------- 1.3$$

or 29 per cent. The remaining 71 per cent of the analyte is in the extracting phase.

A Better Way to Separate Mixtures

The problem with a liquid–liquid extraction is that the separation occurs in one direction only: from the sample to the extracting phase. Let's take a closer look at the liquid–liquid extraction of an analyte and an interferent with distribution ratios of, respectively, 5 and 0.5. Figure A shows that a single extraction using equal volumes of sample and extractant transfers 83 per cent of the analyte and 33 per cent of the interferent to the extracting phase. If the original concentrations of *A* and *I* are identical, then their concentration ratio in the extracting phase after one extraction is

$$\left[\frac{A}{I}\right]A = \frac{0.83}{0.33} = 2.5 ---------1.4$$

A single extraction, therefore, enriches the analyte by a factor of 2.5 × 2.5 ×. After completing a second extraction (Figure A) and combining the two extracting phases, the separation of the analyte and the interferent, surprisingly, is less efficient.

$$\left[\frac{A}{I}\right] = \frac{0.97}{0.55} = 1.8 ----------------1.5$$

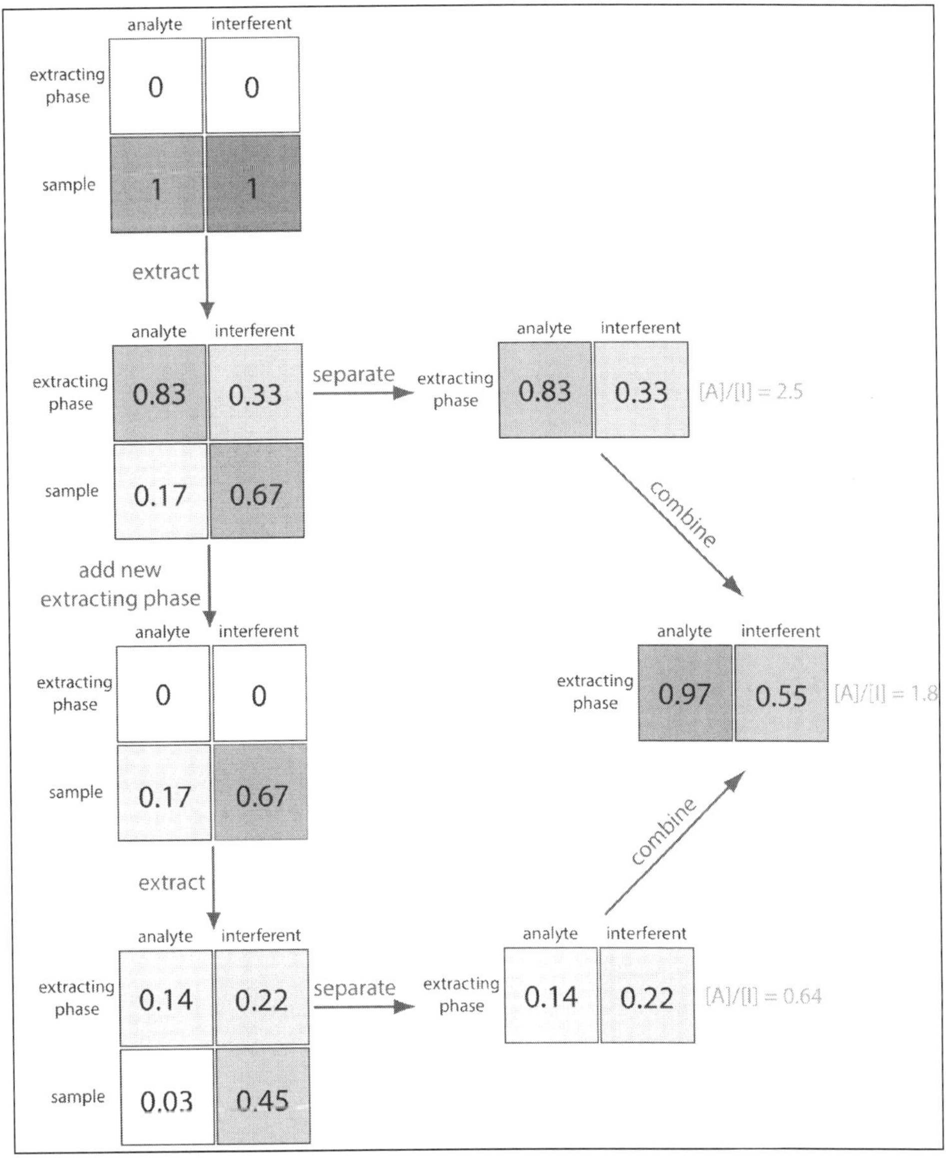

Figure A: Progress of a traditional liquid–liquid extraction using two identical extractions of a single sample using fresh portions of the extractant. The numbers give the fraction of analyte and interferent in each phase assuming equal volumes of sample and extractant and distribution ratios of 5 and 0.5 for the analyte and the interferent, respectively. The opacity of the colors equals the fraction of analyte and interferent present.

We can improve the separation by first extracting the solutes from the sample into the extracting phase and then extracting them back into a fresh portion of solvent that matches the sample's matrix (Figure B). Because the analyte has the larger distribution ratio, more of it moves into the extractant during the first extraction and less of it moves back to the sample phase during the second extraction. In this case the concentration ratio in the extracting phase after two extractions is significantly greater.

$$\left[\frac{A}{I}\right] = \frac{0.96}{0.11} = 6.3 \text{ - - - - - - - - - - - - } 1.6$$

This problem can be avoided by chromatography techniques which will discussed in next section.

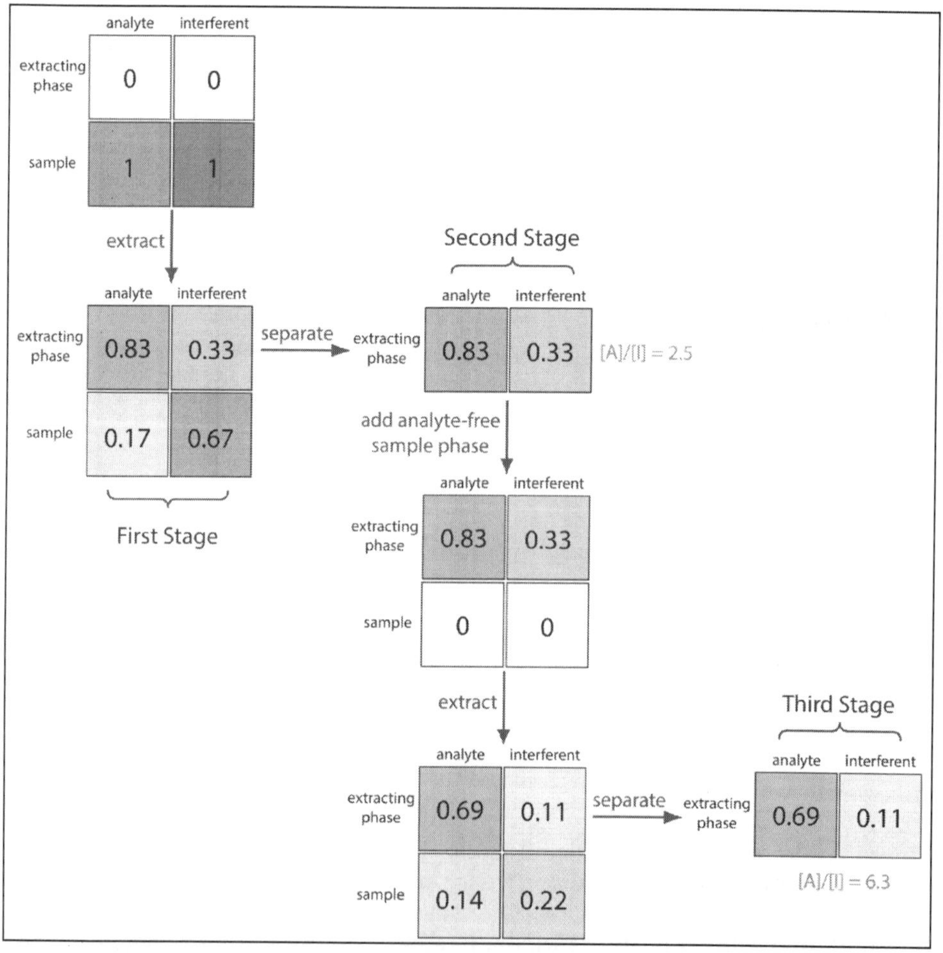

Figure B: Progress of a liquid–liquid extraction in which we first extract the solutes into the extracting phase and then extract them back into an analyte-free portion of the sample's phase. The numbers give the fraction of analyte and interferent in each phase assuming equal volumes of sample and extractant and distribution ratios of 5 and 0.5 for the analyte and the interferent, respectively. The opacity of the colors equals the fraction of analyte and interferent present.

The chromatographic method of separation, in general involves following steps

1. Adsorption or retention of a substance (s) on the stationary phase.
2. Separation of the adsorbed substances by the mobile phase.
3. Recovery of the separated substances by continuous flow of the mobile phase; the method being called elution.
4. Qualitative and Quantitative analysis of eluted substances.

Table 1: Various Types of Chromatographic Technique

Sl.No.	Technique	Stationary Phase	Mobile Phase
1	Column Chromatography or adsorption chromatography	Solid	Liquid
2	Partition Chromatography	Liquid	Liquid
3	Paper Chromatography	Liquid	Liquid
4	Thin layer chromatography (TLC)	Liquid or solid	Liquid
5	Gas liquid chromatography (GLC)	Liquid	Gas
6	Gas-solid chromatography	Solid	Gas
7	Ion exchange chromatography	Solid	Liquid

Table 2: Classification of Chromatographic Methods

Sl.No.	Stationary Phase	Mobile phase	Name
1	Solid	Liquid	Plane chromatography
2	Solid	Liquid	Paper Chromatography
3	Solid	Liquid	Thin Layer Chromatography (TLC)
4	Solid	Liquid	Adsorption column Chromatography
5	Solid	Liquid	High Performance Liquid Chromatography (HPLC)
6	Solid (Ion Exchange Resins)	Liquid	Ion Exchange Chromatography (IEC)
7	Solid	Gas	Gas-Solid Chromatography (GSC)
8	Solid matrix	liquid	Gas permeation chromatography
9	Liquid	Gas	Gas liquid Chromatography
10	Liquid	Liquid	Liquid-liquid Chromatography

1.2 Background History

The concept of chromatography can be traced back to ancient civilizations, such as ancient Egypt and China, where the art of dyeing fabrics and the use of natural pigments involved some principles are similar to modern chromatography. The Russian-Italian botanist Mikhail Tsvet is often credited with the discovery of chromatography in 1901. He used a technique he called "chromatography" to separate plant pigments using a glass tube filled with calcium carbonate. He passed a solution of plant pigments through the tube, and as different pigments interacted differently with the calcium carbonate, they were separated into distinct bands of color. He also used other solid adsorbents like alumina and observed the same principle of separation. The German chemist Richard Kuhn is known for his contributions to chromatography. He used chromatography to investigate the components of carotenoid pigments and later received the Nobel Prize in Chemistry in 1938 for his work on carotenoids and vitamins. The concept of partition chromatography, where components of a mixture partition between two immiscible phases, was developed by Archer John Porter Martin and Richard Laurence Millington Synge in the 1940s. They used liquid-liquid partition chromatography to separate amino acids, which eventually earned them the Nobel Prize in Chemistry in 1952. During the 1940s, Martin and Synge, along with other scientists like Consuelo Gibbs, also pioneered paper chromatography. This technique involved using filter paper or cellulose as the stationary phase and a liquid solvent as the mobile phase, making chromatography more accessible and widely applicable in various fields.

The development of gas chromatography is credited to the work of A.T. James and A.J.P. Martin in the early 1950s. This method uses a gaseous mobile phase and a solid or liquid stationary phase, allowing for the separation of volatile compounds. HPLC is an advancement of liquid chromatography that became popular in the 1970s. It involves the use of high-pressure pumps to improve the efficiency and speed of separation, making it a widely used technique in analytical laboratories.

1.3 Importance and Application of Chromatography

Chromatography is a versatile and powerful analytical technique used in various fields, including chemistry, biochemistry, pharmaceuticals, forensics, environmental science, and more. Its importance lies in its ability to separate and analyze complex mixtures of substances, making it an essential tool for researchers and scientists. Some of the Importance in chromatography are discussed as below

i) Separation of Complex Mixtures: Chromatography can separate a wide range of compounds from complex mixtures based on their physicochemical properties, such as size, charge, polarity, and affinity. This allows researchers to isolate and analyze individual components efficiently.

ii) Sensitivity and Precision: Chromatographic techniques are highly sensitive and capable of detecting trace amounts of compounds in a sample. They also provide precise quantification, making them crucial in quality control and research applications.

iii) Versatility: Chromatography encompasses various methods, each with its own strengths and applications. It can be adapted to analyze a wide range of compounds, from small molecules to large biomolecules like proteins and nucleic acids.

iv) Purification: Chromatography is commonly used for the purification of substances. It enables the isolation of specific compounds from a mixture, which is vital in pharmaceutical, biotechnology, and other industries.

v) Identification: By comparing the retention times and patterns obtained from known standards, chromatography facilitates the identification of unknown compounds in a sample.

Application

i) Analytical Chemistry: Chromatography plays a crucial role in quantitative and qualitative analysis of substances in various fields, such as food safety, environmental monitoring, and drug analysis.\

ii) Pharmaceutical Industry: Chromatographic techniques are extensively used for drug development, quality control, and drug purification processes.

iii) Biochemistry and Proteomics: In proteomics, chromatography is used to separate and analyze proteins, peptides, and amino acids, aiding in the study of protein structure and function.

iv) Environmental Science: Chromatography is employed to detect and quantify pollutants and contaminants in air, water, soil, and other environmental samples.

v) Forensic Science: Chromatography is utilized in forensic laboratories for drug testing, identification of trace evidence, and analysis of toxic substances in biological samples.

vi) Petrochemical Industry: Chromatography is used to characterize crude oil and its fractions, which is essential for refining processes and ensuring product quality.

vii) Clinical Diagnosis: In clinical settings, chromatography is employed to detect and quantify biomarkers and metabolites in body fluids, aiding in the diagnosis and monitoring of diseases.

viii) DNA Sequencing: Chromatographic techniques are used in DNA sequencing, contributing to advancements in genetics and genomics research.

1.4 Theories of Chromatography

The basis of all forms of chromatography is the partition or distribution coefficient;

An analyte is in equilibrium between the two phases; stationary phase and mobile phase. The equilibrium constant, K, is termed the partition coefficient, defined as the molar concentration of analyte in the stationary phase divided by the molar concentration of the analyte in the mobile phase. The time between sample injection and an analyte peak reaching a detector at the end of the column is termed the retention time (t_R). Each analyte in a sample will have a different retention time. The time taken for the mobile phase to pass through the column is called t_M (dead time or void time). A term called the retention factor, k., is often used to describe the migration rate of an analyte on a column. You may also find it called the capacity factor. The retention factor for analyte A is defined as

$$KA' = \frac{(tr - tm)}{tM} \quad 1.7$$

t_R and t_M are easily obtained from a chromatogram. When an analytes retention factor is less than one, elution is so fast that accurate determination of the retention time is very difficult. High reten-tion factors (greater than 20) mean that elution takes a very long time. Ideally, the retention factor for an analyte is between one and five.

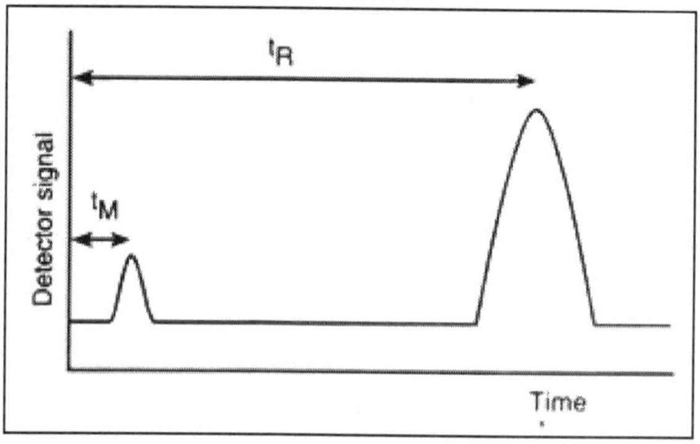

Figure 1.1

The selectivity factor, α, which describes the separation of two species (A and B) on the column

$$\alpha = \frac{kB'}{kA'} \text{ or } tr.B - \frac{tr._B}{tr._A} - t_m \quad 1.8$$

When calculating the selectivity factor, species A elutes faster than species B. The selectivity factor is always greater than one.

1.5 The Theoretical Plate Model of Chromatography

The plate model supposes that the chromatographic column contains a large number of separate layers, called theoretical plates. Separate equilibrations of the sample between the stationary and mobile phase occur in these "plates".

The analyte moves down the column by transfer of equilibrated mobile phase from one plate to the next.

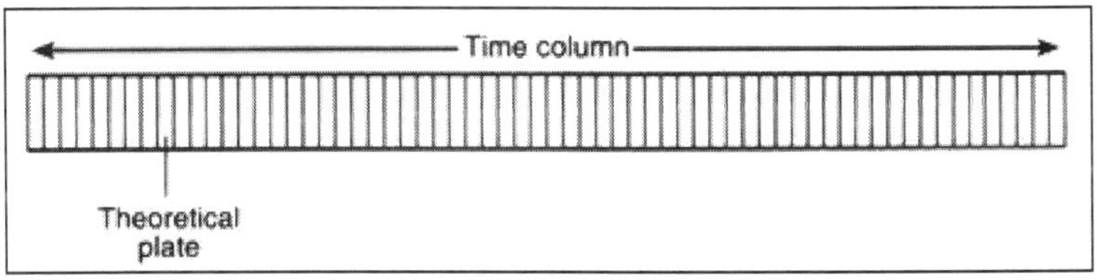

Figure 1.2

It is important to remember that the plates do not really exist; they are a figment of the imagination that helps us understand the processes at work in the column. They also serve as a way of measuring column efficiency, either by stating the number of theoretical plates in a column, N (the more plates the better), or by stating the plate height; the Height is equivalent to a theoretical plate (the smaller the better).

If the length of the column is L, then

$$HETP = \frac{L}{N} \tag{1.9}$$

$$N = \frac{L}{H} \tag{1.10}$$

The number of theoretical plates that a real column possesses can be found by examining a chromatographic peak after elution.

$$N = \frac{5.55t^2 R_{\frac{1}{2}}}{W^2} \tag{1.11}$$

where $W_{1/2}$ is the peak width at half-height.

As can be seen from this equation, columns behave as if they have different numbers of plates for different solutes in a mixture.

1.6 The Rate Theory of Chromatography

The plate model of chromatography assumes that the solute equilibrates instantaneously between the stationary and mobile phases. However, this is not the case in reality. The time it takes for the solute to equilibrate affects the band shape of the chromatographic peak. A slower rate of elution gives the solute more time to equilibrate, resulting in a narrower peak. A faster rate of elution gives the solute less time to equilibrate, resulting in a broader peak.

It is also affected by the different paths available to solute molecules as they travel between particles of stationary phase. If we consider the various mechanisms which contribute to band broadening, we arrive at the Van Deemter equation for plate height.

$$HETP = A + \frac{B}{u} + Cu \tag{1.12}$$

where u is the average velocity of the mobile phase. A is associated with eddy diffusion B is associated with longitudinal diffusion and C is associated with non-equilibrium mass transfer factors which contribute to band broadening. It is discussed in section 4.

1.7 Zone Shapes

The reason of a Gaussian curves for a typical chromatogram is by virtue of the fact that solute molecule during its migration undergoes thousands of transfer between stationary and mobile phase. The time it spends in either phase after a transfer is highly irregular and depends upon its accidental gain of sufficient thermal energy from its environment to accomplish a reverse transfer. Thus in some instances, the residence time in a given phase may be transitory, in others the period may be relatively long. The consequence of these random individual processes is a symmetric spread of velocities around the mean value, which represents the behaviour of the average and most common particles.

The breadth of a Gaussian curve is related to the standard deviation σ approximately 96 per cent of the area under such curve is included plus or minus two standard deviation (± 2σ) of its maximum. Thus, a standard deviation (expressed in units of time or in distance along the length of column) derived from a chromatogram serves as a convenient measure of zone broadening.

From Figure 1.3 explains means for approximating q/ or σ from an experimentally derived chromatogram. Tangents to the two sides of the chromatogram are extended to form a triangle with the abscissa. The area of this triangle corresponds to approximately 96 per cent of the area under the curve. Thus, the intercepts at approximately (± 2 σ from the maximum that is W = 4 q/

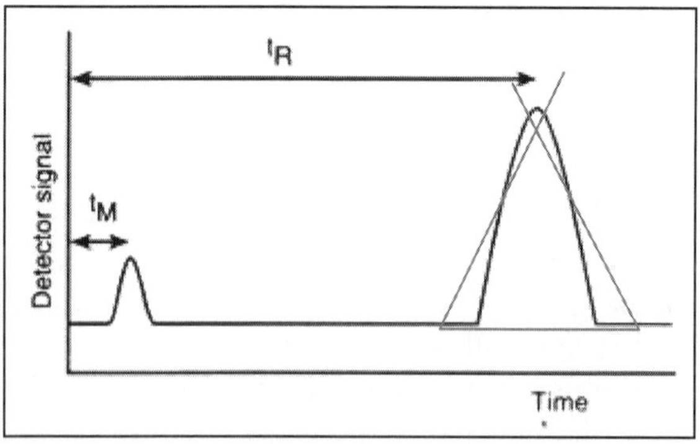

Figure 1.3

1.8 Calculation of H and N from Zone Breadth

The broadening of a chromatographic zone is expressed as the square of the standard deviation (variance) per unit length of column

$$H = \frac{\sigma^2}{L} \tag{1.13}$$

H, is plate height in cm, is the measure of efficiency and L is the length of the column, also in cm [note that σ must carry units of length (cm) to be consistent with H and L.

Substitution of Eq 1.3 we get

$$N = \frac{L^2}{\sigma^2} \tag{1.14}$$

N, number of plates is also given in following expression

$$N = 16\left(\frac{t_R}{W^2}\right) \tag{1.15}$$

t_R is retention time, is the time required after sample injection for the solute peak to appear at the end of column.

1.9 Zone Broadening

1.9.1 Eddy Diffusion

The mobile phase flows through the column, which is packed with stationary phase. Solute molecules diffuse randomly through the stationary phase, taking different paths of different lengths. This causes the solute band to broaden.

The contribution of ***multiple paths*** to the height of a theoretical plate, H_p, is

$$H_p = 2\lambda d_p \tag{1.16}$$

where d_p is the average diameter of the particulate packing material and λ is a constant that accounts for the consistency of the packing. A smaller range of particle sizes and a more consistent packing produce a smaller value for λ. For a column without packing material, H_p is zero and there is no contribution to band broadening from multiple paths.

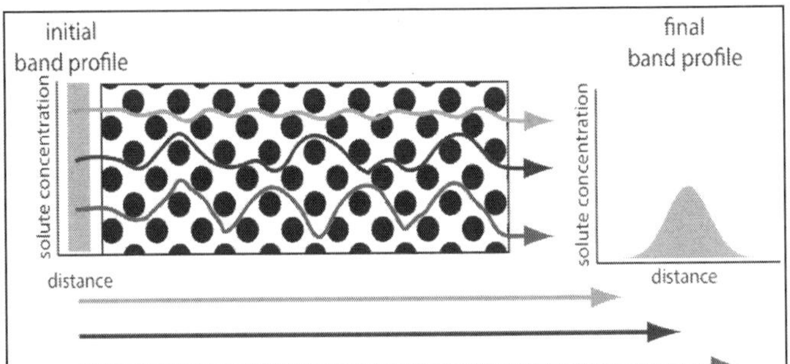

An inconsistent packing creates channels that allow some solute molecules to travel quickly through the column. It also can creates pockets that temporarily trap some solute molecules, slowing their progress through the column. A more uniform packing minimizes these problems.

1.9.2 Longitudinal Diffusion

The concentration of analyte is less at the edges of the band than at the centre. Analyte diffuses out from the centre to the edges. This causes band broadening. If the velocity of the mobile phase is high, then the analyte spends less time on the column, which decreases the effects of longitudinal diffusion. The contribution of longitudinal diffusion to the height of a theoretical plate, H_d, is

$$H_d = \frac{2\gamma D_m}{u} \tag{1.17}$$

here D_m is the solute's diffusion coefficient in the mobile phase, u is the mobile phase's velocity, and γ is a constant related to the efficiency of column packing. Note that the effect of H_d on band broadening is inversely proportional to the mobile phase velocity: a higher velocity provides less time for longitudinal diffusion. Because a solute's diffusion coefficient is larger in the gas phase than in a liquid phase, longitudinal diffusion is a more serious problem in gas chromatography.

1.9.3 Resistance to Mass Transfer

The solute moves between the mobile phase and the stationary phase as it passes through the column. This movement is called mass transfer. Band broadening occurs when the solute's movement within either phase is not fast enough to maintain an equilibrium in its concentration between the two phases. Specifically, a solute molecule in the mobile phase moves down the column before it has a chance to fully transfer to the stationary phase. A solute molecule in the stationary phase, on the other hand, takes longer than expected to move back into the mobile phase. The contributions of mass transfer in the stationary phase, H_s, and mass transfer in the mobile phase, H_m, are given by the following equations

$$H_s = \frac{qkd^2f}{([1+K)]^2 D_s u} \tag{1.18}$$

$$H_m = \frac{f^h\,(d^2p, d^2c)u}{D_m} \tag{1.19}$$

where d_f is the thickness of the stationary phase, d_c is the diameter of the column, D_s and D_m are the diffusion coefficients for the solute in the stationary phase and the mobile phase, k is the solute's retention factor, and q is a constant related to the column packing material. Although the exact form of H_m is not known, it is a function of particle size and column diameter. Note that the effect of H_s and H_m on band broadening is directly proportional to the mobile phase velocity because a smaller velocity provides more time for mass transfer.

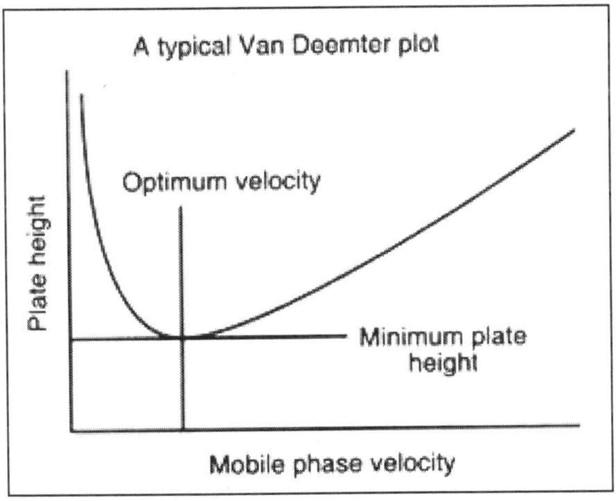

Figure 1.4

1.10 Resolution

Although the selectivity factor, α, describes the separation of band centres, it does not take into account peak widths. Another measure of how well species have been separated is provided by measurement of the resolution. The resolution of two species, A and B, is defined as

$$R = \frac{[(t_R)B - (t_R)A]}{W_A + W_B} \tag{1.20}$$

Baseline resolution is achieved when R = 1.5.

From equation 1.19 we can improve resolution by increasing Δtr, or by decreasing W_A and W_B.

It is useful to relate the resolution to the number of plates in the column, the selectivity factor and the retention factors of the two solutes,

$$R = \sqrt{N}/4(\alpha - 1)/\alpha(1 + k_{B'}/k_{B'}/) \tag{1.21}$$

To obtain high resolution, the three terms must be maximised. An increase in N, the number of theoretical plates, by lengthening the column leads to an increase in retention time and increased band broadening–which may not be desirable. Instead, to increase the number of plates, the height equivalent to a theoretical plate can be reduced by reducing the size of the stationary phase particles.

It is often found that by controlling the capacity factor, k′, separations can be greatly improved. This can be achieved by changing the temperature (in Gas Chromatography) or the composition of the mobile phase (in Liquid Chromatography).

The selectivity factor, α, can also be manipulated to improve separations. When α is close to unity, optimising k′ and increasing N is not sufficient to give good separation in a reasonable time. In these cases, k′ is optimized first, and then α is increased by one of the following procedures:

1. Changing mobile phase composition
2. Changing column temperature
3. Changing composition of stationary phase
4. Using special chemical effects (such as incorporating a species which complexes with one of the solutes into the stationary phase).

1.11 Optimization of Chromatographic Separation

Retention factor (K), Selectivity factor (α) and column efficiency (N) are the factors which affects separation of components in a given mixture. If the two peaks have same retention time, reasonably they will have same peak widths and if we presume that the number of theoretical plates are same for all components in a given mixture then the ratio t_R/W in equation 1.18 is constant. Since the retention time is similar then equation 1.19 becomes

$R_{AB} = 2[(t_R)_B\text{-}(t_R)_A]/W_A+W_B$

$R_{AB} = [(t_R)_B\text{-}(t_R)_A]$ ----1.22{ as retention time for both solutes A and B is similar)

Solving the equation 1.18 for WB and substituting in equation 1.21 we will get

$$R_{AB} = \sqrt{N_B}/4 \times t_{RB} - t_{RA}/t_{RB} \quad (1.23)$$

Rearranging equation I we will get retention time foe solute A and B as

$tr,_A = k_A t_m + t_m$ and $tr,_B = k_B t_m + t_m$

After substituting in equation 1.22 we will get

$$R_{AB} = \sqrt{N_B}/4 \times k_B - k_A/1 + k_B \quad (1.24)$$

Finally retention time for solute A can be eliminated by substituting in equation 1.12. After rearranging, we end up with the following equation for the resolution between the chromatographic peaks for solutes A and B.

$$R_{AB} = \sqrt{N_B}/4 \times \alpha - 1/\alpha \times K_B/1+K_B \quad (1.25)$$

In addition to resolution, another important factor in chromatography is the amount of time needed to elute a pair of solutes, which we can approximate using the retention time for solute *B*.

$$tr,s = 16R^2_{AB/}u \times (\alpha - 1/\alpha)^2 \times (1+KB)^3/K^2_R \quad (1.26)$$

where u is the mobile phase's velocity.

Although Equation 1.24 is useful for considering how a change in N, α, or k qualitatively affects resolution–which suits our purpose here–it is less useful for making accurate quantitative predictions of resolution, particularly for smaller values of N and for larger values of R. For more accurate predictions use the equation

$$R_{AB} = \sqrt{N}/4 \times (\alpha - 1) \times K_B/1+K_{avg} \quad (1.26)$$

Equation 1.25 and Equation 1.26 contain terms that correspond to column efficiency, selectivity, and the solute retention factor. We can vary these terms, more or less independently, to improve resolution and analysis time. The first term, which is a function of the number of theoretical plates (for Equation 1.25) or the height of a theoretical plate (for Equation 1.26), accounts for the effect of column efficiency. The second term is a function of α and accounts for the influence of column selectivity. Finally, the third term in both equations is a function of k_B and accounts for the effect of solute B's retention factor. A discussion of how we can use these parameters to improve resolution will be discussed as given below

a) Retention Factor to Optimize Resolution

One way to improve the resolution of two solutes is to increase the retention factor of solute B. The retention factor is a measure of how strongly a solute interacts with the stationary phase in a chromatography column. A higher retention factor means that the solute will spend more time in the column, and will therefore elute later.

The green curve in Figure 1.5 shows how the resolution of two solutes changes as the retention factor of solute B is increased. The improvement in resolution is greatest when the initial value of the retention factor is small. Once the retention factor exceeds a value of approximately 10, further increases produce only a small improvement in resolution. For example, if the initial retention factor of solute B is 1, increasing it to 10 will improve the resolution by 82 per cent. However, increasing the retention factor to 15 will only improve the resolution by an additional 5.5 per cent. It is important to note that increasing the retention factor also increases the retention time of the solute. The red curve in Figure 1.5 shows how the retention time of solute B changes as a function of its retention factor. The minimum retention time is for a retention factor of 2. Increasing the retention factor from 2 to 10 will approximately double the retention time of solute B. Therefore, increasing the retention factor of solute B can be an effective way to improve the resolution of two solutes. However, it is important to weigh the benefits of increased resolution against the longer retention times that will result.

To increase the retention factor of a solute without changing its selectivity, we can make a general change to the chromatographic conditions that will increase the retention factor of both solutes. In gas chromatography, we can do this by decreasing the column temperature. This is because the vapor pressure of a solute is lower at lower temperatures, so it spends more time in the stationary phase and takes longer to elute. In liquid chromatography, we can increase the retention factor by using a weaker mobile phase. This is because a weaker mobile phase has less solvent strength, so solutes spend more time interacting with the stationary phase and take longer to elute.

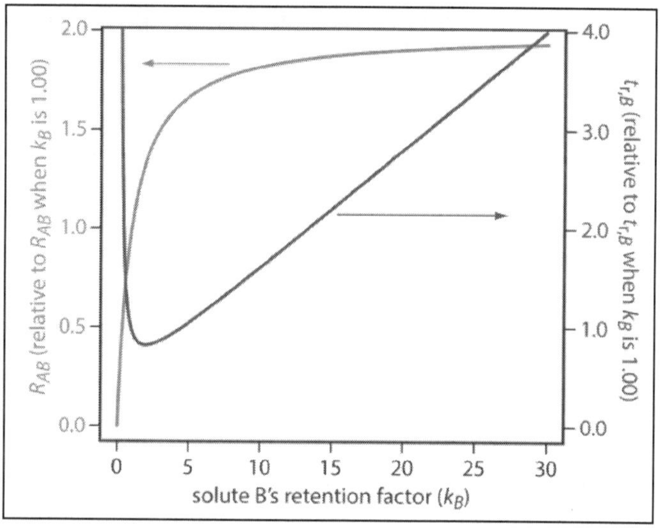

Figure 1.5

Adjusting the retention factor to improve the resolution between one pair of solutes may lead to unacceptably long retention times for other solutes. For example, suppose we need to analyze a four-component mixture with baseline resolution and with a run-time of less than 20 min. Our initial choice of conditions gives the chromatogram in Figure 1.6 a. Although we successfully separate components 3 and 4 within 15 min, we fail to separate components 1 and 2. Adjusting conditions to improve the resolution for the first two components by increasing k_2 provides a good separation of all four components, but the run-time is too long (Figure 1.6 b). This problem of finding a single set of acceptable operating conditions is known as the ***general elution problem***.

One solution to the general elution problem is to make incremental adjustments to the retention factor as the separation takes place. At the beginning of the separation we set the initial chromatographic conditions to optimize the resolution for early eluting solutes. As the separation progresses, we adjust the chromatographic conditions to decrease the retention factor–and, therefore, to decrease the retention time–for each of the later eluting solutes (Figure 1.6 c). In gas chromatography this is accomplished by temperature programming. The column's initial temperature is selected such that the first solutes to elute are resolved fully. The temperature is then increased, either continuously or in steps, to bring off later eluting components with both an acceptable resolution and a reasonable analysis time. In liquid chromatography the same effect is obtained by increasing the solvent's eluting strength. This is known as a gradient elution.

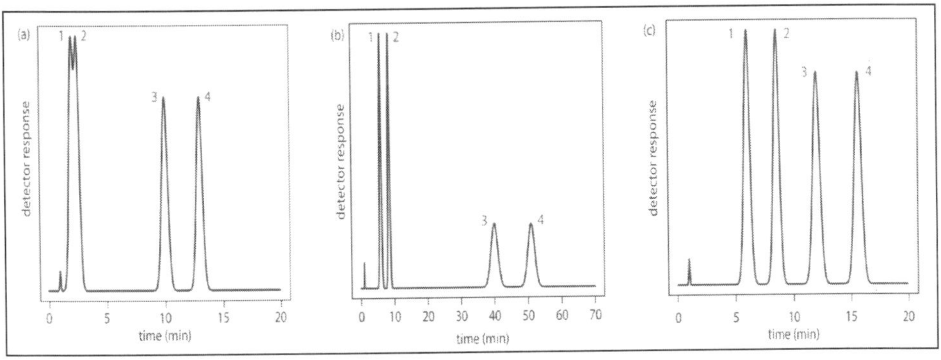

Figure 1.6

b) Selectivity Factor to Optimize Resolution

A second approach to improving resolution is to adjust the selectivity, α. In fact, for $\alpha \approx 1$ usually it is not possible to improve resolution by adjusting the solute retention factor, k_B, or the column efficiency, N. A change in α often has a more dramatic effect on resolution than a change in k_B. For example, changing α from 1.1 to 1.5, while holding constant all other terms, improves resolution by 267 per cent. In gas chromatography, we adjust α by changing the stationary phase; in liquid chromatography, we change the composition of the mobile phase to adjust α.

c) Column Efficiency as a Factor to Optimize Resolution

A third approach to improving resolution is to adjust the column's efficiency by increasing the number of theoretical plates, N. If we have values for k_B and α, then we can use equation 1.24 to calculate the number of theoretical plates for any resolution.

Chapter 2

Paper Chromatography

2.0 Introduction

Paper Chromatography (PC) was first introduced by German scientist Christian Friedrich Schonbein (1865). PC is considered to be the simplest and most widely used of the chromatographic techniques because of its applicability to isolation, identification and quantitative determination of organic and inorganic compounds. Paper chromatography is a type of partition chromatography that uses a piece of paper as the stationary phase and a solvent as the mobile phase. The substances in the mixture are distributed between the two phases, and the ones that are more soluble in the stationary phase will move more slowly than the ones that are more soluble in the mobile phase. This results in the separation of the substances into different spots on the paper. Paper chromatography was originally used to separate mixtures of organic substances, such as dyes and amino acids. However, it can also be used to separate mixtures of inorganic substances, such as cations and anions. To perform paper chromatography, a drop of the test solution is applied to a piece of filter paper. The paper is then placed in a closed chamber with the edge of the paper dipped into a solvent. The solvent will travel up the paper by capillary action, carrying the substances in the test solution with it. The different substances will move at different rates, depending on their solubility in the solvent. Once the solvent has reached the top of the paper, the paper is dried and the spots of the separated substances are visualized using a suitable reagent.

The movement of substance relative to the solvent is expressed in terms of R_F value.

$$R_F = \frac{\textit{Distance travelled by the solute from the origin}}{\textit{Distance travelled by the solvent from the origin}}$$

R_F is function of partition coefficient. It is constant for a given substance, provided the condition of chromatographic system are kept constant with respect to temperature, type of the paper, duration and direction of the development, the amount of the liquid in reservoir and humidity. In some cases, the solvent front runs off the end of filter paper, the movement of a substance in such case is expressed as R_S

$$R_s = \frac{\textit{Dsitance tavelled by the substance from the original line}}{\textit{Distance travelled by the standard from the original line}}$$

R_M: The R_F of chemically related compounds is very close. The influence of individual functional groups was presumed to be added to a rough approximation. According to Bate-Smith and Westall R_M is defined as

$$R_M = Log\left(\frac{1}{R_F} - 1\right) \tag{2.1}$$

The term R_M is additive and is composed of the partial R_M value of the individual functional groups or other groups of atoms in the molecules.

2.1 Types of Paper Chromatography

a) Descending Chromatography

When the development of the paper is done by allowing the solvent to travel down the paper.

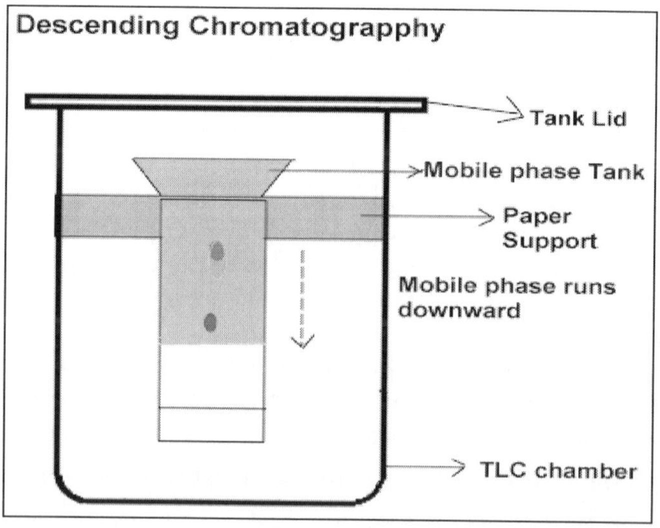

Figure 2.0

The advantage of descending technique is that the development can be continued indefinitely even though the solvents runs off at the other end of the paper.

b) Ascending Chromatography

When the development of the paper is done by allowing the solvent to travel up the paper.

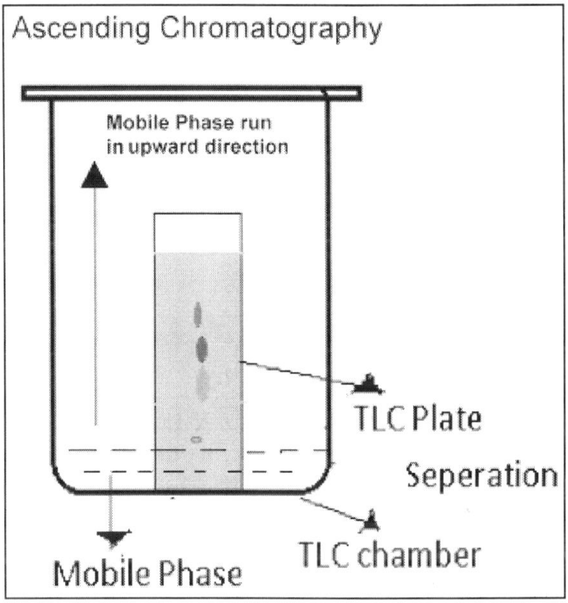

Figure 2.1

c) Ascending-Descending Chromatography

It is the hybrid of the above two techniques. In this technique, the upper part of the ascending chromatography can be folded over a glass rod and allowing the ascending development to change over the descending after crossing the glass rod.

d) Radial Paper Chromatography

This is also known as circular paper chromatography. In this technique the sample is deposited in the centre of the filter paper. After drying the spot the paper is fixed horizontally on the perti-dish possessing the solvent so that the tongue or the wick of the paper dips into the solvent. The solvent rises through the wick or tongue. When the solvent front has moved trough a sufficient large distance, the components get separated in the form of concentric circular zones.

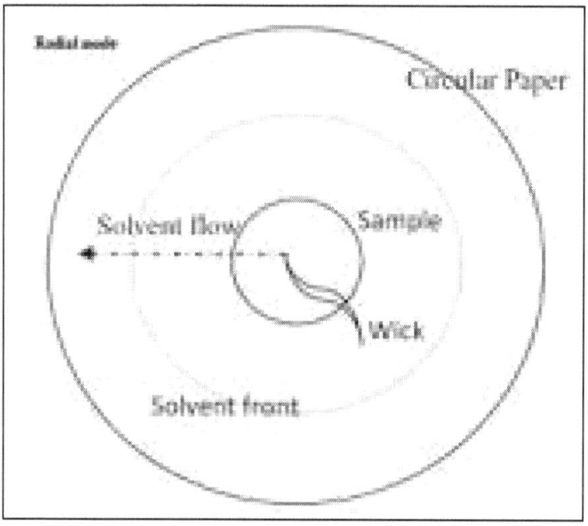

Figure 2.2

e) Two Dimensional Chromatography

In this technique, a square or rectangular paper is used. Then sample is applied to one of the corners. The second development is performed at right angle to direction of the first. This type of chromatography can be carried out with identical solvent system in both the directions or by two solvents.

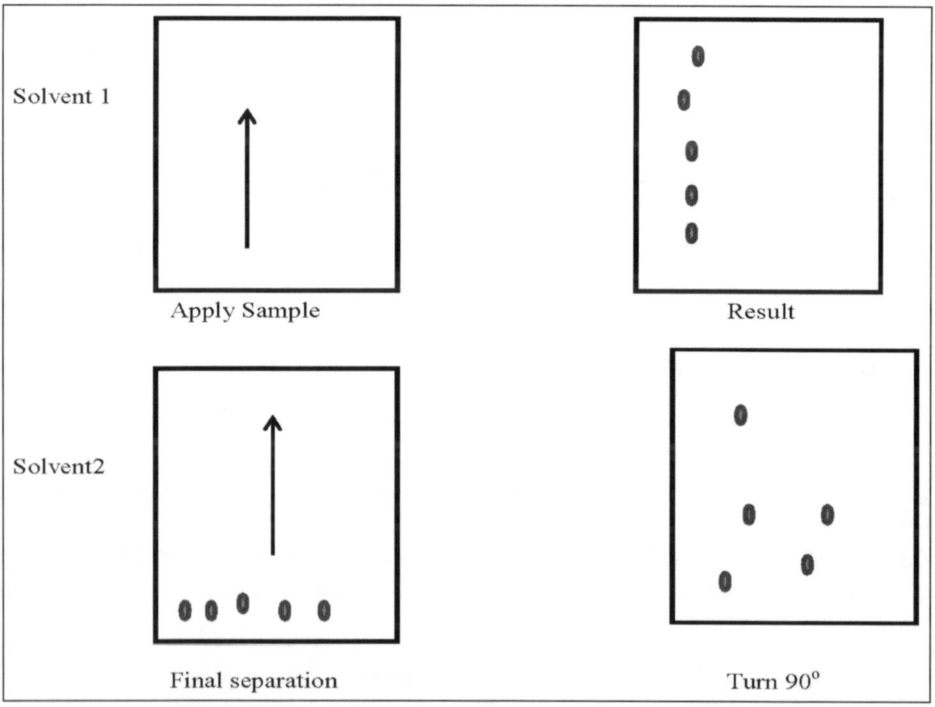

2.2 Requirements for Qualitative Analysis by Paper Chromatography

a)

The first step in paper chromatography is to select the mode of operation. There are five modes: ascending, descending, ascending-descending, radial, and two-dimensional. The choice of mode depends on the nature of the substance to be separated.

b) Choice of Filter Paper

The choice of paper in paper chromatography is governed by the following factors:

- ✰ The purpose of the analysis: Is it qualitative or quantitative?
- ✰ The type of chromatography: Analytical or preparative?
- ✰ The nature of the substances being separated: Hydrophilic, lipophilic, or charged?
- ✰ Coarser and faster papers (*e.g.*, Whatman 31 ET) are used when the substances to be separated have sufficiently wide Rf values.
- ✰ Slower papers (*e.g.*, Whatman 20, Schleicher and Schull 2045 a, Macherey Nagel 260, and Edrol 208) are used for substances with close Rf values.
- ✰ Heavy papers (*e.g.*, Whatman 3MM) are generally used for preparative purposes.
- ✰ The exchange capacity of the paper can be increased by increasing the carboxyl content (1.4 per cent) by partial oxidation. This is done to improve the separation of polar substances.

For cationic separation of protonated amines and amino acid, carboxyl papers are used. For reverse phase chromatography of lipophilic substances like steroids, insecticides and pigments, acetylated papers are used. For separation of low polarity substances such as amines, fatty acids steroids, triglycerides, vitamins and pesticides Kieselghur papers, Alumina papers, Zirconia papers and Silica papers are used.

c) Proper Developing Solvent

The main criterion for selecting the solvents is that the R_F value should be different for different constituents in a given mixture. Following are the general criterion for a good solvent system

i) The R_F values of the sample should lie between 0.05 and 0.85 in the system

ii) The difference between the R_F values of any two components must be 0.05.

iii) The distribution ratio of the components in the solvent system should be independent of the concentration so as to give circular spots.

iv) The solvent should not undergo chemical reaction with any of the component of the sample mixture.

v) The solvent should not interfere with the detection of the spot on the developed chromatogram

vi) The composition of the solvent system should not alter with the time.

Table 2.0: Suitable Solvent System for Paper Chromatography

Stationary Phase	*Mobile Phase*
Water	Isopropanol-ammonia-water (9:1:2)
Water	n-butanol-acetic acid-water (4:1:5)
Water	Phenol saturated with water
Formamide	Choloroform
Formamide	Benzene
Formamide	Benzene-cyclohexane)from 1:9 to 9:1)
Dimethyl Formamide	Cyclohexane
Phenoxy ethanol	Heptane
kerosene	70 per cent iso-propanol
Liquid paraffin	Dimethyl formamide-methanol-water (10:10:1)

2.3 Qualitative Analysis

a) Isolation of Separated Compounds

The simplest way to elute a substance from a spot on a filter paper is to cut out the appropriate part of the paper and soak it in a small amount of solvent. A semi-micron extractor, such as a Soxhlet extractor, can be used to do this. Elution is the process of separating the substance from the spot by washing it with a solvent.

b) Determination

The micro analysis of the resultant eluate can be performed by adopting one or more following techniques

i) Gravimetric estimation

ii) UV Spectrometry

iii) Colorimetery

iv) Polarography

v) Coulometry

vi) Radiometry
vii) Flame photometry

c) *In-situ* Method

i) By visual assessment
ii) By measurement of the area
iii) By desiometer
iv) Potentiometery

2.4 Applications

Paper chromatography has found its application in separation of many organic and biochemical products. For example determination of indole in urine and in the study of barbiturates, antibiotics, carbamyl phosphates, hormones and amino acids. It has also been used in the study of inorganic metal salts and complex ions.

2.5 Paper Chromatography in Pesticide Residues

Reverse phase chromatography is the technique which has generally been applied to pesticide residue studies. In this technique, the paper, in theory but not always iii practice, acts only as a support for an immobile solvent, and the compounds of interest are separated by partitioning between the immobile solvent on the paper and the mobile solvent moving through the paper. Vegetable oils, mineral oils, silicones, propylene glycol, and dimethyl formamide are materials commonly used for immobile phases. This technique is especially applicable to the separation of compounds with very low water solubilities, and thus is useful for compounds such as chlorinated hydrocarbon and organo-phosphorus pesticides. In general, paper chromatography is used to separate pesticide residues without chemical modification of the chromatographic paper. However, fiberglass papers have been used for the reverse phase chromatography of organo-phosphorus compounds and dithiocarbamates. In addition, McKinley and co-workers have used acetylated papers for reverse phase chromatographic separation of organo-phosphorus pesticides.

i) Selection of the Solvent System

When choosing solvents for paper chromatography, several factors should be considered, such as:

- ☆ The solubility and stability of the pesticide in the solvent.
- ☆ The rate at which the solvent moves through the paper.
- ☆ The effect of the solvent on the subsequent detection and quantitative measurement of the residue.

☆ In reverse phase chromatography, the immobile and mobile solvents must not be miscible to any appreciable extent. This means that the two solvents should not mix together easily.

ii) Detection Technique

Generally, the detection technique must be capable of great sensitivity, *i.e.*, detecting a few micrograms of the residue and, in many instances, detecting less than 1 fg of the compound. One widely-used technique for chlorinated pesticide residues involves spraying the developed chromatogram with a solution of silver nitrate and then exposing the sprayed chromatogram to ultraviolet light. The chlorinated compounds appear as black or brown spots on a white background. Papers must be thoroughly washed before chromatography for a satisfactory background. Other techniques used to a limited extent for the detection of chlorinated pesticides on paper chromatograms involve spraying with potassium hydroxide and ferric sulfate, methyl yellow followed by exposure to ultraviolet radiation, and indophenol blue with a weak organic acid. Organophosphorus pesticide residues may be detected on paper chromatograms by several techniques. One of these is the enzymatic technique based on cholinesterase- or carboxylesterase-inhibition by pesticide residues. This is extremely sensitive for most organophosphorus pesticide residues if they are converted to active inhibitors prior to the application of the enzymatic detection procedure. This conversion has most often been accomplished by bromination of the developed chromatogram. Although acetylcholinesterase-inhibition has been used by many workers, probably the most familiar of these techniques is that developed by Getz and Friedman (1963) from the procedure originally outlined by Cook (1955). Human blood plasma is the enzyme source and acetylcholine is the substrate; color is developed on the chromatogram by using an acid-base indicator to detect the acetic acid produced in the chromatogram by the action of the enzyme on acetylcholine. A diazo compound (Azoene Fast Blue RR salt) can be used to detect the 1-naphthol produced over the background of the paper by the action of carboxylesterases on 1-naphthyl acetate. The organophosphorus compounds appeared as white or yellowish spots on a brown background. The enzymatic techniques are extremely sensitive for most organophosphorus compounds; the carboxylesterase procedure can detect less than 1 μg of most of the organophosphates and sometimes as little as 5 ng. Another technique is based on detection of sulfur in the molecule. Since the majority of organophosphorus pesticides contain sulfur, this procedure is applicable to most but not all of the organophosphorus pesticides. Silver nitrate-bromophenol blue spray can be used followed by washing in a citric acid solution to produce blue spots on a yellow background. An iodoplatinate spray can be used as a reagent which apparently detects only those organophosphorus compounds with two or more atoms of sulfur. Other reagents used to detect the sulfur-containing organophosphate pesticides are 2, 6-dibromo-N-chlorop-quinoneimine and metanil yellow, yellow RFS, or methyl yellow after bromination. Spraying the chromatogram first with IV-bromosuccinimide in methyl chloroform, then with a slightly basic, ethanolic solution of fluorescein can be exploited for organophosphate insecticides. This has been used for thiophosphates and some other phosphates. Other spray reagents used for organophosphates are ferric chloride-salicylsulfonic acid reagent.

iii) Quantitative Estimation

Quantitative determination of pesticide residues on paper chromatograms by measurement of spot size has not been widely applied because all stages of the determination for each compound and by each operator must be carefully standardized. Moreover, it is usually applicable only over a very narrow range and, in many chromatographic separations; the spot size appears to be determined largely by factors other than the quantity of the compound present. Probably the best examples of this technique are to be found on the quantitative estimation of chlorinated hydrocarbon pesticide residues in pigeon flesh and soil respectively. In the application of this technique, the separated components are often eluted from the paper for quantitative determination by spectrophotometric or other procedures. The pesticide itself may be eluted or the elution may be carried out after reaction of the separated pesticide to produce a measurable compound. Digestion of the paper chromatographic spot followed by chemical analysis has received considerable attention as a quantitative procedure. One of the major disadvantages of paper chromatographic techniques is the relatively poor spot definition often obtained. Spots on paper chromatograms are usually not as well defined as those on thin layer chromatograms and, therefore, it is more difficult to separate and identify the components of complex mixtures. The characteristics of the paper limit the types of detection techniques. High temperatures or strongly acidic or alkaline solutions cannot be used for paper chromatography because high temperatures often cause severe darkening of the paper and strongly acidic or alkaline solutions usually cause serious degradation of the paper. In reverse phase chromatography, materials suitable for the immobile phase may interfere with the subsequent detection or quantitative measurement of the separated components. Some immobile phases cannot be applied with enzymatic detection techniques because excess bromine is difficult to remove from the chromatogram after oxidation of thionophosphates. Silicones are not suitable for immobile phases on chromatograms to be subjected to oxygen flask combustion prior to quantitative determination of the residues, because the silicones are not digested by this procedure. The application of immobile phases in reverse phase chromatography somewhat limits the use of two-dimensional chromatography for the separation of pesticide residues.

2.6 Disadvantages

- ✰ The spots on paper chromatograms are not as well defined as those on thin layer chromatograms, making it difficult to separate and identify the components of complex mixtures.

- The characteristics of the paper limit the types of detection techniques that can be used. For example, high temperatures or strongly acidic or alkaline solutions cannot be used because they can damage the paper.
- In reverse phase chromatography, the materials used in the immobile phase may interfere with the detection or quantitative measurement of the separated components.
- Some immobile phases cannot be used with enzymatic detection techniques because excess bromine is difficult to remove from the chromatogram after oxidation of thionophosphate.

Silicones are not suitable for immobile phases on chromatograms to be subjected to oxygen flask combustion prior to quantitative determination of the residues, because the silicones are not digested by this procedure. The application of immobile phases in reverse phase chromatography somewhat limits the use of two-dimensional chromatography for the separation of pesticide residues.

2.7 Advantages

The advantages of paper chromatography are the relative simplicity of the operations, the small expense for the required equipment, and the relative ease of varying conditions. Features such as the type of paper, immobile phase, developing solvent, direction and length of development, and detection technique may be readily and rapidly altered in paper chromatography. This factor is particularly valuable in the development of procedures and in the use of specific conditions for specific samples and situations. Paper chromatography has a distinct advantage over thin layer chromatography when the sample is to be digested after chromatographic separation and before quantitative determination since the paper can be readily digested with the sample, whereas it is often necessary to elute the separated components from thin layer chromatograms before digestion. Since paper chromatography has been employed in pesticide residue studies for many years, it currently has the advantage that it is more familiar to many workers than many of the newer chromatographic techniques. This familiarity often results in greater ease and certainty of interpretation of paper chromatographic results.

Chapter 3

Thin Layer Chromatography

3.0 Introduction

The only difference between thin layer chromatography and paper chromatography is that except that a thin (0.25 mm) layer of some inert material, such as Al_2O_3, MgO, and SiO_2 is used as the substrate instead paper. Slurry of these oxides is made in a suitable inert solvent which is uniformly spread over flat surface (glass) and subsequently dried. It may be spread manually or mechanically. The advantage of using inert substrate instead of paper is that more reactive developing reagents such as strong acids, can be used to detect the compounds without destroying the substrate. Thin-layer chromatography (TLC), is a solid-liquid form of chromatography where the stationary phase is normally a polar adsorbent and the mobile phase can be a single solvent or combination of solvents. In thin-layer chromatography, the stationary phase is a polar absorbent, usually finely ground alumina or silica particles. The adsorbent is coated on a glass slide or aluminium sheet or plastic sheet creating a thin layer of the particular stationary phase. Almost all mixtures of solvents can be used as the mobile phase. By manipulating the mobile phase, organic compounds can be separated. In TLC, molecules are continuously moving back and forth between the free and adsorbed states with millions of molecules adsorbing and millions of other molecules desorbing each second. The equilibrium between the free and adsorbed states depends on three factors:

1. The polarity and size of the molecule
2. The polarity of the stationary phase
3. The polarity of the solvent.

The separation and analysis of high molecular weight biochemical compounds can be appreciated in thin layer chromatography by using reagents like H_2SO_4 which otherwise is not possible in paper chromatography. Thin layer chromatography can be used for broad spectrum of compounds like amino acids, dyes, food coloring, drugs, sugars, natural products and insecticides may be separated and identified. TLC has also been applied to micro analysis of inorganic ions and salts (mixture of Ru, Rh, Pd, Ir and Au). The technique of thin-layer chromatography was first discovered by Izmailov and Shraiber in 1938. TLC is often named by other names such as drop, strip, spread layer, surface chromatography and open column chromatography.

3.1 Advantages of TLC over Paper Chromatography

i) Simple equipment: Low cost effective equipment's are required.

ii) Wide choice of stationary phase: The method may be developed for adsorption, partition or ion exchange chromatography

iii) Short displacement: For good separation on inorganic adsorbent layers development time, in case of TLC is about 1 hr while as paper chromatography require several hours or days

iv) Separation effects: TLC has better separation effect

v) Early recovery of separated components: The spots or zones which are quite discernible on the powdery coating of the plates can be easily removed quantitatively by scraping with knife and required compounds dissolved in a suitable solvent are determined by calorimetric or spectrophotometric analysis.

vi) Easy visualization of separated compounds: Detection of fluorescence compounds under UV light is easier than on paper because the inorganic background does not fluoresce.

vii) Sensitivity: Extremely sharp delineated spots are obtained in TLC (about 10 to 100 folds)

viii) Chemically inert stationary phase: TLC by virtue of its inert stationary phase provides a better window of opportunity in affording vigorous means of detection including application of strong heat or of corroding reagent such as conc. H_2SO_4

ix) Variable thickness of thin layers: The method used in TLC analysis may also be applied for the preparative separation with the aid of thicker layer as well as to most separation by column chromatography.

3.2 Experimental Techniques

Although the techniques employed in TLC is similar to paper chromatography, however the main differences are given as under

i) Coating material: Popular coating materials used in TLC is highlighted as given in Table 3.1

Table 3.1: Coating Material Used in TLC

Adsorbent	*Acidic or Basic*	*Activity*	*Separatory Mechanism*	*Components to be Separated*
Silica gel	acidic	active	Adsorption partition	Acidic and neutral substances
Alumina	base	active	Adsorption partition	Basic and neutral
Kieslegur	neutral	Inactive	partition	Strongly hydrophilic substances
Cellulose powder	neutral	none	partition	Water soluble compounds

ii) Preparation of thin layers in plates: The various methods of preparing layers in thin layers are very important and different applicators are available in the market which is used for coating the glass plates with different materials. The various methods includes as follows.

a) Pouring: In this method, a known amount of slurry is put on a given size plate which is kept on a level surface. The plate is then tipped back and forth to spread the slurry uniformly over the surface. Although this method is simple although not used widely.

b) Dipping: This method was first proposed by Peifer (1962). In this method, plates are prepared by dipping them at a time, back to back, in chloroform-methanol slurries of the adsorbent. This method is also not of much use.

c) Spraying: This method was proposed by Reitsema. He employs a small point sprayer for distribution of the slurry on the glass plate.

d) Spreading: The slurry is placed in applicator. This is either moved over the stationary plate or it is held stationary and plate is pushed through.

An improved TLC spread was developed by DESAGA (West Germany). By this the layer thickness can be adjusted from 0 to 2.0 mm.

3.3 Activation of the Adsorbent

The adsorbents are activated by drying the thin layer plate for 30 min in air and then in oven for next 30 min at 110oC. For Alumina and silica gel plates the activation is done by heating at 150oC for 4 hours.

3.4 Purification of Silica Gel "G" Layers

Iron is main impurity in silica Gel which can lead to distortions in chromatogram and the iron can be removed by done by preliminary development with methanol-conc HCl (9:1 v/v). The iron gets migrated with the solvent front to upper edge of the plate. The pates are dried again and activated at 110°C.

3.5 Sample Preparation

For quantitative work Agla micro-syringe is generally used for transferring the sample solution to the thin layers and for qualitative work capillary tubes may be used. The range of sample volume and concentration is far greater than with the paper chromatography. Solvents used for sample preparation should be volatile and non-polar as possible.

3.6 Solvent Systems

For selecting suitable solvent Stahl's triangle may be sued if nature of solute is well known in advance otherwise trial and error method is used for selecting solvent if we do not know the components of mixture. The triangle is inter relating adsorbent activity, nature of the solute and the nature of the solvent. Table 3.1 gives list of selected one and two components solvent systems in order of increasing eluting power (i.e in order of increasing polarity). With the mixture of solvent, it is possible to obtain intermediate elution power behaviour. Mixtures of two or more solvents of different polarity often give better separation than chemically homogenous solvents.

Table 3.2: Elutropic Series of Selected One and Two Component Solvent Systems with Increasing Polarity

High boiling parafins	Benzene, ethanol (9:1)
Cyclohexane	Chloroform, acetone (85:15)
Benzene	Benzene, ethyl acetate (1:4)
Cyclohexane, ethyl acetate (95:5)	Diethyl etherEthyl acetate
Benzene, ethyl acetate (95:5)	Ethyl acetate, methanol (99:1)
Chloroform	Benzene, acetone (1:1)
Cyclohexane, ethyl acetate (85:15)	Acetone
Benzene, ethyl acetate (85:15)	Methanol
Chloroform, acetone (9:1)	Dimethyl Formamide
Benzene, methanol (95:5)	Dimethyl sulphoxide
Cyclohexane, ethyl acetate (1:1)	Water

3.7 Detection of Components

Coloured substances can be physically visualized while as colourless components can be obtained under UV light or by treating with the plate with a visualizing reagent like chromic acid/sulphuric acid.

3.8 Evaluation of Chromatogram

Solutes on the plates are fixed by their position by marking them and after detecting the their evaluation may be either qualitative or quantitative.

a) *Qualitative*: This is same as with paper chromatograms except to say that R_F values in TLC shows inferior reproducibility.

b) *Quantitative*: It can be carried out by two ways

   i) Direct method

   ii) Indirect method

3.8.1 Direct Method

a) *Visual assessment of chromatograms*: It is semi-quantitative and has many shortcomings.

b) *Determination by measurement of spot area*: Mathematical relationship between spot area and amount of substance present is invoked however random errors in this procedure are generally high as result of the spot shapes during the separation.

c) *Quantitative TLC incorporating densiometry*: Densiometry is a method whereby the intensity of the colour of a substance is measured directly on a chromatogram

d) *Direct spectrophotometry on thin layer chromatograms*: Quantitative measurements are obtained by evaluation with the wavelength of maximum absorption of the substances. Characterization of the separated chromatograms zones by reading the absorption or fluorescence curves directly from TLC plates is done by chromatogram spectrophotometer introduced by Zeiss, Stahl and Jork. Infra-red, reflectance spectroscopy, spark chamber method have also been used for the direct evaluation of chromatogram.

3.8.2 Indirect Method

Indirect method involves chemical constituents of compounds of interest, their recovery by elution and subsequent determination with a suitable instrument. These methods are known as elution techniques.

In this method area containing the adsorbed compound after visualization is marked and scooped with the help of vacuum cleaner. This method does not involve any loss while transferring the chromatographed species. Simple agitation with suitable solvent followed by the removal of adsorbent by Elution of the solute from the adsorbent is done by simple agitation with solvent followed by the removal of adsorbent by centrifugation. The micro analysis of the resultant eluate can be performed by adopting one or more of the following techniques:

a) Gravimetric
b) UV spectrophotometry
c) Colorimetry
d) Polarography
e) Coulometry
f) Radiometry
g) Flame photometry
h) Vapour-Phase Chromatography

In the assessment of the merits of any procedure, the following information is required

i) The nature of the substance to be assessed
ii) The scientific equipment available and its sensitivity
iii) The time available
iv) The alternative method available.

3.9 Application of TLC

TLC is used extensively for the quantitative determination of high molecular weight compounds, particularly in medical and biological research. Qualitative identification can be based on the rate at which a compound flows along the plate. A compound can be identified by direct comparison with known compound. This requires guessing the identity of the components in order to use pertinent "known compounds". A better method involves measuring the RF value of the sample. For any particular system of substrate, solvent and sample, the R_F value is constant (at constant temperature). A unknown compound is then compared to compounds with the same R_F value, which are relatively few in number and subsequent identification is then much easier.

The various application of TLC are:

i) *As a check on processes*. Checking of the other separation procedures and purification processes. Can be achieved by using TLC techniques. It has been used for checking of distillation fractions and form checking the progress of purification by molecular distillation.
ii) *In organic chemistry*: In isolation and separation of individual components of mixtures. The main reason for the popularity of TLC as an analytical and preparation methods as follows:

a) TLC can be used for most of chemical compounds
b) High speed of separation
c) High selectivity.

iii) *For checking the purity of sample:* TLC is mainly used as a check on samples obtained by other methods. In this respect, its high sensitivity often enables impurities to be observed in so-called pure samples.

iv) *As a purification process:* Elemental analysis and other physical methods are required for characterisation of pure organic substance. The conventional means of purification can be tedious however using TLC technique sufficient pure material can be obtained on 2 cm plates. The incorporation of 25 ppm of sodium fluorescein during the preparation of the plates will make compounds (aromatic ones) visible under UV light (254 nm).

v) *Examination of reaction:* TLC has enabled the chemist to assess that a reaction is complete and it has followed the expected outcome. When the reaction did not proceed as expected an examination of the behaviour of the spot with standard reagent will sometimes give information for the rapid identification of the product.

vi) *For identifying organic compounds:* TLC is being increasingly employed as a rapid method for isolation, purification, and identification of organic substances. The main requirements are that the reaction must be separable and sufficiently non-volatile to remain on then plate during the development and also unaffected by the adsorbent and developing solvent.

vii) *Characterization of organic compounds:* TLC has been successfully used for characterizing acids, alcohols, glycols, alkaloids, amines, amino acids, antibiotics *etc.*

Table 3.3: Analysis of Components of Foodstuffs by TLC

Food	*Components Determined*	*Stationary Phase*	*Mobile Phase*	*Visualizing Reagent*
Fruits	Amino acids	Silica gel 0.25	Chloroform, 1-Pentanol acetic acid (30:30:1)	Ninhydrin
Milk	Amino acids	Cellulose 0.25	Methanol, chloroform, conc NH3 Soln (2:2:1)	Ninhydrin
Orange juice	Amino acid(as adulteration)	Kiesel Gel-G(as adulteration)	Phenol,water (3:1)	Ninhydrin
Apple, Pear	Sorbitol	Kiesel Gel-G	Isopropanol, ethylacetoperiodate/ acetic acid (54:7:1.2)	Benzidine

Food	*Components Determined*	*Stationary Phase*	*Mobile Phase*	*Visualizing Reagent*
Wine	Glucose, sorbitol (as adulteration)	Kiesel Gel-G	Butanol-acetone, water (4:5:1)	Benzidine
Milk (cow) Sheep goats	Glycerides	Silica gel	Light petroleum (b.p 30-60oC), diethyl ether, acetic acid (80:20)	Iodine vapour
Butter	Glycerides (as adulteration)	Silica gel	Chloroform, acetic acid	Rhodamine B

3.10 Application of TLC for Separation of Inorganic Ions

Recently TLC has been used for separating cationic, anionic, purely from covalent species and also some organic derivatives of the metals. The first inorganic work on thin layers was reported by Meynard and Hall in 1949 that separated Zn and iron by radial chromatography on layers of mixture of aluminium oxide and celite bound with starch.

The various separations of cations and anions have been summarized in Table 3.4.

Table 3.4: Separation of Metal Ions by TLC

Metal Ion	*Solvent Systems*	*Colour with PTC*	*R x 100*
Ag(I), Hg(I)	n-butanol-3 dm mol HCl-3 moltartaric acid (10:1:1:V/V)	Light pink, black pinkish red	0.08,88.8
Cu(II), Pb(II) Bi(III), Cd(II), Hg(II)	n-butanol-1, 5 mol dm 3HCl-Me,Et, ketone(75:15:2 v/v)	Brown, pinkish red darkBrown, yellowish black	23.0,38.060.0,66.0
As(III), Sb (III), SN((II)	n-butanol-3-mol dm-3HCl-1M-tartaric acid (10:1:, v/v)	Yellow, orange, brown	55.0,61.077.0
Fe(III), Cr(VI) Al(III)	Acetone-3-mol dm HCl-1M-tartaric acid (10:1:1, v/v)	Black, green, red	73.8,39.022.7
Zn(II), Co(II) Ni(II)	—do—	Cream/pink with dithiozone, black yellow	54.0,39.0,022.7
Cu(II), Ni(ii), Co(II)	Acetone, MIBK (3:1,v/v)	Brown, yellowish, orange, green	77.8, 66.6, 22.7
Pb(IV), V(V), Rh(II), Hg(II)	Iso-amylacetate conc HCl (30:2,v/v)	Light brown, dark brown, pink, grey golden yellow	60.7,49.2,0.0,67.6

3.10.1 Application of TLC in Quantitative Analysis

Estimating the area of the spot and based on prior calibration, the amount of material can be made for the quantitative analysis of compounds as separated by thin-layer chromatography. More accurate determination can be made by carefully removing the entire sample spot (and the substrate) from the plate. The sample is then removed from the substrate by dissolving it in a known volume of suitable solvent. It can be determined by UV absorption, IR absorption or some other suitable analytical method.

Two basic approach which are generally employed for the quantitative analysis of substances separated on chromatograms are

i) Analysis of fractions on a plate

ii) Determination of fraction after elution from the coating material

A subsequent chromatographic method is employed for assessing the composition of fraction resolved on plates, thereby yielding quantitative analysis on total in an indirect way

iii) *Analysis of fractions on a plate* This technique is quite simple and analysis of fractions on chromatoplates can be carried out by a) Visual comparison by standard b) The precise determination of spot area or spot size c) photo densitometry

iv) Radioactive substances can be determined as a) photo-desitometry of X-ray prints b) By liquid scintillation counting in suspension c) Direct radio scanning of chromatoplates

The size of spot can be found by the amount of substance applied to the plate. The size shape and colour of a spot are influenced by nature of solvent used for dissolving the sample, process of putting the solution on the layer, composition and thickness of the layer, comparison of developing solvents, condition of environment in chromatojar.

A Graphical Method, in which a calibration curve is obtained by plotting the area of spots (formed by reference material) against their weights and the amount of unknown fractions is determined by relating its area to the calibration curve. According to Purdy (1962) the spot area is a linear function of logarithmic of the weight (W) of material present in adsorption as well as in partition chromatography and thus straight line calibration curve are obtained by plotting log of weights (W) against square root of areas (A)

$$Log\ W - Log\ W_s + \left[\frac{\sqrt{A} - \sqrt{A_s}}{\sqrt{A_d} - \sqrt{A}}\right] \tag{3.1}$$

Where A = Spot are of the unknown, A_d = Area of correspondence spot in dilute sample, d = dilution factor, As = Spot are of standard W_s = Weight of standard.

The second method consists in chromatographing equal volumes of a solution of the sample, a dilute solution of the sample and a dilute solution of the same concentration as the first but containing in addition a certain amount of the material to be determined. The weight of the fraction to be analysed is then calculated from the relation given as

$$Log\left[W_d + \frac{a}{W}\right] = \frac{\left[\sqrt{A_s} - \sqrt{A}\right]\log d}{\sqrt{A_d} - \sqrt{A}} \tag{3.2}$$

where A = spot area of unknown

A_d = Area of the corresponding spot in the first dilute sample

D = Dilution factor

A_s = Spot are of the dilute sample containing the reference material

Some quantitative methods are like Spectrophotometric measurement, Fluorimetric methods, Spectral reflectance, Visual comparison, spot are and weight, spot Densitometer, Vapour phase chromatography, Radioactive method, Volumetric analysis

3.10.2 Application of Thin Layer Chromatography for Pesticides

3.10.2.1 Sample Preparation

Although traditional sample preparation methods such as liquid-liquid and Soxhlet extraction together with large column chromatographic clean-up on adsorbents such as Florisil are still widely used prior to TLC analysis, modern extraction and clean-up methods are being used to an increasing degree. For example, pyrethrins and piperonyl butoxide in soil extracts can be purified and concentrated using Isolute C_{18} solid phase extraction (SPE) cartridges eluted with methanol prior to reversed phase (RP)-TLC analysis. Matrix solid phase dispersion (MSPD) can be used in the determination of the organophosphorus (OP) pesticides like diazinon and ethion. Samples (with oils) is blended with C_{18} adsorbent and packed into a column on top of lanthanum silicate ion exchanger as a co-column. After washing the column with hexane, the purified residues were extracted with dichloromethane–ethyl acetate (1 + 1) before TLC analysis. Additional sample preparation methods are described in some of the applications cited in the sections below.

3.10.2.2 Thin-Layer Plates, Mobile Phase Selection, and Plate Development

Most pesticide determinations are performed on silica gel TLC or HPTLC commercially precoated plates. The layers almost always contain a gypsum binder (designated as G layers) or an organic binder. Silica gel H layers are especially hard and rugged and permit the use of high concentrations of water in mobile phases without loss of adherence. Layers often contain a fluorescent indicator or phosphor to facilitate detection of compounds that absorb 254 nm UV light as dark zones on a fluorescent background. Cellulose, alumina and kieselguhr layers are also used occasionally for pesticide TLC.

Chemically bonded silica gel phases are being used with increasing frequency for TLC analysis of pesticides, most notably cyano, amino, and diol normal phase (NP) layers and C18 (octadecyl as discussed in section) and C18W (water wettable octadecyl) reversed phase (RP) layers. Impregnated layers are also sometimes applied for pesticide analysis, *e.g.*, silica gel impregnated with mineral oil for the RP separation of pyrethrins and piperonyl butoxide [7]. Lanthanum silicate or lanthanum tungstate ion exchange layers are being used for the separation and detection of diazinon and ethion residues recovered from nuts by MSPD (matrix solid phase dispersion). Mobile phases were methanol– 10 per cent ammonia (95: 5) and methanol–dichloromethane–10 per cent ammonia (65: 31: 4), respectively, and visualization was obtained with palladium chloride reagent.

3.10.2.3 Detection, Identification, and Documentation of Zones

Pesticide zones can be detected on layers in daylight as colored zones or under 366 nm UV light as fluorescent zones in their natural state, or after post chromatographic derivatization with a chromogenic or fluorogenic reagent applied by spraying or dipping. In the case of chlorinated pesticides, silver nitrate detection reagent can be impregnated into a silica gel layer prior to development, followed by exposure to UV light for about 10 min to allow observation of dark zones on a light brown background; this method isused to detect the pyrethroid insecticide fenvalerate after development with hexane–acetone (3: 1). Many pesticides absorb shortwave UV light and can be detected on F layers as dark zones on a bright fluorescent background (green fluorescence with most commercial layers) by inspection in a viewing box under illumination from a 254 nm lamp; this detection method is based on quenching of the layer fluorescence by the pesticides.

Chapter 4

Liquid-Liquid Partition Chromatography

4.1 Introduction

In normal phase partition chromatography, the stationary phase is water that is supported by a matrix. The matrix can be cellulose, starch, or silicic acid for column chromatography. For thin layer chromatography, the matrix is most commonly cellulose, but it also exhibits adsorption effects. In paper partition chromatography, the water that is naturally present in the paper acts as the stationary phase. However, in column and thin layer chromatography, the correct amount of water must be added to the support to achieve the correct partitioning. Some supports can be prepared to contain as much as 50 per cent water by weight and still be free-flowing powders that can be packed into columns or coated onto plates.

The mobile solvent phase is generally an organic solvent that is immiscible with water, or a mixture of aqueous organic solvents that have a dielectric constant less than that of water. In some cases, the solvent may be miscible with water. However, in these cases, the solute is not expected to partition between the stationary and mobile phases because it forms an insoluble complex with the supporting matrix and is effectively immiscible with the mobile phase.

In reversed-phase partition chromatography, the stationary phase is a non-polar compound, such as liquid paraffin, that is supported by a suitable matrix. The matrix is similar to those used in other types of chromatography. The stationary phase is prepared by treating the matrix with a solution of the mobile phase in a suitable non-polar and volatile solvent. The solvent is then removed by evaporation.

Liquid-liquid chromatography was first developed by Martin and Synge in 1941. They used this technique to separate amino acids, which was a major step forward in biochemistry. They used a liquid stationary phase that was mutually insoluble with the mobile phase and the stationary phase. The stationary liquid phase was adsorbed onto the surface of the substrate, forming a thin layer. The mobile liquid phase could then flow over the stationary phase without removing it from the column.

A sample injected into the column in the normal fashion and distributed itself between stationary phase and mobile phase according to distribution coefficient. The sample then moved through the column in the mobile phase until it reached the detector. The separation of two samples was achieved when their distribution coefficients were different, which resulted in different retention times.

The factors that affect peak broadening are similar in gas-liquid chromatography and liquid-solid chromatography. In gas-liquid chromatography, the sample and the mobile phase are both gases. However, in liquid-liquid chromatography, the sample is a liquid and the mobile phase is another liquid. The solubility of one liquid in another is a physical property of the liquid-liquid system. The solubility is usually limited, and it may be below or high but is not usually above infinite. This is particularly true for solvents selected for liquid-liquid chromatography, where a limited solubility leads to control of chromatographic separation.

4.2 Theory: Consider a column packed with a finely divided inert solid, with each particle coated with a thin layer of liquid. This liquid layer is the stationary phase in liquid-liquid chromatography. When a solution containing the components A and B is poured into the column, each component will distribute itself between the stationary phase and the mobile phase (the liquid that is flowing through the column). The relative affinity of each component for the stationary phase is determined by its partition coefficient. the partition coefficient K_A, of the solute A is defined as

$$K_A = \frac{Concentration\ of\ A\ in\ the\ stationary\ phase}{Concentration\ of\ A\ in\ the\ mobile\ phase} \tag{4.1}$$

and that of B

$$K_B = \frac{concentration\ of\ B\ in\ the\ stationary\ phase}{concentration\ of\ B\ in\ mobile\ phase} \tag{4.2}$$

A comparison of partition coefficient yields a method of predicting the efficiency of a separation process. For example, when $K_A = 25$ and $K_B = 15$, a larger fraction of component A than component B will be dissolved in stationary phase at any given time.

In partition chromatography one solvent usually water, is held on the stationary supporting phase which is in the form of a column of film of inert material. The other phase consist of a mobile phase, water saturated, organic liquid that flows past the stationary phase. The components of a mixture are separated if their partition coefficient between the solvents is different.

If a sample is partially soluble in two immiscible solvents, it will come to equilibrium and distribute itself between these solvents with X_m, in mobile phase and X_s in stationary phase, where X_s and X_m are the quantities of sample compound involved, from which we can derive the distribution coefficient K as

$$K_s = \frac{X_s}{X_m} \tag{4.3}$$

If K_s is high, the sample is mostly in stationary phase and vice versa.

If we include the volume of mobile and stationary phase in the calculation of distribution, we have a new constant, the capacity factor Ks where

$$\mathbf{Ks}\square = \frac{total\ moes\ of\ X\ in\ stationary\ phase}{total\ moles\ of\ X\ in\ mobile\ phase} \tag{4.4}$$

$$\frac{V_x}{V_m} * \frac{X_s}{X_m} = \frac{V_s}{V_m} * K_s \tag{4.5}$$

whereas V_s = volume of stationary phase in column, V_m = Volume of mobile phase.

When a peak maximum passes the detector, half of the component has eluted in the retention volume V_r, half is left in the column distributed between stationary and mobile phase. Therefore

$$V_r X_m = X_m V_m + V_s X_s \tag{4.6}$$

$$V_r = V_m + V_s X_s \frac{\square}{\square} X_m$$

$$= Vm + VsKs \tag{4.7}$$

$$= V_m \left[1 + \frac{V_s}{V_m K_s}\right]$$

$$= V_m\ [1 + \mathbf{KsI}]$$

$$= \frac{V_r}{V_m} - 1 = \mathbf{Ks}\square \text{ ------------------- } 4.8$$

V_r = Flow rate of sample × retention time ft_r

t_r = retention time of sample in column

V_m = Value of mobile phase in the column

From equation 4.2 we can deduce that

$$t_r - t_m \qquad t_r \tag{4.9}$$

This equation is fundamental to both gas and liquid chromatography. It illustrates the dependence of the retention time on the relative solubility of the sample in the stationary phase and the mobile phase. This basic relationship holds in liquid-liquid and gas –liquid chromatography.

4.2 Solid Support

The supports most commonly used include silica gel diatomaceous earth (kiselguhr, celite *etc.*) and cellulose. As it is not possible to achieve complete coverage of the support, adsorption also plays some role in the separations. Silica gel used for the purpose is in the form of a fine white powder which is having 30 per cent water which act as stationary phase. Cellulose as well as other support material also contains water. Some columns and their application has been given in Table 4.1.

Table 4.1: Stationary Support Materials for Liquid-Liquid Partition Chromatography

Support	*Stationary Phase*	*Mobile Phase*	*Applications*
Celite	Water	$CHCl_3$ or CCl_4	C_1-C_4 alcohols
Silica gel	Water	$CHCl_3$/Butanol	C2-C8 fatty acids
Silica gel	Water	Iso-octane	Phenols
Cellulose	Water	Acetone, HCl	Inorganic ions, phenols
Cellulose	Water	Methanol/butanol $CHCl_3$	Inorganic ions, phenols
Starch	Water	propanol/Butanol HCl	Amino acids

4.3 Selection of Mobile and Stationary Phase

In liquid-liquid chromatography, it is important to choose two liquids that are virtually insoluble in each other. Water and hexane are two suitable liquids. When water is used as the stationary phase and hexane as the mobile phase, the sample is injected into the hexane and separation takes place as usual. However, it is important to note that water is slightly soluble in hexane. Over time, the hexane will dissolve the stationary water phase and it will be lost from the column. To prevent this, the hexane can be pre-saturated with water. This means that the hexane will already contain a certain amount of water before it is used in the chromatography experiment. The pre-saturated hexane will not dissolve any more water from the stationary phase, which will remain intact throughout the separation process.

Liquid-liquid chromatography is limited by the need for the stationary phase to be a good solvent for the sample but not soluble in the mobile phase. This means that only samples with comparatively low capacitance values can be separated by liquid-liquid chromatography. Changing the mobile phase is also difficult, and great care must be taken in selecting solvents that will meet the requirements for separation. Even if the stationary phase is only slightly soluble in the mobile phase, it can still be lost over time. This is because the stationary phase is slowly dissolved from the column during use. This can lead to the loss of substrate material and the breakdown of the equipment. The problem can be significantly reduced by pre-saturating the mobile phase with the stationary phase material. This prevents the stationary phase from being dissolved from the column during operation.

4.4 Solvent System

The stationary and mobile phases in liquid chromatography are not independent of each other. The optimum combination of the two phases is often difficult to predict, but water is typically used as the stationary phase and an organic solvent as the mobile phase. Two immiscible organic solvents of different polarities may also be used. However, practical problems can arise from factors such as complexation, pH, mixed-solvent effects, and non-equilibrium. A number of the more common solvent combinations are listed in Table 4.2.

Table 4.2: Suggested Solvent Systems for Column Partition Chromatography

Stationary Phase	*Mobile Phase*
Water	Alcohol (n-butanol, isobutanol)
Water plus acid	Hydrocarbons (Benzene, cyclohexane, hexane)
Water plus alkali	Chloroform
Water plus buffer components	Ethyl acetate
Aqueous Alcohol (MeOH, EtOH)	Ethylene glycol, monomethyl ether
Formamide	Methyl ethyl ketone
Glycols (Ethylene, Propylene, Glycerol)	Pyridine

Reverse-Phase Chromatography

In this system, a substrate such as glass or silica is used as a base. Organic material is then used on this base to create an organic surface that acts as the stationary phase. The mobile phase can be aqueous or any other solvent with a high dielectric constant. The most common stationary phases used to date are silica gels that have been esterified with alcohols to form silicate esters. The SiOH groups on silica gel can also be silanized with organochlorosilicates, which creates surfaces onto which the substrate has been chemically bonded. This method has been successfully used for reversed-phase chromatography.

When bonded organic surfaces are used as the stationary phase, the mobile phase can be changed. One way to do this is to use the initial mobile phase for a period of time, and then switch to methanol as the mobile phase. However, a more popular and efficient method is to use gradient elution. In gradient elution, the composition of the mobile phase is gradually changed over time. The initial mobile phase is typically water, but the percentage of methanol is gradually increased until the final mobile phase is pure methanol. Other pairs of solvents that can be used for gradient elution include water-acetonitrile and tetrahydrofuran-dioxane.

Table 4.3: Solvent Systems for Reversed Phase Chromatography

Stationary Phase	*Mobile Phase*
n-Butanol	Water
Octanol	Water plus acid
Chloroform	Water plus alkali
Chlorosilanes and Silicones	Water Plus alkali
Mineral oil	Water plus buffer components
Paraffin	Aqueous alcohols (MeOH, EtOH) Formamides, Glycols

4.5 Application of Partition Chromatography

Liquid-liquid chromatography (LLC), especially in its high-performance form (HPLC), is a technique used to separate a wide range of biological compounds. In its reverse phase, it is particularly effective for separating polar compounds. A variety of polystyrene and acylic ester macro-porous polymeric adsorbents with different polarities are commercially available under the trade names Bio-Bead SM (Bio Rad) and XAD (Rohm and Haas). Some common applications of LLC include:

i) Detergent removal from protein solution
ii) Separation of steroids, bile acids and mycotoxins
iii) Removal of pesticides, phenols, insecticides
iv) Trace metal concentration from aqueous solutions.

4.6 Comparison between Thin Layer Chromatography and Paper Chromatography

Thin Layer Chromatography	*Paper Chromatography*
Principle of separation is adsorption. Principle of separation is partition	Minimum quantity of sample is required
Consuming less time (15-45 min.)	Consuming more time (1-3 h)
Plates may be heated in oven for longtime	Paper cannot be heated for long time
Very sharp separation	Less sharp separation

Thin Layer Chromatography	*Paper Chromatography*
Physical strength of the plate is high (Ascending/Descending)	Physical strength of the paper is low(Descending/Small paper only ascending)
Plates are prepared as thin coat with adsorbents.	Paper is used as such since coating can't be done on paper
Corrosive reagent can be used	Corrosive agents can't be used since it. Destroys it.
Compound can be detected under UV	Can't be detected under UV
Sensitivity of detection is high	Less sensitivity of detection
Spots are less diffused	Spots are easily diffused.

Chapter 5

Adsorption Chromatography

5.1 Introduction

Liquids can be separated into their components by passing them over a solid surface, in a process similar to how gases are separated by passing them over a liquid surface. The liquid solvent is the mobile phase, and the solid surface is the stationary phase. Just like in gas chromatography, the different components of the mixture will flow at different rates, depending on how much time they spend in the stationary phase and how much time they spend in the mobile phase.

In liquid-solid chromatography, a suitable inert solvent is used as the mobile phase, just like an inert gas is used in gas chromatography. The ratio of the time that a sample molecule spends in the solvent and on the surface of the stationary phase depends on the solubility of the molecule in the solvent and the adsorption effects of the stationary phase on the sample molecules. The Van Deemter equation, which describes the factors that affect the efficiency of a chromatographic separation, is also applicable to liquid-solid chromatography. These factors include the geometry of the column, diffusion, mass transfer, and flow rate. An adsorbent is a solid that has a high surface area and can attract and hold molecules on its surface. This adsorption can be selective, meaning that one solute may be adsorbed more strongly than another. The separation of the components of a mixture by liquid-solid chromatography depends on the differences in the degree of adsorption of the solutes by the adsorbent and their solubility in the solvent.

5.2 Theory

Compounds are adsorbed on to the column and an equilibrium set up between molecules bound to column and those free in solution. The extent of the binding is governed by the charge, van der Waal's force, dipole interaction, hydrogen bonding and steric factors and depends on the structure of the compounds. The mass of the solute adsorbed per unit weight of adsorbent (*m*) depends on the concentration of the solute (*s*) and the Langmiur derived an equation on the basis that a) only o monolayer is adsorbed; b) only a proportion of the molecules in collision will result in adsorption. This is known as Langmuir adsorption isotherm

$$m = \frac{K_1 K_2 C}{1 + K_2 C} \qquad (5.1)$$

K_1 is a measure of number of active adsorption sites per unit weight of adsorbent and depends upon nature of adsorbent; K_2 is a measure of affinity of solute for the adsorption and is affected by all the compounds of the system.

Langmuir assumed only one binding site, but in practice there are number of different sites on the surface of adsorbent, each with a different affinity, thus giving a series of Langmuir type isotherm

Hinshelwood therefore suggested that a equation

$$m = \frac{\Sigma K_1 K_2 C}{1 + K_2 C} \qquad (5.2)$$

Equation 5.2 gives a more accurate picture of the situation, and thus indeed approximation of Freundlich adsorption isotherm

$$m = Kc^2 \qquad (5.3)$$

This mixture of binding sites of different affinities is the cause of tailing observed in the elution profile of compound. This tailing can be overcome by eluting with suitable gradient of pH, ionic strength, polarity so that more strongly adsorbed molecules meet a higher concentration of displacing compounds than the more weakly bound molecules.

5.3 Adsorbent

Stationary phases in chromatography can be made of a variety of compounds, including silicic acid (silica gel), aluminum oxide, calcium carbonate, magnesium carbonate, zinc carbonate, magnesium oxide, and cellulose. Hydroxyapatite (calcium phosphate) is a commonly used stationary phase for the separation of proteins, nucleic acids, and viruses. It has ion-exchange properties that help to separate these molecules. In nucleic acid work, hydroxyapatite is particularly valuable because it can bind double-stranded DNA but not single-stranded

DNA. This is because has a negatively charged phosphate backbone, which can interact with the positively charged calcium ions on the surface of the hydroxyapatite.

Care is needed in then choice of adsorbents, since occasionally some adsorbents may degrade certain compounds during evaporation. Other adsorbents have a tendency to take up water from the atmosphere during storage and this may adversely affect their properties. In such case it may be necessary to activate the adsorbent by keeping it for a period of time at 110oC to remove water. Separation on the adsorbent may be enhanced by the presence of water in the adsorbent as this may result in both adsorption and partition effects.

Adsorbents for thin layer chromatography are often impregnated with various ions during plate preparation and this may enhance the separation. The incorporation of silver nitrate into the adsorbent enhances the separation of compounds differentiating in the number and position of double bonds. This is referred as argentation. TLC, poly-E- caprolactum TLC sheets are particularly versatile and do not require activation. It is possible to determine the role of adsorbent by several factors like surface area, chemical composition and geometrical arrangements of atoms or groups at the surface. The stronger adsorbent naturally has more polar structure. In addition their capacity to take part in hydrogen bonding also plays an important role. It has been found that the strength of adsorption of polar groups on polar compounds increase in order of –CH=CH > OCH_3 > COOR > =C=O.CHO > SH > NH > OH > COOH.

Weekly polar compounds are chromatographed on a polar adsorbent and vice versa.

In addition to silica and alumina, a wide variety of material of varying activity find use is adsorbents. Some are regarded as acidic (silica) while other are considered as basic (alumina) which can be prepared both in acidic and basic forms.

Table 5.1: Stationary Phase Materials for Adsorption Chromatographs

Material	*Nature of Active Sites*	*Separation Applications*
Alumina	Acidic and basic	Sterols, dyestuffs, vitamins, esters inorganic compounds
Silica gel	Acidic	sterols Amino acids
Carbon	Neutral and acidic	Peptide, carbohydrate and amino acid
Magnesia	Basic	Similar to alumina
Magnesium silicate	Acidic	Sterols esters glycerides alkaloids
Calcium carbonate	Neutral	Carotenoids xanthophylls
Aluminium silicates	Acidic	Sterols
Starch	Neutral	Enzymes

5.4 Solvents

The mobile phase in chromatography can be any organic solvent that is pure enough. The choice of solvent depends on the polarity of the compounds to be separated and their distribution coefficient. Low polarity solvents (such as hexane, heptane, toluene, acetonitrile, diethyl ether, and chloroform) are commonly used for compounds that are non-polar or weakly polar. Intermediate polarity solvents (such as ethanoic acid, dichloromethane, and pyridine) are used for compounds with intermediate polarity. Highly polar solvents (such as propan-1-ol, butan-1-ol, acetone, ethanol, and methanol) are used for compounds that are highly polar. In gradient elution, a mixture of miscible solvents with gradually increasing polarity is used. This allows for better separation of compounds with a wide range of polarities.

For silica, the eluting power of a series of solvent decreases in the order

Pure water > methanol > ethanol > n-propanol > acetone > ethyl acetate > diethyl ether > Chloroform > dichloroethylene > benzene > toluene > trichloroethylene > carbon tetrachloride > cyclohexane > hexane

5.5 Limitations: The phenomenons of adsorption sites are not simple or haphazard. The sample molecules are not adsorbed by all parts of the surface at "active sites". The true nature of active sites is not clearly understood. We do not know that they are dependent on local charge created by the surface molecules themselves. They are also affected by the dimensions of the crystal lattice that make up the surface. In general, the closer the fit between the crystal lattice size and dimensions of sample molecules, the greater will be the affinity between the two and adsorption effects will be increased. The active sites, however attract not only the sample molecules but also solvent (mobile phase) molecules. Consequently, there is competition between the solvent molecules and the sample molecules for the active sites. Generally there is a large excess of solvent molecules present and because of mass action effect, the solvent molecules are able to displace the sample molecules from the active sites. By choosing a solvent with low affinity for the active sites, the sample molecules are permitted to remain on the surface of stationary phase for an accepted length of time, but if the solvent has high affinity for active sites, the sample molecules are quickly displaced and flow rapidly through the column. As can be imagined, the correct choice of solvent and solid substrate is critical in liquid–liquid chromatography.

Another problem peculiar to solid substrate is their loss of capacity with time to handle sample of suitable size. The capacity of the stationary phase is directly proportional to the number of active sites on its surface rather than the gross surface area. When the number is high, the capacity is high and vice versa. In practice there are often low concentrations of impurities present in the solvent that have a high affinity for the active sites. They are not normally displaced once they become adsorbed. The active sites lost to the system and gradually as all the active site become covered with impurities, the surface become ineffective as a stationary phase. As a result chromatographic separation se[eration are no longer possible.

5.5 Application of Adsorption Chromatography

Liquid-solid chromatography is a useful technique for separating high molecular weight compounds. This includes organic reaction residues, distillation residues, and compounds that may isomerize or decompose in gas chromatography. These compounds cannot be analyzed by gas chromatography because their boiling points are too high or they are unstable. To establish the identity or non-identity of two substances, the two substances can be mixed and their solution can be poured over a critical adsorbent column. The different compounds will be separated based on their affinity for the adsorbent, and the different bands of compounds can be identified.

5.5.1 Application of Adsorption Chromatography in Pesticide Residue

The cleanup procedures play a vital role in estimating pesticide residues in a matrix where adsorption chromatography is used. The extraction method, cleanup method, and measurement technique must all be considered in any pesticide residue analysis. The nature of the material being extracted and the pesticide(s) being sought must also be taken into account. Anhydrous sodium sulfate is commonly used in cleanup columns.

Tables 5.2 list the adsorbents which have been used most extensively for residue analysis in combination with the pesticides. The compounds marked (x) are those which have been recovered in amounts of at least 70 per cent from a fortified control extract.

Table 5.2: Suitability of Six Adsorbents for Clean Up of Extracts Containing Organo-Chlorine Pesticides[a]

Pesticide	*$A1_2O_3$*	*$A1_2O_3$-PE[b]*	*SiO_2*	*Florisil*	*C*	*MgO*
Aldrin	x		x	x	x	x
Chlorbenside				x		
Chlordane	x	x	x	x	x	x
DDD	x	x	x	x	x	x
DDE	x	x	x	x	x	x
DDT	x	x	x	x	x	x
Dichloran				x		
Dieldrin	x	x	x	x	x	x
Dyrene	x					

Pesticide	$A1_2O_3$	$A1_2O_3$-PE[b]	SiO_2	Florisil	C	MgO
Endosulfan			x	x	x	
Endrin	x	x	x	x	x	
Heptachlor	x		x	x	x	x
Heptachlor epoxide	x		x	x	x	
Lindane		x	x	x	x	x

a: The adsorbent is listed as suitable (x) only if at least 70 per cent recoveries were obtained from fortified control extracts; b: Polyethylene-coated alumina.

Table 5.3: Suitability of Six Adsorbents for Clean-up of Extracts Containing Organo-phosphorus Pesticides[b]

Pesticide	$A1_2O_3$	$A1_2O_3$-PE[b]	SiO_2	Florisil	C	MgO
Azinphos-methyl	x	x			x	x
Carbophenothion				x		x
Chlorthion	x	x				x
Cidial	x					
Coumaphos	x	x				
Demeton-O	x				x	
Demeton-S		x			x	
Diazinon	x	x				
Dichlorvos	x	x			x	
Dimethoate		x	x		x	x
Ethion	x	x				
Malathion	x	x		x	x	
Parathion	x		x	x	x	
Phosphamidon		x				

a: The adsorbent is listed as suitable (x) only if at least 70 per cent recoveries were obtained from fortified control extracts; b: Polyethylene-coated alumina.

Chapter 6

Column Chromatography

6.1 Introduction

In chemistry, Column chromatography is a technique which is used to separate a single chemical compound from a mixture dissolved in a fluid. It separates substances based on differential adsorption of compounds to the adsorbent as the compounds move through the column at different rates which allow them to get separated in fractions. This technique can be used on small scale as well as large scale to purify materials that can be used in future experiments. This method is a type of adsorption chromatography technique. Early chromatographic columns were large glass tubes with a diameter of 40 to 50 centimeters and a length of 50 to 500 centimeters. They were packed with solid particles of the stationary phase, which were typically larger than 150 to 200 micrometers in diameter. This was done to ensure that the flow rate of the mobile phase was reasonable. The mobile phase was typically forced down the column by the head of liquid above the packing. Flow rates were typically a few tenths of a millimeter per minute, which made separations time-consuming.

It was developed by the American Petroleum chemist D.T. Day in 1900, M.S. Tswett, the polish botanist, in 1960 used adsorption columns in his investigations of plant pigments. It was not until about 1930 that the method was used extensively by chemists. Column chromatography is also known as adsorption chromatography.

6.2 Principle

The rate at which different materials are adsorbed by a given adsorbent is known to vary. This principle of selective adsorption is used in column chromatography. In column chromatography, the mixture to be separated is dissolved in a solvent and passed through a column filled with the adsorbent. The components of the mixture are adsorbed to the adsorbent at different rates, depending on their affinity for the adsorbent. The component with the strongest affinity for the adsorbent is adsorbed at the top of the column, and the component with the weakest affinity is adsorbed at the bottom of the column. The initial separation of the components can be improved by passing a different solvent through the column. The solvent will elute the components from the column at different rates, further separating them. The bands of the components in the column will become more distinct as the solvent elutes them. The column with the adsorbed components is called a chromatogram. The operation of developing the chromatogram is the process of passing the solvent through the column to elute the components. The portion of the column occupied by a particular component is called a zone. The narrower the zone, the more concentrated the component is in the eluate.

In order to separate or to estimate the various components, two procedures may be adopted

i) After the development, the column of adsorbents may be pushed out of the tube, the various zones are cut with a knife at boundaries and the substance present in zones extracted with a suitable solvent. This process of recovery of constituents from the chromatogram is known as elution.

ii) After development, the column may be washed with more solvent, now termed as eluent, and each component is collected as it reaches the end of column and is released.

When the mobile phase along with the mixture that needs to be separated is introduced from the top of the column, the movement of the individual components of the mixture is at different rates. The components with lower adsorption and affinity to stationary phase travel faster when compared to the greater adsorption and affinity with the stationary phase. The components that move fast are removed first whereas the components that move slow are eluted out last. The adsorption of solute molecules to the column occurs in a reversible manner. The rate of the movement of the components is expressed as:

Rf = the distance travelled by solute/the distance travelled by solvent Rf is the retardation factor.

6.3 Experimental Details

a) Apparatus

A simple straight glass tube tapered at the bottom and often fitted with a top is still commonly used. The ratio of length to diameter should be greater than 20:1; a ratio of 40:1 can be taken as standard. This tube is about 20-30 cm long and 2-3 cm in diameter. This may hold 50-100 g of adsorbent and may retain several grams of adsorbate. The adsorbent is supported on a plug of cotton or glass wool may be employed Figure 6.1. Long and barrow tubes are used for difficult separations. Columns fitted with cooling jackets are used when constant temperature conditions are required. Separating columns with detachable adapters are being used. In these, the adsorbent is packed between two sintered glass discs and small bore Teflon tubings is used for liquid input and output.

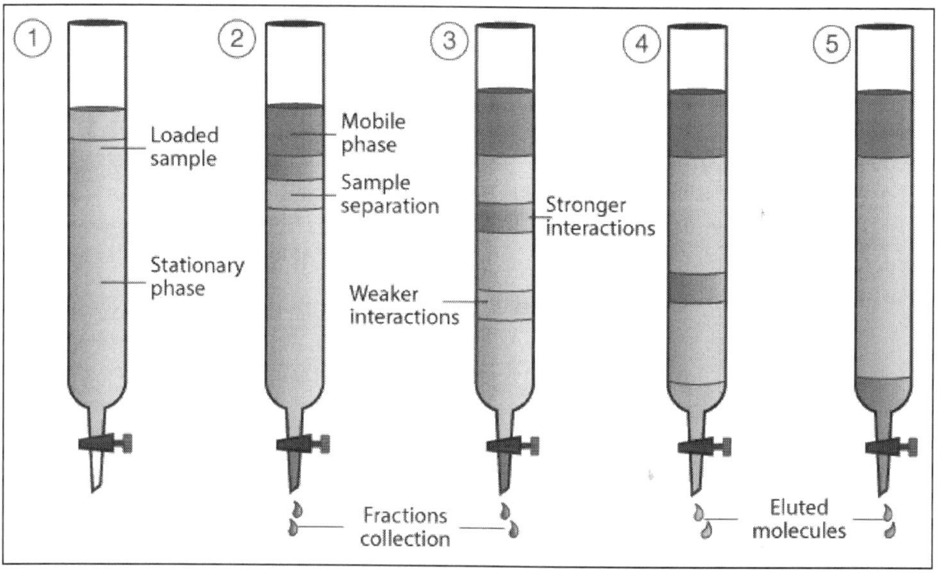

Figure 6.1: Column Chromatography.

b) Adsorbents

General requirements. In order to obtain columns with good filtration properties, the adsorbents should have following characteristics.

i) Their particles should be spherical in shape and uniform in size
ii) Their mechanical stability must be great enough to prevent the formation of fine dust which might be deposited in the channels of packing's
iii) They should not react chemically, either with the eluting solvents or with the sample components
iv) They should contain as small amount of soluble component as possible
v) They should be categorically inactive and as a rule have neutral surface. However exceptions are ion exchangers.

Table 6.1: Grades Series of Adsorbents

1. Sucrose, Starch	8. Lime
2. Inulin	9. Activated silicic acid
3. Tale	10. Activated Magnesium silicate
4. Sodium carbonate	11. Activated alumina
5. Calcium carbonate	12. Fuller's earth
6. Magnesium carbonate	13. Calcium phosphate
7. Magnesia mixtures	

c) Preparation of Adsorption Columns

The glass wool or cotton plug is used as a support for the column. It is first kept in position in tube. The tube is then clamped vertically. Now the adsorbent is added in portions so that the tube is packed uniformly with then adsorbent. Whenever any portion is added, it is pressed from above with a flattened glass before the next portion is added. This is continued nearly two-third of it is filled.

An alternative procedure involves the preparation of slurry of the adsorbent in a suitable liquid medium. The latter may be petroleum ether or sometimes as the same solvent which nis being used in chromatographic process. Now the slurry is added to the tube gradually, aliitle at time. The best method is to connect the tube to suction pump through stop-cock. Suction is applied after each addition. This process is continued till a column of desired height is obtained. It is important to remember that the surface of the column should remain covered with petroleum ether all along.

d) Solvents Used

The choice of solvents will naturally depend on the first palce upon the solubility. The solvents generally employed possess boiling points between 40oC and 85oC. The most widely used medium is light petroleum (b.p not aove 80oC), others are cyclohexane, carbon dioxide, chloroform, carbon tetrachloride, methylene chloride, ethyl acetate, ethyl alcohol, acetone ether and acetic acid. The solvents used as mobile phase can be arranged according to their eluting powers (Table 6.2). Thus eluotropic series are obtained which vary with type of adsorbent used. Mixtures of solvents are included in the term mobile phase.

Table 6.2: Solvents Used as Mobile Phase

Solvents	$€_{20}$
Petroleum ether	1.90
Cyclohexane	2.02
Carbon tetrachloride	2.24
Trichloroethylene	3.40

Solvents	ϵ_{20}
Benzene	2028
Chloroform	4.81
Absolute alcohol	4.34
Ethyl acetate	6.11
Pyridine	12.40
Acetone	21.40
n-Propanol	21.80
Ethanol	25.80
Methanol	33.60
Water	80.40
Formamide	>84.0

The solvents used in chromatography have three functions to perform

i) They serve to introduce the mixture to the columns

ii) They effect the process of development by which the zones of chromatogram are separated to their fullest extent. When used for this purpose, the solvent are termed as developers

iii) They are also used to remove the required content of each zone from the mechanically separated parts of the column or from the column as whole after it is partly developed. The solvents for this purpose are called as eluants. A good eluent must be liquid which can dissolve readily the various components from the column. While the same solvent may serve throughout. It is often necessary to use different solvents at different stages of the chromatographic separation.

e) Successive Elution

For complex samples, it may be necessary to use a series of solvents with increasing polarity to elute the different compounds. A solvent with low polarity will dissolve and elute only the least polar compounds, such as hydrocarbons. The more polar compounds will remain undissolved and close to the top of the column. After the least polar compounds have been eluted, a more polar solvent can be used to elute the next group of compounds, and so on. For example, the first solvent used in the analysis of a complex mixture of compounds might be heptane. This will remove the paraffins from the sample. The paraffins can then be further separated by gas chromatography or another suitable technique. A different solvent such as benzene can then be used to elute the aromatic compounds from the sample. After all the aromatics have been removed, benzene can be replaced by a more polar solvent such as diethyl ether. Each change in solvent will elute a different group of compounds from the sample, based on their polarity. The different fractions of compounds that are eluted from the column can then be studied further by chromatography, spectroscopy, or other analytical techniques. This allows for the qualitative and quantitative analysis of the compounds in the sample. The polarity may be expressed as solvent strength. A list of common solvent and their strength on alumina is listed in Table 6.3.

Table 6.3: Solvent Strength of Common Solvents

Solvent	*Solvent Strength Parameter*
n-heptane	0.00
Cyclohexane	0.04
Carbon tetrachloride	0.18
Toluene	0.29
Benzene	0.32
Ethyl ether	0.38
Chloroform	0.40
Tetrahydrofuran	0.45
Acetone	0.56
Methyl acetate	0.60
Acetonitrile	0.65
i-Propanol	0.82
Ethanol	0.88
Methanol	0.95
Formic acid	large

f) Gradient Elution

In traditional chromatography, the mobile phase is changed abruptly from one solvent to another. This can lead to problems, such as difficulty in collecting fractions near the interface between the two solvents. Gradient elution is a technique that solves this problem by gradually changing the composition of the mobile phase over time. This is done by mixing two solvents that are miscible, but have different dielectric constants. The first solvent, which has lower solvent strength, is used as the mobile phase initially. After a suitable period of time, a small amount of the second solvent is introduced. The concentration of the second solvent in the mobile phase then gradually increases until it reaches 100 per cent. This gradient in solvent composition leads to a progressive change in the solvent strength and solubility of the sample components. This allows for better separation of the components, even those with very different properties. Any two miscible solvents can be used for gradient elution, but solvents with different properties are often selected to achieve the best separation. Some popular pairs of solvents include water-methanol, water-acetonitrile, and dioxane-tetrahydrofuran.

g) Detectors

These are used to monitor and determine quantitatively the dissolved substances emerging from the column. The various detectors are as follows

1) Optical detector. These are old type of flow analysers which are small cell made from glass or quartz and are used for continuous photometric analysis with the visible or UV light of appropriate wavelengths. If the sample do not exhibit appreciable absorption in the cited region, reagents can be added to produce colour reactions which can be measured photometrically.
2) Differential refractometers: Claesson has employed the refraction of emerging power for the detection. This method has been improved to a differential refractometer having a sensitivity (limit of detection 10^6 g)
3) Detector based on heat of absorption: These detectors are known as micro-absorption detectors. In these, liquid emerging from a separating column is passed through two cells, one located on the top of the other, while the lower cell is filled with a adsorbent. In the centre of packing of each of the cell, the glass-covering measuring point of small thermistor is located. The total deflection (+ve and- ve peak) is proportional to the concentration at least to a range of 10^2.
4) Flame ionization Detectors: In these detectors there is metal wire which is to column exit. The decomposition product of the substances, which are transported by the wire, are led to flame ionization or argon detector.
5) Conductivity detector: These are suitable for ionized substances in aqueous solution. The effluent is passed through the measuring cell of the detector which contains two or three platinum electrodes within a wheatstone bridge circuit and is operated by A.C.

6.4 Theory of Development

If a solution of concentration "C" is passed through an adsorbent column in such a manner that *n* grams get adsorbed per gram of the adsorbent, then n may be given as follows

$$n = f(C) \tag{6.1}$$

When only one gram column is considered, then only pure solvent will be flowing until n gram of the solution get adsorbed on the column. This volume of sovent is known as retention volume R and is given by

$$R = \frac{f(C)}{C} \tag{6.2}$$

The retention volume has been found to be independent of the concentration of the solution

i) Column Efficiency

When a solute travels down a column, its band broadens due to kinetic and flow effects. This results in a Gaussian concentration profile within the band. The width of the base of the peak is typically measured as four standard deviations (4σ), and the extent of peak broadening is quantified by the variance (σ^2). For a symmetrical Gaussian peak, the standard deviation σ is equal to half the width of the peak at the inflection point (0.607). However, in some cases, the peak may be asymmetrical, with either a fronting (extended front) or tailing (extended tail) appearance. Peak asymmetry can be caused by a number of factors, including overloading the column, poor column packing, poor sample application, or solute-support interactions.

The success of any chromatographic procedure is measured by its ability to separate completely (resolve) one compound from a mixture of similar compounds. Peak resolution (R_s) is related to properties of the peaks

$$Rs = \frac{2(t_R - t_{Rs})}{W_a + W_b} \tag{6.3}$$

t_R and t_{Rs} = retention time of compounds A and B respectively

w_a, w_b = the base width of the peaks for A and B

It can be shown that when R_s= 1.5 the separation of the two peaks is 99.7 per cent complete. In more practical cases, R_s value of 1.0 corresponding to 98 per cent separation is satisfactory. According to definition of resolution R_{s-1} peak fit between any two peaks in question. Peak resolution is determined by three factors, *Selectivity* which measures the discriminatory power of the system, *retention* which measures the retentive power of the system and which is related to the partition coefficient, K_d of the compounds and efficiency which measures the relative narrowness of the peaks in terms of the ratio of retention time to peak width.

Not uncommonly in chromatography it is not possible to completely resolve two peaks. As a result they produce fused peaks and for this analysis the assumption has to be made that the peak characteristics of the two unresolved compounds do not affect each other.

The number of theoretical plates (Plate number) (N) involved in the elution of a particular compound is given by

$$N = 16\left(\frac{t_r}{W}\right)2 \tag{6.4}$$

$$N = 5.54\left(\frac{t_r}{W_b}\right)2 \tag{6.5}$$

W = the peak base width and is equal to 4 σ

W_h = the peak width at half height and is equal to 2.35 σ

The value of N can therefore vary easily calculated from the data. The plate number can be increased by simply increasing the column length (N) but there is limit to this since retention time and peak width increases proportionally with L whilst the peak height deceases as square root of N. Whilst N is measure of efficiency of column, the plate height, which is also called height equivalent to a theoretical plate (HETP) is useful for comparative purposes such as operating the column under different conditions. It can be shown that

$$HETP = \frac{L}{N} = H \tag{6.6}$$

The maximum number of peaks that can be separated by a specific chromatographic system is called peak capacity (n). It is related to the retention volume of the first and last peaks (Vα and V_w respectively) and to plate number

$$n = \frac{1 + N}{16\left(\frac{\ln V_w}{V_\alpha}\right)} \tag{6.7}$$

Peak capacity is determined by acceptable separation times and by detection sensitivity. In practice it can be increased either by procedure of gradient elution as in liquid chromatography or by temperature programming as in gas-liquid chromatography.

Effect of particle size of packing and flow rate upon plate height H for solute N, N-diethyl p-aminobenzene and mobile phase which is mixture of hexane, methylene chloride and isopropyl alcohol.

Factor Affecting Column Efficiency

i) *Nature of Solvent*: Solvent of low viscosities are generally used for high efficiency separations. The reason for this is that rate of flow is inversely proportional to viscosity and hence it becomes necessary to select a solvent of lowest viscosity and proper elution strength.

ii) *Dimension of columns*: It is possible ti improve the column efficiency by increasing the length width ratio of the column. For common preparative separations, sample/column packing ratios have found to range from 1:20 and 1:100

iii) *Particle size of column packing*: It is possible to increase column efficiency by decreasing particle size of the adsorbent. The usual particle size ranges from 100 to 200 mesh.

iv) *Pore diameter of column packing*: Generally polar adsorbents possess a pore diameter of ≤20 Å. According to Kleen a decrease in average pore diameter from 170-200 Å does not affect efficiency

v) *Temperature of column*: Difficult soluble samples are generally separated at higher temperature while other samples are separated at room temperature.

6.5 Application of Column Chromatography

a) Analytical Uses

Glass or copper capillaries with an internal diameter of 0.05 to 2 mm and a length of 1 to 20 m are used for analytical purposes. The narrow tubing's inner surface serves as a support for the liquid phase or as an adsorbent. Glass capillaries whose inner surfaces have been treated with concentrated ammonia at 300 degrees Celsius can be used to analyze amino acids using a mobile phase of butane, pyridine, and dilute acetic acid (in a 5:5:1 ratio), or xylose, glucose, and maltose using a mobile phase of butanone, acetic acid, and water (in a 3:1:6 ratio). Vestergaard and Sayegh could operate seven urinary steroids within 5 hours which requires 36 hours on normal column. They have used narrow Teflon packed with aluminium oxide silica gel. An optical detector used for the analysis of steroids by gradient elution using acetone in chloroform

b) Separation of Geometrical Isomer

Winterstein was the first to chromatographically separate cis/trans isomers of bixin and crocetein dimethyl ether. Later, Zechmeister separated cis/trans isomers of carotenoids on calcium carbonate and alumina columns, including trans-cis-trans and trans-cis-cis-trans isomers. Cis/trans isomers of carboxylic acids have also been separated on charcoal and silica gel columns. The separation of cis/trans isomers is based on steric factors. Isomers whose functional groups can more easily approach the surface of the adsorbent are more strongly adsorbed.

c) Separation of Diasteromers

In some cases, a derivative having an optically active partner cannot be separated from the latter. However such application have been realized by column chromatography on various adsorbents. For example, the separation of diasterometric 7-chloro-azibiccyclo(4:1-heptane has been done on silica gel using heptane/diethyl ether as solvent

d) Separation of Tautomeric Mixtures

As the separation of tautomeric mixtures requires high temperature gas chromatography is rarely used. However the separation of thsese cpuld be done by column chromatography. For example the keto and enol-forms of p-hydroxyl-phenyl pyruvic acid and indol pyruvic acid could be separated in the liquid phase. The separation is done in weakly acidic medium. The enol form appears in elute before the keto form

e) Separation of Racemates

The first successful seperations of racemates using organic solvents were achieved on lactose the flamous separation of enantiomers of Trogers base.

Chapter 7

Gel Chromatography or Gel Permeation Chromatography

7.1 Introduction

Gel Chromatography (also known as gel permeation, molecular sieving or size exclusion chromatography) is a chromatographic technique in which the separation of components is based on the difference of molecular weight or size, and is one of the effective methods used to isolate and analyze the bio-macromolecular substances. Gel permeation chromatography or exclusion chromatography uses porous material as the stationary phase and a liquid as a mobile phase. The diameter of the porous of the porous material are the order of 50-3000 Å which is similar to the size of many molecules. The latter penetrate the pores according to their size. Small molecules penetrate more rapidly than larger molecules, which frequently are excluded from the smaller pores present. This results in a difference in the rates at which the molecules pass down the column, the smaller molecules travelling faster than the than the larger one. Selectivity based on steric factors between the adsorbate and the adsorbent is utilized in achieving specific separations through gel chromatography. The technique is known as gel filtration, molecular sieve filtration and exclusion chromatography. Gel chromatography is a simple and reliable method for separating molecules according to size. Its versatility makes it generally applicable to the purification of all, classes of biological substances, including giant macromolecules not readily fractionated by other technique. Good separation and high activity yields are easily obtained.

In gel permeation separation takes place according to size of the solute molecule. Exclusion takes place according to molecular size as depicted in Fig 7.1. In molecular filtration separation are carried out on natural and synthetic zeolite (metal-aluminium silicates) called molecular sieve. A typical zeolite or molecular sieves have a general formula $M_{2/n}$. Al_2O_3 x SiO_2. yH_2O where M is cation of valency *n*. The molecular sieve contains permanent cavities and channels in their structure. The sieve properties of the zeolite is determined by the size of these cavities, while adsorptive properties of the zeolite are determined by the large surface area and presence of M, AL, Si and O in the zeolite structure.

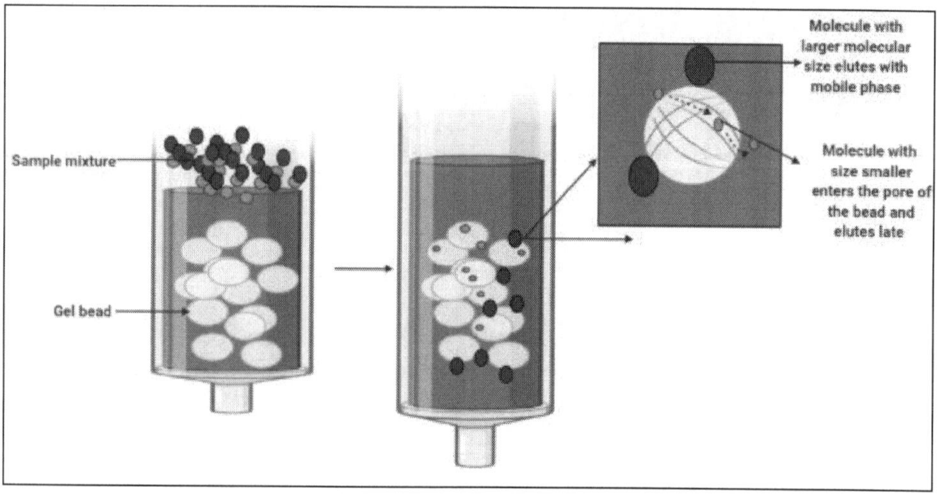

Figure 7.1: Gel-Permeation Chromatography.

7.2 Principle

Size exclusion chromatography separates molecules based on their size and shape using porous materials, such as gels or porous glass. This technique is also known as gel filtration or controlled-pore glass chromatography. Gels and porous glass have a network of pores of different sizes. When a sample is passed through a column of these materials, larger molecules are excluded from the pores and elute from the column first. Smaller molecules, on the other hand, can enter the pores and are retained for a longer time, eluting from the column later. The separation of molecules by size exclusion chromatography is based on the following principles:

- Large molecules are excluded from the pores. Molecules that are larger than the pores of the stationary phase cannot enter the pores and are eluted from the column first.
- Smaller molecules enter the pores. Molecules that are smaller than the pores of the stationary phase can enter the pores and are retained for a longer time, eluting from the column later.

- ☆ Molecules are distributed between the mobile phase and the pores. Molecules that enter the pores are distributed between the mobile phase inside the pores and the mobile phase outside the pores. This distribution is dependent on the size of the molecule and the size of the pores. Size exclusion chromatography is a versatile technique that can be used to separate a wide range of molecules, including proteins, polymers, and carbohydrates. It is a mild technique, so it does not damage the molecules being separated. For a given type of gel, the distribution coefficient, K_d, of a particular solute between inner and outer solvent is a function of its molecular size. If the solute is larger and completely excluded from the solvent within the gel, $K_d = 0$, whereas if the solvent is sufficiently small to gain complete accessibility to the inner solvent, $K_d = 1$. Due to variation in pore size for a given gel, there is some inner solvent which will be available and some which will not be available to m solute of intermediate size, hence Kd values vary between 0 and 1. It is this complete variation of K_d between the two limits which makes possible separation of the solute within a narrow molecular size range on a given gel. The elution volume, V_e of a given solute depends on the void volume, V_o, the distribution coefficient, and the volume inside the gel matrix itself Vi

$$V_\alpha = V_\alpha + K_d V_i \tag{7.1}$$

The inner volume, V_i can be calculated from the known dry weight of the gel, α and the water regain value W_r

$$V_i = \alpha W_r \tag{7.2}$$

The numerical value of V_e for a given solute will vary with the size of the column, whereas K_d is a characteristic value for this solute and is independent of the geometry of the gel bed.

For two substances of different molecular weights and K_d values, $K_{d.}$ and $K_{d..}$ the difference in their effluent volume, V_s is given by

(7.3)

Therefore

$$V_s = (K_{d'} - K_{d''})V_i \tag{7.4}$$

To completely separate two substances using gel filtration, the sample volume must be smaller than the void volume (Vs). However, in practice, it is advisable to reduce the sample volume even further below Vs to account for deviations from ideal behavior, such as poor column packing. This is because the ratio of sample volume to gel volume affects the sharpness of the separation and the degree of dilution of the sample. Equations 7.2 and 7.4 can be used to calculate the optimal bed volume for a given purification. Gel filtration can also be performed in the thin layer mode (TLG). TLG is similar to thin layer chromatography (TLC), but there are some key differences. In TLG, a layer of swollen gel is spread onto a glass plate. The gel beads adhere to the plate without the need for a fixative and form the stationary phase. The interstitial fluid forms the mobile phase. Unlike TLC, the TLG plate must not be dried, so there is no solvent front. The TLG plate is placed in an airtight container and connected to reservoirs at either end by filter paper bridges. The plate is tilted at an angle of 20 degrees to the horizontal to facilitate the movement of the mobile phase through the layer. Equilibrium must be established for at least 12 hours before the sample can be applied. The main purpose of this equilibration is to normalize the ratio of stationary to mobile phase volumes. The sample is applied to the plate as spots or bands and then developed for a suitable period of time. The separated spots are detected using an appropriate method. Whereas TLC is mainly used for the separation of amino acids, sugars and oligosaccharides, alkaloids, steroids and lyophilic substances in general TLG is used for separation of hydrophilic substance requiring mild conditions such as proteins, peptides, nucleic acids, ie higher molecular weight biological material. The great advantage of TLG over column gel filtration is that a number of samples can be chromotographed at the same time, under identical condition. In addition very small amounts of the sample can be used.

7.3 Materials

Gel filtration chromatography uses gels to separate molecules based on their size. The most common gels used are cross-linked dextrans (Sephadex), agarose, polyacrylamide, and polystyrenes. Sephadex gels are made by cross-linking the polysaccharide dextran with epichlorhydrin. They are hydrophilic and swell rapidly in aqueous media, forming gel particles. Sephadex gels are available in a range of porosities, which makes them useful for separating molecules of different sizes. Agarose gels are made from agar, which is a linear polysaccharide consisting of alternating residues of D-galactose and 3,6-anhydro-galactose units. Agarose gels are also hydrophilic and cause very little denaturation or adsorption of sensitive biochemical compounds. They are more porous than Sephadex gels, and can be used to separate larger molecules, such as viruses, nucleic acids, and polysaccharides. Polyacrylamide gels are made by polymerizing acrylamide and methylene bisacrylamide. The porosity of polyacrylamide gels can be varied by changing the relative proportions of the two monomers. Polyacrylamide gels have similar characteristics to Sephadex and agarose gels, and can be used to separate molecules with a molecular weight range of 1800 to 400,000 daltons. Gel filtration chromatography is a versatile technique that can be used for a variety of purposes, including:

- Separation of proteins and peptides: Gel filtration chromatography is a commonly used technique for separating proteins and peptides based on their size.
- Size-exclusion reaction chromatography (SEC): SEC is a type of gel filtration chromatography that is used to determine the molecular weight of polymers.
- Separation of nucleic acids and nucleotides: Gel filtration chromatography can also be used to separate nucleic acids and nucleotides based on their size.
- Virus particle separation: Gel filtration chromatography can be used to separate virus particles from other components in a sample.
- Endotoxin removal: Gel filtration chromatography can be used to remove endotoxin, a type of bacterial toxin, from samples.

Different types of gels are used for different applications:

- Sephadex gels are commonly used for the separation of globular proteins and randomly coiled polymers.
- Agarose gels are commonly used for the separation of viruses, nucleic acids, and polysaccharides.
- Polyacrylamide gels are commonly used for the separation of proteins and peptides.

When choosing a gel, it is important to consider the following factors:

- Molecular weight range of the molecules to be separated: The porosity of the gel should be appropriate for the size of the molecules to be separated.
- Sample volume: The capacity of the gel should be sufficient to accommodate the sample volume.
- Resolution: Finer gels provide better resolution, but they also have slower flow rates.

Gel filtration chromatography is a powerful technique that can be used to separate and purify a wide variety of molecule.

Some commonly used gels are listed in Table 7.1. The Sephadex and poly acrylamide gels must be converted to their swollen form before use whereas the Sephacryl and agarose gels are supplied in the pre-swollen forms. Many gels are available in a range of size: superfine, fine, medium and coarse. The coarser the bed, the better column flow rate but poor resolution it gives. Thus the fine and superfine beads are preferred for analytical work and the coarser ones for preparative purposes. The capacity of a particular gel is a measure of the weight of a solute that can penetrate a particular weight of gel. It therefore gives an indication of the amount of a solute that can be separated by a particular column or gel.

Table 7.1: Some Commonly Used Gels for Exclusion Chromatography

Polymer	*Trade Name*		*Fractionation Range (Daltons)*	*Bed Volume (cm^3 g^{-1} dry gel)*
Dextran	Sephadex	G10	<700	2-3
		G25	1×10^3 to 5×10^3	2-3
		G50	1.5×10^3 to 3×10^4	4-6
		G100	4×10^3 to 1.5×10^5	15-20
		G200	5×10^3 to 6×10^6	30-40
	Sepacryl	S300	1×10^4 to 1.5×10^6	---
		S400	2×10^4 to 8×10^6	---
Agarose	Sepharose	2B	1×10^4 to 4×10^6	---
		4B	6×10^4 to 2×10^7	---
		6B	7×10^4 to $4\ 10^7$	---
	Bio-Gel	A5m	1×10^4 to 5×10^6	---
		A15m	4×10^4 to 1.5×10^7	
		A50m	1×10^5 to 5×10^7	---
		A150m	1×10^6 to 1.5×10^8	---
Polyacrylamide	Bio-Gel	P2	1×10^2 to 1.8×10^3	---
		P6	1×10^3 to 6×10^3	---
		P30	2.5×10^3 to 4×10^4	---
		P100	5×10^3 to 1×10^5	---
		P300	6×10^4 to 4×10^5	---

7.4 Gel Preparation, Column Packing and Detectors

The methods which are generally used for the preparation of gel are

1. The dry powder of the gel is allowed to swell in the solvent which is to be used as eluent. In actual practice a weighed amount of the dry powder of the gel is mixed with solvents and allowed to swell. The mixture is kept until the equilibrium is established.
2. In the second method gel slurry is armed at about 100°C in a water bath. As a result the gel swells in a few days. The Slurry is cooled and packed in the column

There are generally two types of column packing's; porous glasses or silica's and the porous cross linked organic gels such as dextrans, methacarylate-based gels, polyvinyl alcohol based gels and hydroxethyl cellulose gels. The detectors used are based on UV fluorescence, UV absorption, or change in refractive index.

7.5 Applications

Gel Chromatography has been used with great success in the separation of the sugars, polypeptides, proteins, liquids, asphalts, butyl rubbers, polystyrenes, silicon polymers and others. The extensive analytical applications of GPC cover both organic and inorganic materials. Gel permeation chromatography (GPC) has been applied to studies of complex, biochemical or highly polymerised molecules.

The main application of gel chromatography is the separation and characterization of molecules of different of molecular weights. Very often it becomes possible to separate molecules of similar molecular weights by appropriate gel and column length. Another important application is the separation of large molecules of the biological origin from inorganic and ionisable species. This termed as desalting. For example, employing a column of Sephadex gel, haemoglobins can be separated from sodium chloride. Fractionation of mixtures of biopolymers is the most widespread applications of gel chromatography. The method can also be used for preparative purposes. However the main application of gel chromatography is as follows

a) The main application of exclusion chromatography is in the purification of biological macromolecules. Viruses, proteins, enzymes, hormones, antibodies, nucleic acids, and polysaccharides have all been separated and purified by the use of appropriate gels or glass granules. Mixtures of lower molecular weight compounds may also be separated. The amino acids can be separated from peptides, and peptides obtained from the partial hydrolysis of a protein can be fractionated as can oligonucleotides from nucleic acid hydrolysates. Low molecular weight dextrans in such mixture as corn syrup oil can also be separated.

b) Molecular weight determination. The effluent volumes of globular proteins are largely determined by their molecular weights. It has been shown that over a considerable weight ranges, the effluent volumes is approximately a linear function of the logarithm of the molecular weight. Hence the construction of the calibration curve, with protein of a similar shapes and known molecular weight, enables of a molecular weight of other protein, even in crude protein preparations to be estimated.

c) Solution Concentration: Solutions of high molecular weight substances can be concentrated by the addition of dry Sephadex G-25 (coarse). Water and low molecular weight substances remain in solution. After ten minutes the gel is removed by centrifugation, leaving the high molecular by weight material in solution whose concentration has increased but whose pH and ionic strength are unaltered.

d) Desalting: By use of a column Sephadex G-25 solution of high molecular weight compounds may be desalted. The high molecular weight substances move with the void volume while the low molecular weight components are distributed between mobile and stationary phase and hence move slowly. This method of desalting is faster and more efficient than dialysis. Applications include removal of phenols from nucleic acid preparations, ammonium sulphate from protein preparation as well as monosaccharaides from polysaccharides and amino from proteins

e) Protein-Building studies: Exclusion chromatography is on one of a number of methods commonly used to study the reversible binding of a ligand to a macromolecules such as proteins including receptor proteins. A sample of the protein/ligand mixture is applied to a column of a suitable gel (G-25) which has previously been equilibrated with a solution of the ligand of the same concentration of ligand and protein in the effluent determined. The early fraction will contain unbound ligand, but the subsequent appearance of the protein will result in an increase in the total amount of ligand (bound plus unbound). By repeating the experiment at a series of ligand concentrations the appropriate binding constant can be calculated.

The chromatogram obtained results from a separation based on the size of the sample molecules, which are directly related to molecular weights of the components. If the sample is single polymer, the chromatogram represents the molecular weight distribution. This method is very valuable for determining the molecular weight distribution of polymers up to very high molecular weights. The actual calculations are usually based on a comparison with standard polymer compounds such as proteins. Gel permeation conditions, mixture of polymer can be characterized with respect to the percentages and weight range of each polymer present.

7.6 Pesticide Residue Analysis using Gel Permeation Chromatography

Extraction of pesticide residues from samples produces complex mixtures that often require sample purification and preparation steps to isolate the targeted pesticides for analysis. In addition, a multi-residue analytical strategy is often necessary to facilitate the quantitative determination of individual pesticide residues due to the differences among their chemical and physical properties and incompatible detection techniques. The cost of labour and materials, and long turnaround times could be significantly reduced if sample preparation and cleanup procedures are performed by automated methods. Many traditional procedures used for extraction of fatty samples are time consuming and solvent intensive. The more widely-used extraction techniques for pesticides are Soxhlet extraction (SOX), shake-flask, liquid–liquid extraction (LLE), solid-phase extraction (SPE), solid-phase micro-extraction (SPME), matrix solid-phase dispersion (MSPD), supercritical fluid extraction

(SFE), accelerated solvent extraction (ASE) and microwave-assisted extraction (MAE). ASE is an extraction technique that speeds up the extraction process and reduces the total amount of solvent used. The system uses conventional solvents at elevated temperatures and pressures, which results in improved extraction kinetics to achieve quantitative extraction from solid and semisolid samples in a short time with a small amount of solvent. This method has been accepted as the standard method for solid sample extraction by U.S. EPA Method 3545. The extraction step is usually followed by a cleanup procedure. The gel permeation chromatography (GPC), first used in 1972 by Tindle and Stalling to separate pesticides from fish lipids using polystyrene gel, Bio-Beads SX-2, and cyclohexane solvent. Another system was further developed in 1974, using Bio-Beads SX-3 and toluene–ethyl acetate for quantitative recovery of non-ionic chlorinated pesticides and polychlorinated biphenyl compounds. Gel Permeation Chromatography (GPC) appears to be the best suited technique to multi-residue analysis as it affords cleanup of both polar and non-polar pesticides with a single injection on a fully automated system. Therefore, GPC is now considered a universal cleanup procedure for separating pesticides from lipids, waxes and other low volatile large non-polar co-extractives. It offers the best protection for GC capillary columns. The volume of organic eluents used is reduced considerably by miniaturization. Acetonitrile is now adopted as the extracting solvent and, after salting-out, cleanupis being performed by GPC with a graphitised carbon column, silica gel and Florisil cartridges in series to detect 87 pesticides in agricultural products by GC–MS.

The pesticides extracted by supercritical fluid extraction (SFE) with collection using Extrelut, defatted with Extrelut + C_{18} and cleaned up by using gel permeation chromatography (GPC) and a Sep-Pak Florisil cartridge. The test solution are subjected to HPLC with ultraviolet (UV) and fluorescence (FL) detection, GC with flame photometric detection (FPD), hydrogen flame ionization detection (FID) or electron-capture detection (ECD) and GC/MS selected ion monitoring (SIM) detection. It has been proposed that simultaneous and consecutive analytical methods for 63 organophosphorus pesticide residues in vegetable and fruit. Extraction of the sample with acetonitrile followed salting-out. Co-extractives were removed by GPC and the pesticide eluate was separated to two fractions. Firstly, the second fraction (85–125 mL) will be passed through a silica-gel mini-column. After putting a Florisil mini-column before the silica-gel mini-column, the first fraction (55–85 mL) is passed through the tandem mini-column. The eluate are being subjected to dual-column GC with dual FPD and nitrogen-phosphorus detection (NPD) system. Some investigators use the method for determining 55 organophosphorus, organochlorinated compounds, and pyrethroids commonly used in crop protection. Pesticide residues are extracted from samples with a mixture of ethyl acetate and sodium sulfate. No additional clean-up steps were necessary. Analysis is performed by gas chromatography by using a combination of positive chemical ionization and electron impact ionization modes and tandem mass spectrometry. There are few methods used for the determination of 251 pesticide and degradation product residues in fruit and vegetable samples. Extraction of the sample with acetonitrile is followed by a salting-out step. Co-extractives were removed by passing a portion of the acetonitrile extract through an octadecyl (C_{18}) solid-phase extraction cleanup cartridge and then, in a second cleanup, through a carbon cartridge coupled to an amino propyl cartridge. Determination is by gas chromatography with mass-selective detection and by liquid chromatography with post-column reaction and fluorescence detection for *N*-methylcarbamates. Some Scientists proposed analytical methods for 385 pesticide residues in vegetable and fruit. Extraction of the sample with acetone is followed by salting-out. Sample extracts are cleaned up with modified multimethod S19. The test solution is subjected to gas chromatography with atomic emission detection or gas chromatography with mass spectrometric detection. Some workers used C_{18} and Florisil double solid-phase extraction (SPE) columns for cleanup and GC with electron-capture detector for determination of pesticide residues of 21 organochlorines, 6 pyrethroids, and 7 PCBs. It has been proposed that C_{18} and active carbon double SPE columns be used for respective determination of multi-organophosphorus and multi-organochlorines in milk and eggs. Alcohol + ethyl acetate (5 + 95) can be adopted for extracting pesticide residues in milk samples, C_{18} and Sep-Pak Aminopropyl series SPE cartridges for clean up and GC-ECD, -FPD, and -MS and LC-fluorescence detection for the determination of 59 pesticide residues of organochlorine compounds, organophosphorus compounds, and carbamates. Few investigators put forward a proposal that matrix dispersive technique and GC-ECD can be utilized for determination of 22 organocholorine compounds and 6 PCBs in milk. Alcohol + ethyl acetate (5 + 95) mixing solvent for extraction and GPC for cleanup of 43 organophosphorus compounds, 17 organochlorine compounds, and 11 *N*-methylcarbamates in animal tissues, the former two of which were determined respectively by GC-FPD and GC-ECD and the latter one by LC-fluorescence detection. The Scientific literature introduced pressurized liquid extraction (PLE), GPC and silica-gel mini-coulumn cleanup and GC-FPD, -ECD, -NPD, and LC-fluorescence detection for the determination of 220 pesticide residues in plant original foods and animal original foods. Acetone + light petroleum (1 + 1) can be used for mixing solvent for extraction, Florisil double columns cleanup and GC-ECD determination of nine pyrethroids in animal tissues.

It has been established a new method for the determination of 437 pesticide residues in animal tissues by gas chromatography–mass spectrometry (GC–MS) and liquid chromatography–tandem mass spectrometry (LC–MS–MS). The samples can be extracted with cyclohexane + ethyl acetate (1 + 1), GPC cleanup and compared to the methods in the literature; the pesticide varieties for determination have been added in the new method. The analytical procedures have been simplified and modes of detection have been changed so that the accuracy and reliability of the method can be improved.

Chapter 8

Ion Exchange Chromatography

8.1 Introduction

For many years, scientists have known that clays contain metal ions and that these ions can exchange with other metal ions in their vicinity. This observation was important to understanding soil chemistry. Zeolites, which are sodium aluminium silicates, have some similar properties to clays. Because of this, natural and artificial zeolites can be used to remove calcium and magnesium ions from water. The calcium and magnesium ions dissolved in the water are exchanged for sodium ions from the zeolite, producing soft water. Ion exchangers can also be created by combining a polymer with a functional group. These ion exchangers have properties very similar to those of zeolites. There are two types of ion exchangers: cationic and anionic. Both types use a resin as the insoluble inert support, but they use different functional groups to provide the desired type of exchanger. The phenomenon of ion exchange was first reported by two British agricultural chemists in 1850. They proved that soil can remove potassium or ammonium salts from water by releasing an equivalent amount of calcium salts. After the work of Gans in 1913, natural and synthetic inorganic cations with short chains containing ionisable functional groups were developed. The fundamental work on chromatographic separation by use of ion exchange resin is mainly derived from the studies carried out in connection with the work on atomic energy in the USA.

8.2 Definition

Ion exchange is a reversible process in which ions of similar charge are exchanged between a solid ion exchanger and a solution. The ion exchanger is typically made of an insoluble organic polymer with charged groups attached to it. The most common type of ion exchanger is made of a styrene-divinyl benzene copolymer. Copolymers of acrylic acid derivatives and divinylbenzene are also commonly used. For use with biological macromolecules, charged groups are often introduced into cellular fibers. Cross-linked dextran (Sephadex) has also been used as a carrier material for several years.

The most common properties of all ion exchangers as follows

They are almost insoluble in water and organic solvents such as benzene, carbon tetrachloride, ether, *etc.*

They are complex in nature i.e in fact they are polymeric

They have active or counter ions that will exchange reversibly with the other ions in a surrounding solution without any substantial change in the material

8.3 Principle

The principal feature underlying this form of chromatography is the attraction between oppositely charged particles. Many biological materials, like amino acids and proteins have ionisable group and the fact that they carry a net positive or negative charge can be utilized in separating mixtures of such compounds. The net charge exhibited by such compounds is dependent on their pK_a and on the pH of the solution in accordance with the Henderson – Hasselbach equation.

Ion-exchange separations are typically done in a column filled with an ion exchanger. There are two types of ion exchangers: cation and anion exchangers. Cation exchangers have negatively charged groups that attract positively charged molecules. These exchangers are also called acidic ion exchange materials because their negative charge comes from the ionization of acidic groups. Anion exchangers have positively charged groups that attract negatively charged molecules. These exchangers are also called basic ion exchange materials because their positive charge typically comes from the association of protons with basic groups.

The actual ion exchange mechanism is thought to be composed from five distinct steps:

i) Diffusion of ion through exchange surface. This occur very quickly in homogenous solutions

ii) Diffusion of ion through matrix structure of the exchanger to the exchange sites. This is dependent upon the degree of cross linkage of the exchanger and the concentration of the solution. This process is thought to be the feature which controls the rate of the whole ion exchange process.

iii) Exchange of ions at exchange sites to occur instantaneously and is in equilibrium process

Cation Exchanger

RSO^{-}_{3}....... Na^{+} + $N^{+}H_{3}R^{\cdot}$ ⇌ RSO_{3}^{-}.......$N^{+}H_{3}R^{\cdot}$ + Na^{+}

Exchanger Counter Ion Charged molecule to be exchanged Bound molecular ion

Anion Exchanger

$(R)_4N^+,Cl^- + {}^-OOCR \rightleftharpoons (R)_4N^+..........{}^-OOCR + Cl^-$

The more highly charged the molecule to be exchanged, the tighter it binds to the exchanger and the less readily it is displaced by the other ions;

iv) Diffusion of the exchanged ion through the exchanger to the surface

v) Selective desorption of the eluent and diffusion of the molecule into the external solution. The selective desorption of the bound molecule is achieved by changes in pH and or/ionic concentration or by the affinity elution, in which case an ion which has great affinity for the exchanger than has the bound molecule is introduced in to the system.

The rate of ion exchange is controlled by the law of mass action. For example, in the case illustrated alone Na^+ from solution will replace H^+ from the resin. However if every strong acid is washed through the resin the high concentration of H^+ in the acid will by mass action reverse the process and replace Na^+ from the resin.

At all times the metal ion present are in competition for the functional groups. In general at all concentration the functional group sites are taken by the metal with the greatest affinity for that group. The affinity is controlled as follows

The ion with the highest charge has the affinity, that is $Na^+ < Ca^{2+} < Al^{3+}$

The ion with greatest size (and Charge) has the higher affinity that $Li^+ < Na^+ < K^+ < Cs^+ < Be^{2+} < Mg^{2+} < Cu^{2+} < Sn^{2+}$

If the ion becomes too large, distortion of the resin takes place and rules breaks down

8.4 Cation Exchangers

A cation exchanger is a high molecular weight, cross linked polymer having sulphonic, carboxylic, phenolic *etc.* groups as an integral part of the resin and an equivalent amount of cations. Thus, a cation exchanger is nothing but a polymeric anion to which active cations are attached. In these cation exchangers, the hydrogen ion s are mobile and exchangeable with other cations,. The anions (-COO^-, SO^{2-}_{3} and –O^-) remain attached to the resin network.

When a cation exchanger is kept in a solution of a salt, some of the H+ ions of the resins enter the solution and an equivalent of the cations of the salt get attached to the resin. This reaction may be represented as follows

$$\underset{\text{(Resin)}}{H_nR} + \underset{\text{(Solution)}}{nNA^+} \rightleftharpoons \underset{\text{(resin)}}{NanR} + \underset{\text{(Solution)}}{nH^+}$$

Some commercially available cations ion exchangers are given in Table 8.1

The resin having sodium ions, produced in then above reaction can exchange these ions with other cations. The reaction with calcium ions may be represented as follows

$$2Na_nR + nCa^{2+} \rightleftharpoons Ca_nR_2 = 2nNa^+$$

Table 8.1: Some Commercial Cation Exchangers

Trade Name	*Functional Group*	*Framework Material*
Amberlite IR-120	$-SO_3H$	Styrene/divinyl benzene (copolymer homoporous)
Dowex	SO_3H	–do–
Zerolit	SO_3H	–do–
Amberlite-200	SO_3H	Same framework but heteroporous
SE cellulose	$-C_2H_4-SO_3H$	Cellulose
Amberlite IRC-50	-COOH	Methacrylic acid/divinyl
CM	$-CH_2COOH$	Cellulose, fibrous

8.5 Anion Exchangers

An anion exchanger is a polymer amine or quaternary ammonium group as integral part of the resin and an equivalent amount of anions such as Cl^-, SO_4^{2-}, OH^-. The anion exchange behaviour of these materials may be represented as follows

$$\underset{\text{(Resin)}}{2RCl_n} + \underset{\text{(Solution)}}{nSO_4^{2-}} \rightleftharpoons \underset{\text{(resin)}}{R_2(SO_4)_n} + \underset{\text{(Solution)}}{2nCl^-}$$

$$\underset{\text{(Resin)}}{RCl_n} + \underset{\text{(Solution)}}{nOH^-} \rightleftharpoons \underset{\text{(resin)}}{R(OH)_n} + \underset{\text{(Solution)}}{nCl^-}$$

Some commercially available anion exchangers are given in Table 8.2.

8.6 Regeneration

Anion exchangers are typically supplied as salts, especially amines, because they are only stable in this form. Cation exchangers can be converted to the

H+ form by treating them with an aqueous acid, followed by washing out the ammonium base with hydroxyl groups from strongly basic anion exchangers using sodium hydroxide. Weakly basic anion exchangers are converted to free amines. Regardless of the form in which the resin will be used, it should always be subjected to a full cycle of charging and discharging before use. This step improves its effective capacity.

Table 8.2: Some Commercially available Anion Exchanger

Trade Name	*Functional Group*	*Framework Material*
Amberlite IRA-400	-CH2-N+(CH3)2	Styrene/divinyl benzene(copolymer homoporous)
Zerolit FF-IP	–do–	Same as above (but heteroporous)
AmberliteIRA-410	$-CH_2-N^+-CH_2-$ CH_2 / \ CH_3 CH_2	Styrene/divinyl benzene(copolymer homoporous)
Zerolit N-Ip	–do–	Same as above but isoporous
QAE Sephadex A-25	–N–	Dextran

8.7 Ion Exchange Column used in Chromatographic Separation

Ion-exchange chromatography works because different cations (positively charged ions) or anions (negatively charged ions) have different abilities to swap places with ions on the surface of a special material called an ion exchanger. This ability to swap places depends on the charge and size of the hydrated ion in solution. Ions with a higher charge or a larger hydrated size will bind more strongly to the ion exchanger. In other words, ion-exchange chromatography separates ions based on their charge and size. Ions with a higher charge or a larger hydrated size will elute (come off the column) later than ions with a lower charge or a smaller hydrated size.

If we compare other hydrated ions of the same size it is observed that the ionic charge plays an important role to determine their capacity to undergo exchange reaction. According to this view, the capacity amongst cations has been found to decrease in the following order.

$$Th^4+ > Al^{3+} > Ca^{2+} > Na^+$$

Similarly, the capacity amongst anions of the same size has been found to decrease in the following order

$$PO_4^{3-} > SO_4^{2-} > I^-$$

If we compare the ions carrying the same charge the size of hydrated ion plays an important role to determine their capacity to undergo exchange reactions. Thus amongsr the univalent cations, the capacity has been found to decrease in the following order

$$H_+ > Cs_+ > Rb_+ > K_+ > NH_{4+} > Na^+ > Li^+$$

Amongst the doubly charged cations, the capacity has been found to decrease in the following order

$$Ba^{2+} > Pb^{2+} > Sr^{2+} > Ca^{2+} > Ni^{2+} > Cu^{2+}\ Co^{2+} > Zn^{2+}\ Mg^{2+}$$

If the active polyvalent ion in a resin is to be exchanged with an ion of lower valency, the exchange has been found to be favourable per unit weight of material. Greater the number of ion, greater the capacity of the resin for the exchange process. The efficiency of the resin has been found to depend upon the degree of cross-linking *i.e.* the greater the crosss linking, the higher the efficiency of resin.

8.8 Selection of Suitable Systems

It is useful to classify ion exchangers according to the pK values of their ion groups. However Table 8.3 gives a simplified survey of their working ranges.

Table 8.3: Operating Ranges of Commonly Used Ion Exchangers

Specification of Exchanger	*Ionic Groups*	*Suitable Chromatographic Medium*
Strong acid	$-SO_4^-$	Acidic and alkaline
Weak acid	$-COO^-$	Only alkaline
Strong base	$-N^+R_3$	Acidic and alkaline
Weal base	$-NHR_2$	Only acidic

Ion exchange resins can only remove dissolved materials from a solution if the materials are in their dissociated form. This means that the molecules must have split into their charged ions. Therefore, it is important to work at a pH where the material to be removed is dissociated, while the other components of the solution are not. The concentration of the oppositely charged ions in the solution also affects the ion exchange process. The higher the concentration of these ions, the more difficult it will be for the resin to remove the desired ions. Once the desired ions have been retained by the resin, they can be eluted (removed) by using a solution with a high concentration of cations or anions. This will displace the desired ions from the resin without changing the pH of the solution, which would affect the dissociation relationships.

8.9 Ion Exchange Capacity

The ion exchange capacity (IEC) of a resin is the number of ion-active groups per gram of resin. It is expressed in milliequivalents per gram (meq/g). The IEC of a cation exchange resin is typically determined in the laboratory by measuring the amount of sodium (Na+) ions that can be adsorbed by 1 gram of the resin in the hydrogen (H+) form. To do this, a solution of sodium chloride (NaCl) of known concentration is passed through a column containing a known weight of the resin. The acid that elutes from the column is then collected and titrated with a standard solution of an alkali. The amount of Na+ ions adsorbed by the resin is calculated from the titration results. The IEC of a cation exchange resin is an important property that determines its effectiveness in removing positively charged ions from solution. A resin with a high IEC will have more ion-active groups available to exchange ions, and therefore will be able to remove more ions from solution. In V ml of an alkali solution of strength N in milliequivalents per litre is required to titrate all the acid eluted from the resin weighting W g then

$$Capacity = V \times \frac{N}{W} \tag{8.1}$$

$$\text{Capacity} = V \times N/W \tag{8.1}$$

Similarly, the exchange capacity of an anion exchanger is determined by passing a solution of sodium nitrate of known concentration through a column. The concentration ions or the hydroxide ions in the solution eluting from the column is found by titration with a standard solution of silver nitrate or acid respectively.

(i) Ion Exchange Techniques

There are two main techniques for contacting ion exchange resins: batch and column methods.

Batch Method

In the batch method, the resin and solution are mixed together in a vessel until equilibrium is reached. The solution is then filtered off. The extent to which the ions from the solution are exchanged for those on the resin depends on the selectivity coefficient. In this type of single-step equilibrium, only a small portion of the exchange capacity of the resin is utilized.

The batch method is used for softening water and producing deionized or demineralized water. For water softening, the calcium and magnesium ions that cause hardness are exchanged for sodium ions. The sodium form of a sulfuric acid resin is generally used.

Column Method

In the column method, the resin is packed into a column and the solution is passed through it. The ions in the solution are exchanged for the ions on the resin as the solution flows through the column. The column is regenerated periodically by backwashing with a salt brine solution.

The column method is more efficient than the batch method because it uses the full exchange capacity of the resin. It is also more continuous, which makes it more suitable for large-scale applications.

$$RSO_3^- \ldots\ldots Na^+ + Ca^{2+} \rightleftharpoons (RSO_3^-)_2 + 2Na^+$$

When the resin is exhausted, a 15 per cent solution of sodium chloride is added to the resin. This removes the calcium or the magnesium ions from the resin by exchange with sodium ions.

Deionized or demineralized water is water from which all cations and anions of electrolytes have been removed. It is prepared by treating water with a cation exchanger in the acid or the hydrogen form. All the cations of the electrolyte are replaced by hydrogen ions

$$RSO_3^- \ldots\ldots H^+ + M^+ \rightleftharpoons RSO_3^- \ldots\ldots M^+ + H^+$$

The water is then treated with an anion exchanger in the basic or the hydroxide form

$$R^+OH^- + Cl^- \rightleftharpoons R^+Cl^- + OH^-$$

Combination of H^+ and OH^- ions forms water almost completely free from ions of any dissolved electrolytes. Deionized water finds a large number of applications in the laboratory. Its conductivity equals that of conductivity water prepared by distillation, which is a time consuming process. Deionized water can be used in conductance experiments and for preparing solution of silver nitrate. Deionized water is also used in biological studies. Biologists have very often to prepare culture media of precise composition. If the solutions are prepared even in distilled water, the ions present can introduce source of uncertainty into the experiments. Use of deionized water eliminates this uncertainty and errors. Deionized water is also used in lead storage batteries and industrial boilers. Commercial units to prepare deionized water for use in laboratories are now available.

ii) Column Method Ion Exchange Chromatography

Ion exchange chromatography is a technique that separates different components of a mixture based on their different attraction to a resin. This resin has charged groups on its surface that can attract ions with the opposite charge. The different components of the mixture have different affinities for the resin, so they move through the column at different speeds. There are three main types of ion exchange chromatography:

1. Frontal analysis: In frontal analysis, the sample mixture is passed through the column continuously. The components of the mixture that

have the strongest affinity for the resin will be retained by the column for the longest time. The components with the weakest affinity for the resin will elute from the column first.

2. Elution analysis: In elution analysis, the sample mixture is passed through the column once, followed by a buffer solution that has a higher ionic strength than the sample mixture. The buffer solution displaces the components of the mixture from the resin, and they elute from the column in order of decreasing affinity for the resin.
3. Displacement development: In displacement development, a solution containing a displacing ion is passed through the column after the sample mixture has been loaded. The displacing ion has a stronger affinity for the resin than the components of the mixture, so it displaces them from the resin and they elute from the column in order of decreasing affinity for the resin.
4. Ion exchange chromatography apparatus: The apparatus used in ion exchange chromatography consists of a glass column packed with the resin. The column is fitted with a glass wool plug or sintered glass disc at the bottom to prevent the resin from falling out.

The resin should be packed into the column as a slurry. To do this, the resin is mixed with distilled water to form a slurry. The slurry is then poured into the column slowly, taking care to avoid forming air bubbles. Once the column is packed with the resin, it is backwashed with distilled water to remove any air bubbles or fine particles.

8.10 Application of Ion Chromatography

Because of the particular selectivity forces of material, separation with ion exchange is used mostly in inorganic chemistry. Organic ions which form salts with oppositely charged ions are another phase can be separated. Almost all ion exchanger reactions require aqueous solvents. Some applications of ion exchangers are as follows:

i) Separations of similar Ions from one another: Ion exchange chromatography is being used in separate similar ions from one another because the different ions undergo exchange reactions to different anions. For instance, a mixture of Li^+, Na^+, and K^+ ions can be separated by passing their solution through a cation exchanger. Subsequently, 0.1 N HCl has been used as eluent. Similarly ion exchange chromatography has been used to separate a mixture containing Cl^-, Br^- and I^-. The mixture solution is passed through a basic anion exchanger. Sodium nitrate solution has been used as eluent. When 0.5 N sodium nitrate is used, Cl^- ion will come first in the elute. If the concentration of sodium nitrate is raised, Br- will be eluted and by further increasing then concentration of sodium nitrate, I^- will be elute.

ii) Removal of interfering Radicals: In the estimation of Ca^{2+} or Ba^{2+} ions by then oxalate or sulphate method, phosphate ion is found in interfere. Therefore, its removal become necessary which is achieved by passing a solution of Ca^{2+} or Ba^{2+} ions having phosphate ion through sulphonic acid cation exchanger. The Ca^{2+} or Ba^{2+} get exchanged with H+ ions while phosphate ions will pass as such through the column. The process has to repeated so that the phosphate ions will completely removed. Now Ca^{2+} and Ba^{2+} ions held by resin will be removed by using suitable eluent. Finally, these ions are estimated by the usual methods.

iii) Softening of hard water: We are aware of the fact that the hardness of water is due to the presence of Ca^{2+}, Mg^{2+} and other divalentions. These ions may be removed by passing hard water through cation exchangers charged with Na+ when the following reaction takes place

$$\underset{\text{(Resin)}}{2Na_nR} + \underset{\text{(Hard water)}}{nCa^{2+}} \rightleftharpoons \underset{\text{(Resin)}}{Ca_nR^{2+}} + \underset{\text{(Solution)}}{2nNa^+}$$

The Ca^{2+} and Mg^{2+} ions are retained in the column whereas the Na+ ions pass into the solution. These Na+ ions are harmless for washing purpose. After using the ions exchanger for a long time, it become inactive. Its activity can be revived by percolating through a concentrated solution of NaCl when the following reverse reaction takes place

$$\underset{\text{(Resin)}}{Ca_nR_2^+} + \underset{\text{(Solution)}}{2nNa^+} \rightleftharpoons \underset{\text{(Resin)}}{nCa^{2+}} + \underset{\text{(Regenerated)}}{2Na_nR}$$

v) Complete demineralization of water: Ion exchange is a water purification process that removes cations and anions. It is commonly used to soften water and demineralize water. To soften water, the water is first passed through a cation exchanger. The cation exchanger exchanges metallic cations (such as Na^+, Ca^{2+}, and Mg^{2+}) for H^+ ions. The water from the cation exchanger is then passed through an anion exchanger. The anion exchanger exchanges anions commonly present in water (such as Cl^-, NO_3^-, and SO_4^{2-}) for OH^- ions. The H^+ and OH^- ions combine to form unionized water. Cation exchangers are typically made of sulphonic acid resin, while anion exchangers are typically made of strong basic resin. To regenerate mixed bed ion exchange resins, the resins are separated by flotation. One resin is made of lower specific gravity than the other, so it floats on top. Each resin is then given its specific regeneration treatment. Ion exchange chromatography is a powerful technique for separating lanthanides and actinides.

vi) Separation of Lanthanides: During World War II, the separation of fission products was a major challenge for analytical chemists, especially for separating lanthanides (La, Sm, Eu, Y, Pr, Nd, and Pm) and mixtures of calcium (Ca), rubidium (Rb), strontium (Sr), zirconium (Zr), and niobium (Nb) in trace amounts. Synthetic resins were used in combination with ion exchange chromatography to separate these elements. When a solution containing a mixture of lanthanides is passed through a column packed with ion exchange resin, the cations in the solution exchange with hydrogen or other cations present on the resin. The cation with the strongest affinity for the resin is retained at the top of the column, while the other cations are retained lower down in the column in order of decreasing affinity. When the column is eluted with a citrate buffer solution, the lanthanide ion with the strongest affinity for the eluent comes out first, followed by the other lanthanides in decreasing order of affinity. Improved separation of lanthanides in kilogram quantities has been achieved using other complexing agents, such as H_4EDTA and a variety of hydroxyl and amino acids.

vii) Separation of Actinides: Ion exchange chromatography has also played a unique role in the discovery of transplutonium elements in the actinide series. In the actinide series, the elution order is reversed due to actinide contraction. This proved to be the only way to identify these elements, some of which were produced in quantities as small as a few atoms. It can be said that their separation would have been impossible using any other method.

viii) Purification of organic compounds extracted in water: Many natural products extracted in water have been found to be contaminated with ions originally present in water. Those ions can be removed by using ion exchange processes.

viii) Separation of sugars: This method was developed by Khym and Zill (1951). The sugars are first of all converted into borate complexes. Separation of borate complexes of sugars have been achieved on 11 x 0.9 cm column 200- 400 mesh Dowex 1 resin, using a loading of 5- 10 cm long borate complexes and flow rate of 0.5 – 1 ml/min. Quantitative recovery of sugars is possible after separation of the borate complexes. Similarly, disaccharides can be separated from the monosaccharaides and the individual compounds of hexose and pentose mixture resolved.

ix) Separation of amino acids. Ion exchange chromatography has been used to separate complex mixture of 18 amino acids obtained by the acid hydrolysis of proteins. The mixture of amino acids is first introduced into a very short column at pH 2 and eluted with 0.35 N sodium citrate buffer at pH 5.25 with which the column has been equilibrated. Acidic nas neutral amino acids at first leave the column un-separated and are followed by lysine, histidine, NH_3 and arginine. Now a second sample chromatographed into second column with 0.2 N sodium citrate buffer at pH 3.28 and subsequently eluted with 0.88 N citrate buffer at pH 4.25.

8.11 Application of Ion Chromatography in Pesticide Residue Analysis

Certain pesticides, known as anionic polar pesticides, are difficult to analyze because they cannot be extracted using common methods and are not retained well by reversed-phase liquid chromatography. As a result, not all of these pesticides are tested for in all UK and EU food monitoring programs. Instead, specific pesticides are targeted in specific foods as needed. For example, in 2018, the EU only required ethephon to be tested for in sweet peppers, wheat, and table grapes, and glyphosate to be tested for in table grapes and wheat. In contrast, in 2019, 2020, and 2021, glyphosate must be tested for in all foods. The Quick Polar Pesticides Method for products of Plant Origin (QuPPe-PO) is a sample extraction and preparation method that allows for the simultaneous analysis of several polar pesticides in a wide range of plant-based foods. This method underpins the use of a variety of liquid chromatography tandem mass spectrometry (LC-MS/MS) based methods. The researchers investigated the use of ion chromatography coupled with tandem mass spectrometry (IC-MS/MS) as an alternative to HPLC-MS/MS for the determination of anionic polar pesticides in fruit and vegetables. They developed a multi-residue method for the simultaneous determination of 9 analytes, including the metabolites of glyphosate, with limits of detection low enough to comply with EU maximum residue levels (MRLs).

Ion chromatography (IC) is a technique for separating ions and polar molecules. It was introduced in 1975 and has since evolved to include more efficient columns and a wider range of detectors. IC is particularly well-suited for analyzing polar pesticides, which are challenging to analyze using other techniques such as reversed-phase LC. There are two main approaches to analyzing polar pesticides using IC: a) Derivatization of the analyte to make it more amenable for retention on a reversed-phase column. b) Using a different type of chromatography, such as normal-phase chromatography or IC. Derivatization involves adding a functional group to the analyte to change its properties. This can make the analyte more hydrophobic, so it will retain better on a reversed-phase column. Normal-phase chromatography is a type of chromatography that separates compounds based on their polarity. Hydrophilic interaction LC (HILIC) and zwitterionic columns are two types of normal-phase columns that have been shown to be effective for analyzing polar pesticides.

IC is a good choice for analyzing polar pesticides because it can separate ions and polar molecules even in the presence of high levels of non-volatile salts. The application and evaluation of IC coupled to high-resolution MS using a quadrupole (Q)-Orbitrap MS instrument for the multi-residue detection of polar pesticides in fruits and vegetables provides a good platform for analysis of these pesticides. The validated method is fast and straightforward in order to be implemented in routine laboratories. It provides an appropriate LOQ relative to EU maximum residue limits in all cases, and the linearity, retention time, precision, mass accuracy, and repeatability cover the most stringent criteria of QC performance for routine food control.

8.12 Instrumentation and Principles

Ion chromatography (IC) is a popular analytical method for detecting and measuring inorganic and organic ions in various samples. It is a relatively simple and inexpensive method with high efficiency and good validation metrics, such as sensitivity, selectivity, precision, robustness, low detection limit, and quantification. IC can easily replace conventional wet chemistry methods, which typically require more steps and generate more waste. IC is widely used in the food, pharmaceutical, and environmental monitoring industries, as well as in academic research laboratories. Ion chromatography (IC) is a popular technique for measuring organic and inorganic ions. It has been used in environmental monitoring, food analysis, criminal forensics, water quality control, and the pharmaceutical industry since its development by Small in 1975. IC is a robust and sensitive technique that can separate ions in a sample and quantify inorganic anions and cations, as well as charged complexes and organic substances that can exist in ionic form, such as the pesticide glyphosate and its metabolite AMPA. IC is based on electrostatic interactions between charged fragments on the surface of molecules and oppositely charged functional groups attached to a stationary phase. It is a surface process that occurs when an ionic solid comes into contact with a solution. IC can replace traditional chemical analysis methods such as gravimetry, volumetry, and endpoint detection. These methods can be time-consuming and difficult to automate, especially when analyzing a large number of samples. They can also be less sensitive and more susceptible to interference and sample contamination from analyst handling. Due to its advantages and rapid development, IC is expected to be used in new areas in the near future, especially when coupled with other techniques to improve performance. The various components of ion chromatography system are discussed as under

i) Mobile Phase

The eluent is the solvent that moves the sample through the chromatography column. It can have a fixed or variable concentration during the run, depending on the type of chromatography being used. For some types of chromatography, the eluent is generated by an eluent generator cartridge. This cartridge contains a concentrated electrolyte solution, which is diluted to the desired concentration by the mobile phase. The sample is initially dissolved in the eluent, and then the mixture is introduced to the chromatography column. The different components of the sample interact with the stationary phase and the eluent to different degrees, which causes them to separate. The eluent is chosen based on the type of sample being analyzed. For example, anion analysis typically uses a dilute buffer solution containing sodium bicarbonate and sodium carbonate, while cation analysis typically uses a solution of dilute acid like sulfuric, acetic, citric, sulfonic and carboxylic acids.

ii) Degasser

The quality of the eluent has a significant impact on the performance and wear of a chromatograph. A degasifier is a high-pressure device that removes dissolved gases from the eluent, preventing the formation of bubbles that can interfere with the chromatographic process. Pump systems typically use double piston pumps to reduce noise in the chromatogram baseline and provide a constant, pulse-free flow of eluent, which is necessary for sensitive detectors such as conductivity, UV/VIS, and amperometry. Electronic circuits and pulse shock absorbers are used to further reduce residual pulse. Some pumps only operate in isocratic mode, while others can also operate in gradient mode. Gradient elution allows for the separation and analysis of a wider range of ions than isocratic elution, but it is more complex to set up and operate.

iii) Injector

The sample is injected into the system through a valve injector. The injection valve typically contains six ports in a two-position change format. One position is used to load the sample in the injection circuit (Figure 8.1) and, after that, there is a change of the injection valve, this being the second position, where the sample is transported by the mobile phase to the guard column and, in then to the separation column Figure 8.2.

iv) Column or Stationary Phase

Separation columns or chromatographic columns, also known as stationary phase, are supports packed in cylindrical tubes of different diameters and lengths, which have inlet and outlet at their ends. The internal contents of these columns are generally porous and solid substrate particles with positively charged or negatively charged ionic functional groups on their surface. The ion exchange resin must be stable and insoluble in common solvents, and can be organic or inorganic. Organic resins are the most used ion exchangers. Those with a silica base have greater efficiency in chromatographic separations. The most common are polystyrene, which easily forms an addition polymer with divinylbenzene. There are also resins with inorganic characteristics, whose ion exchange properties are favorable. This resin offers greater thermal stability. The most used inorganic ion exchange are hydrated oxides, acid salts of polyvalent metals, salts of heteropoly acids, insoluble ferrocyanides and aluminosilicates.

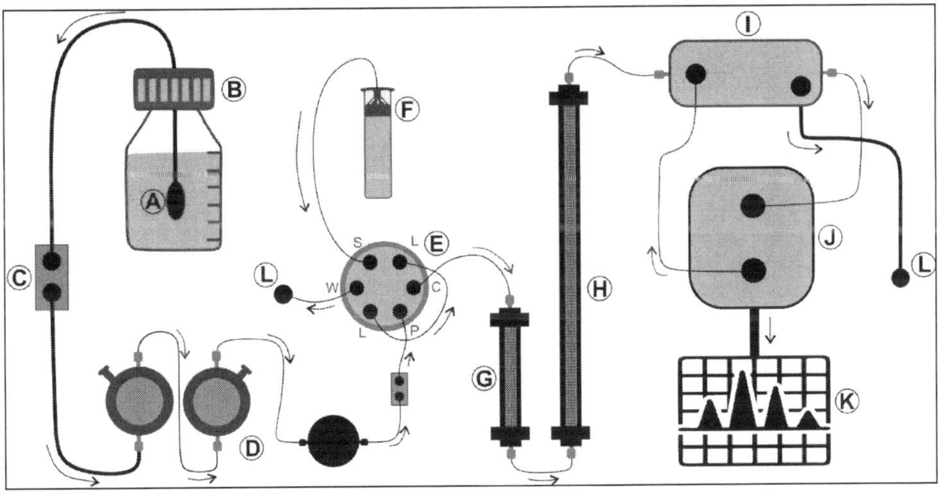

Figure 8.1: Schematic of an Ion Chromatography System.

A-Filter; B-Bottle with eluent; C-Degasser; D-Pump system; E-Injection System; F-Vial Sample; G-Guard column; H-Separation Column; I-Chemical suppressor; J-Detector; K-Chromatogram; L- Waste.

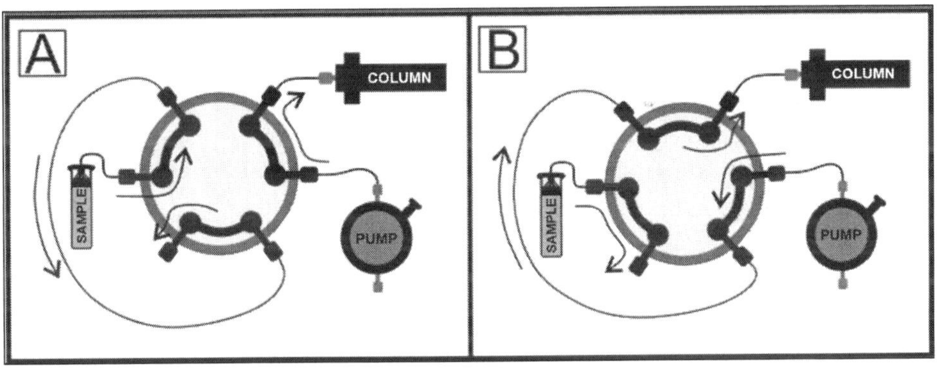

Figure 8.2: Sample Injection System.

Loop loading with sample (A) and drag the sample through the eluent into the separation column (B).

v) Ion exchangers

Functional groups have three different types of ion exchangers. In cation exchange chromatography, negatively charged ligands bond to positively charged molecules whereas the opposite occurs for anion exchange chromatography, whereas Zwitterionic ligands have both charges in a single column.

1. Anion exchanger : Anion exchangers are resins that are capable of capturing anions. There are strong base resins that are derived from the reaction of tertiary amines with styrene-DVB and weak base resins that are functionalized with primary and secondary amine groups. They can be applied in the adsorption of strong acids, but the kinetics is prolonged. Amine resins have good yield and are regenerated by NaOH, NH_4OH or $NaCO_3$.
2. Cation Exchanger : Cationic ion exchangers are composed of a strong or weak organic polymeric matrix. Strong acid resins are produced by sulfonation, whose functional group is sulfonic acid. These resins work in a good pH range, but require a greater amount of regenerant. This is the most chosen for almost all water analyses. Weak character exchangers are carboxylic resins that have exchange capacity and are used to treat larger organic ions.
3. Zwitterionic Exchanger : Zwitterionic ion exchangers are those stationary phases used in the multi-selective retention mechanism. These exchangers have both positive and negative charges, so they carry a net zero charge. Zwitterionic stationary phases accumulate equal amounts of oppositely charged groups, fixed close to the surface or within the volume of the stationary phase. Ideally, the zwitterionic phase can be considered as a phase containing an equivalent amount of strongly acidic and strongly basic groups

vi) Suppressors

After sample separation and before detection there is a suppressor cell. The main function of the electrochemical suppressor system as part of the detection unit is to reduce the background signal of the mobile phases, increasing the detectability of sample ions. Thus, the high background conductivity of the electrolytes in the eluent is reduced through a selective exchange membrane, lowering the baseline and enabling detection at low concentrations.

vii) Detectors

Passing through the suppressor, the analyte is conducted to the detector of the apparatus. There are several types of detectors; they can be conductivity, amperometric, potentiometric, spectrophotometric or fluorescence. Conductivity cells are the most used for IC, where anions and cations are determined through electrical conductance, producing a signal based on the physical or chemical properties of the analyte, generating the chromatogram when processed by the computer system. UV-VIS detection is one of the most popular modes for high performance liquid chromatography, but its application in ion chromatography is limited. The fluorescence detector is rarely used in IC, as only some ions have a fluorescent appearance. Amperometric detection can be used for samples with pK values greater than 7. Refractive index detection is rarely used in IC. One of the most efficient detection methods used is mass spectrometry (MS). However, it is not popularly found coupled to the IC due to the high cost compared to the other detectors mentioned

Chapter 9

High Performance (Pressure) Liquid Chromatography

9.1 Introduction

Early liquid chromatography was slow, taking days to complete a single analysis. This was due to the use of gravity-fed systems, which allowed the mobile phase to flow through the column at a very slow rate. This long retention time led to poor resolution, as the sample components had more time to diffuse and blend together. High-performance liquid chromatography (HPLC) overcomes this problem by using a pump to force the mobile phase through the column at high pressure. This reduces the retention time and improves resolution. All of the factors that affect separation in liquid chromatography also apply to HPLC, including plate height, sample distribution between phases, and the selection of stationary and mobile phases. However, the main advantage of HPLC is its speed. By reducing the retention time, HPLC reduces diffusion and improves resolution. HPLC also uses smaller particles in the stationary phase than traditional liquid chromatography. This further improves resolution by reducing the geometric factor A in the Van Deemter equation. Typical HPLC pressures range from 30 to 200 atmospheres, depending on the type of column used. The pressure is adjusted to achieve the optimum linear flow rate for the mobile phase, which produces the smallest theoretical plate height. This is determined experimentally, similar to the H/V relationship in gas chromatography.

9.2 Principle

Chromatographic columns with longer lengths and more theoretical plates per unit length have better resolving power. However, there is a limit to column length due to peak broadening. The number of theoretical plates is related to the surface area of the stationary phase, so smaller particle sizes improve resolution. However, smaller particles also increase resistance to eluent flow. All previous forms of column chromatography relied on gravity or low-pressure pumping systems to deliver eluent to the column. This resulted in relatively low flow rates, which gave more time for band broadening due to diffusion. Faster flow rates were not possible because they would create back-pressure that could damage the stationary phase and impair resolution. In the past decade, there have been dramatic developments in column chromatography technology. New stationary phases have been developed that can withstand high pressures, and pumping systems that can deliver reliable flow rates. These developments have resulted in faster and better resolution, and explain why high performance liquid chromatography (HPLC) has emerged as the most popular, powerful, and versatile form of chromatography. HPLC was originally referred to as high pressure liquid chromatography, but the term high performance liquid chromatography is now preferred because it better describes the characteristics of the chromatography and avoids the impression that high pressures are necessary for high performance. This is now known not to be the case, and the term medium pressure liquid chromatography has been coined for some separations. The new technology in stationary phases has also been applied to thin-layer chromatography, giving rise to high performance thin-layer chromatography (HPTLC). However, the impact of new technology has not been as great in thin-layer chromatography as it has been in column chromatography. All factors affecting separations on liquid column chromatography apply to HPTLC as well, such as plate height, sample distribution between stationary and liquid phases, and the selection of the stationary and liquid phases. Various methods of development of the chromatograms (elution, gradient elution) can also be used in HPTLC.

9.3 Instrumentation

General instrumentation for HPLC has following components.

i) One or more solvent reservoirs for the mobile phase.

ii) A pump to deliver the mobile phase with varying range of pressures up to several hundred atmospheres to achieve reasonable flow rates.

iii) Sampling valves or loops where the sample may be injected into the flowing mobile phase. Sample may be dissolved in mobile phase.

iv) A guard column or an on-line filter to prevent contamination of the main column.

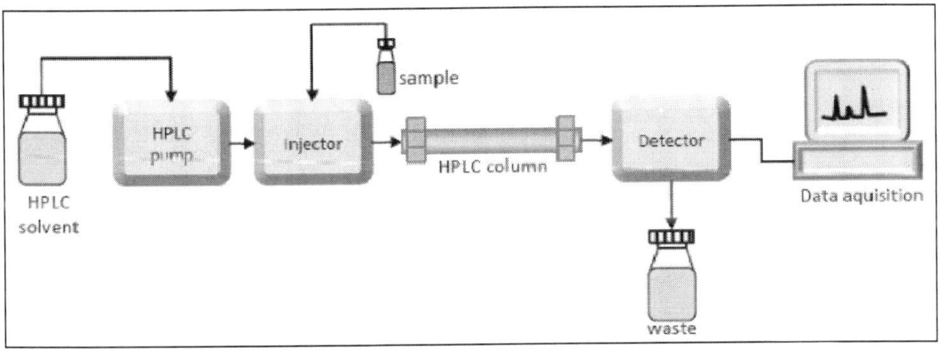

Figure 9.1: Basic Components of HPLC.

v) A pressure gauge, inserted in front of the separation column, to measure column inlet pressure.

vi) Separation column containing packing to accomplish desired separation. These may be modified silica gel, ion-exchange resin, gel or some other unique packing.

vii) A detector capable enough of measuring the solute concentrations.

viii) Display and recording device for plotting time vs peak intensity.

9.3.1 Sample Injection System

It is a limiting factor in the precision of HPLC measurement because of reproducibility with which samples may be introduced onto the column packing. Insertion of the sample into the column must be through a narrow plug so that peak broadening is minimized and the system should have no dead volume by itself. Generally, samples are dissolved in a mobile phase solvent to avoid solvent peak and 10 to 50 μL is introduced through micro sampling valves. These devices form an integral part of liquid chromatography equipment having interchangeable loops with a choice of sample size from 5 to 500 μL. The most widely used method of introduction is based on sampling loop as shown in Figure 9.2. It is filled by thoroughly flushing it using a sample solution by means of a micro-syringe at pressures up to 7000 psi. A rotation of the valve rotor places the sample filled loop into the high pressure mobile phase stream whereby the sample is sent to the column. The system can be located within a temperature controlled oven if handling at elevated temperatures is required. Many HPLC instruments incorporate an auto sampler with an automatic injector that can inject variable volumes as per requirement. In stopped flow injection method, pump is turned off till atmospheric pressure is attained, syringe is inserted and the sample injected. The flow of sample can be brought to zero and rapidly resumed by diverting the mobile phase using a three way valve placed before the injector. This method is especially very useful for very high pressures. For best results, a two to fivefold excess of sample should be passed through the loop to ensure that previous sample has been purged thoroughly.

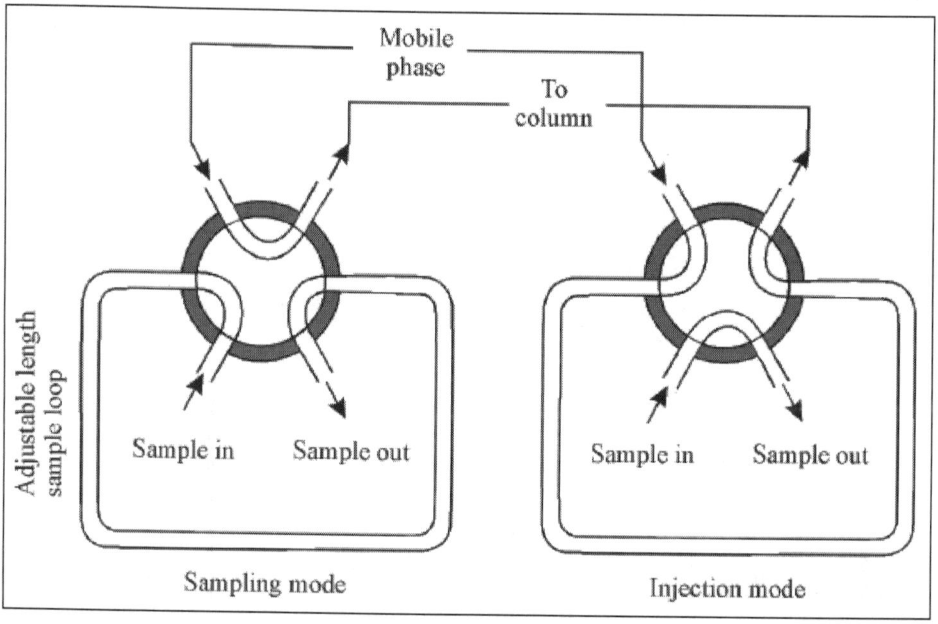

Figure 9.2: Schematic of Injector Valve with External Sample Loop in a Micro-Volume Sampler.

9.3.2 Columns

The HPLC separation column is the most important part of the instrument, where the actual separation of the sample takes place. It is typically made of a heavy-walled, glass-lined metal or 316-grade stainless steel tube that can withstand high pressure and resist chemical corrosion from the mobile phase. The inside of the tube must be smooth and have a uniform bore diameter. Straight columns that can be operated vertically are preferred. Some typical tubing materials used in HPLC columns given in Table 9.1. The best columns are precision-bored with an internal mirror finish, which allows for efficient packing of the column. Porous plugs of stainless steel or Teflon are used at the ends of the columns to retain the packing material. These plugs must be homogeneous to ensure a uniform flow of solvents through the column. Column temperature may need to be thermostatically controlled during analysis for some separations involving liquid partition and ion exchange. Column fittings and connectors must be designed to minimize void volume and avoid unswept corners. Column length ranges from 10 to 30 cm with an inner diameter of 2 to 5 mm, providing 40,000 to 60,000 plates per inch. Shorter columns of 3 to 8 cm can also be used for fast separations, but in such cases, the sample size will be limited. The length of the column can affect both the resolution and speed of separation. Longer columns provide better resolution, but they also take longer to run. Standard column lengths vary with the manufacturer, but the most common values are 30, 25, 15, 12.5, 10, and 7.5 cm. Shorter columns are often described as high-speed columns. Columns packed with finer particles are more expensive than the standard 5 μm packing, but they also provide better resolution.

Table 9.1: Column Tubing Materials and its Uses

Material	*Use*
Stainless steel (SS)	General utility material, good for high pressure system
Poly (ether-ether) ketone (PEEK)	Inert to most organic solvents except methylene chloride, THF, DMSO and conc. Sulphuric and nitric acids. Holds pressuresup to 5000 psi (34MPa).Good for metal-free biological systems.
Tefzel	Withstands pressures up to 5000 psi,corrosion resistant; expensive
Fused silica Glass	Used for capillary LC.
Glass	Limited pressure range.
Glass-lined SS	Inert, withstands pressures but difficult to know when the glass is broken.

Guard column: In order to increase the life of analytical column, a short guard column, also called pre-column, is placed before the main column. It removes contamination from the solvent. Guard column serves to saturate the mobile phase with the stationary phase so that losses of stationary phase in the column are minimized. However, it is essential that the composition of the guard column should be similar to that of the analytical column but its particle size may be larger to minimize the pressure drop.

9.3.2.1 Column Packing

The packing used in modern HPLC consists of small, rigid particles having a narrow particle size distribution. The most common packing material used for LC is prepared from silica (acidic) and alumina (basic) particles which are synthesized by agglomerating submicron size particles under conditions that lead to larger particle diameter. These are often coated with thin organic films which are physically or chemically bonded to the surface. For nearly all HPLC applications, chemically modified or unmodified micro particulate silicas of 3, 5 or 10 μm diameter are preferred. This form of LLC, in which both monomeric and polymeric phases have been bonded to a wide range of support materials, is called bonded phase chromatography (BPC). Characteristics of typical packing materials used in HPLC are listed in Table 9.2. The particles used in HPLC, which are totally porous (macro-porous) or superficially porous (pellicular) support, may be spherical or irregular in shape but it is essential that the size range is as narrow as possible to ensure high column efficiency and permeability. These adsorbent packing's retain solute molecules almost exclusively on the internal surface of the pores, thus, separating these from others. micro-porous support where micro-porous ramify through the particles which are generally 5-10 μm in diameter. Porous particles are made up silica, polystyrene-divinyl-benzene synthetic resin, alumina, or other type of ion-exchange resin. Silica is the most common type of porous particle packing material. The characteristics of various types of bonded phases are described below:

i) Spherical bonded phase: These spherical packing's consist of a solid, spherical nonporous core (usually a glass bead) with a layer of attached functional groups forming an outer shell containing unmodified or modified silica gel, resin, polyamide, *etc.* Various functional groups are used depending on the nature of solutes to be separated.

ii) Porous layer beads: A porous or pellicular layer bead type packing material consists of a solid, spherical with an average particle diameter 30-40 μm coated with a thin porous outer shell, typically of 1-3 μm thick. It may be a silica gel layer, a network of small spherical particles bonded to the solid core. It may also be monomeric or polymeric organic phase. Surface areas of the porous layer beads range from 5 to 15 m^2/g. These materials are easy to be packed because of its dense core but suffer from limited sample capacity due to small surface areas. Porous layer packings exhibit good efficiency because of improved mass transfer within the stationary phase. Longer columns are possible because the pressure drop is lower due to larger particle size of porous layer supports. Thicker coatings give rise to slower mass transfer but have increased sample capacity.

iii) Porous particles: Totally porous particles have a large surface area in the range 100 to 860 m^2/g with average being 400 m^2/g. The mean pore diameter is inversely related to the specific surface area where small molecules enter the pores. The particles can be packed into the HPLC column of efficiencies up to 800 theoretical plates per centimetre if 5 μm particle sizes are used. However, larger particles give proportionately small number of theoretical plates whence the efficiency of separation goes down.

iv) Macro-porous particles: These are recently introduced graphitized carbon and styrene-divinyl-benzene polymers having large channels besides micro-pores. The rigid, porous polymeric macro-porous beads do not swell or shrink with changes in the ionic strength of the mobile phase (can be used over an extended pH range of 1 and 13) or deform at high velocities and are most suited for separations in non-aqueous media. These materials have increased the choice of stationary phases and the scope of HPLC, particularly for highly polar and basic substances.

Table 9.2: Characteristics of some Commercial HPLC Column Packing Materials

Type	*Pellicular*	*Microprous*
Silica	Corasil (37-50μm) Vydac (30-40μm)	Lichrosorb (5,10 and 20 μm) Micropak (5 and 10 μm) Porasil (15 and 20 μm) Spherisorb (5 μm)
Alumina	Perisorb PA	Micropak Al (5 and 10 μm) Spherisorb Al

9.3.2.2 Chemistry of Bonded Phase

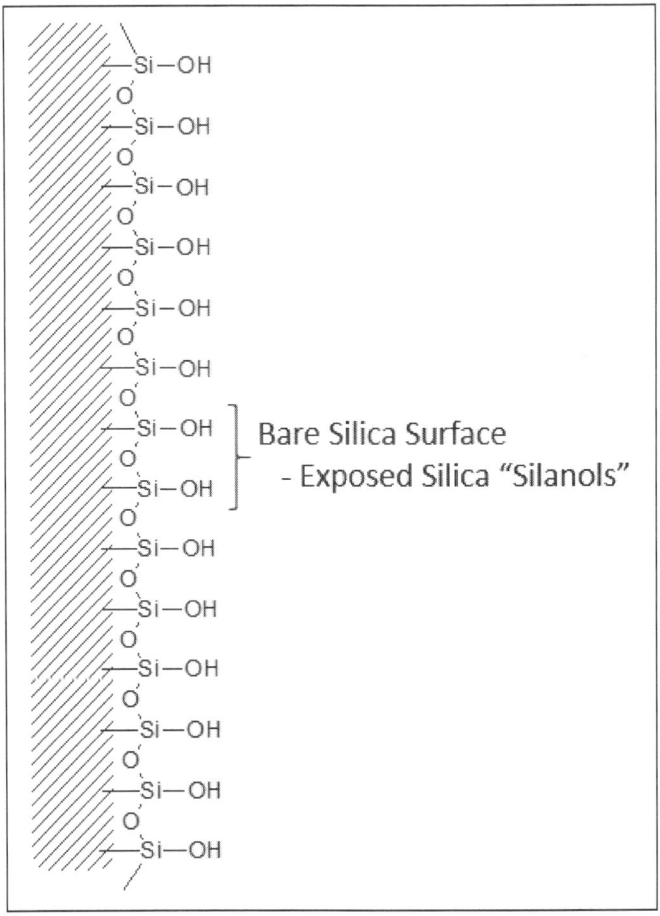

Through chemical reaction, attachment of the primary bonded phase is achieved.

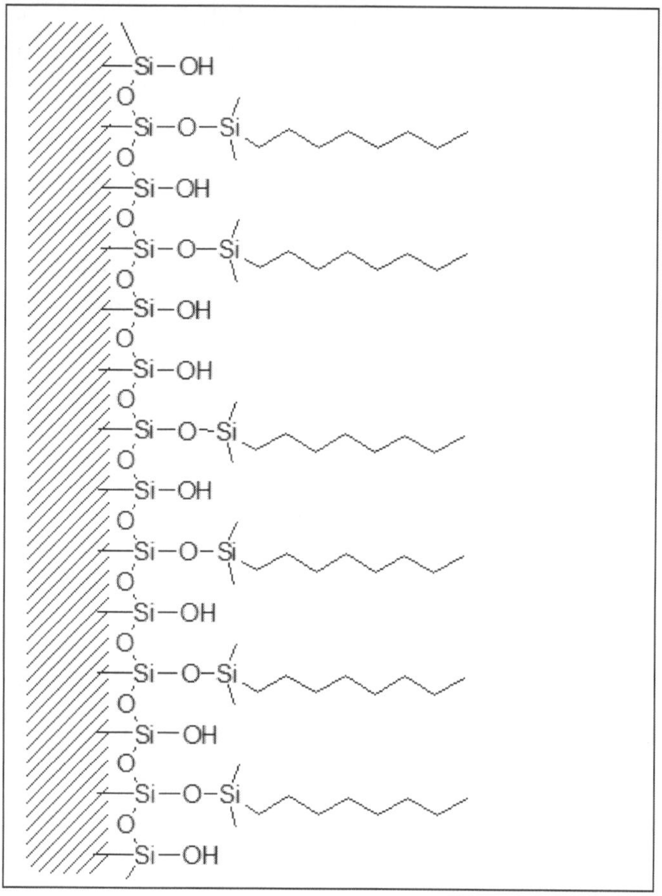

Manufacturers refer to the amount of bonded phase coverage in terms of per cent Carbon load.

Much of the silica surface remains unreacted – some surface silanol groups are still exposed.

Frequently, a secondary bonding step introduces smaller "end-cap" molecules between the primary bonded phase strands. These endcaps may be polar or non-polar.

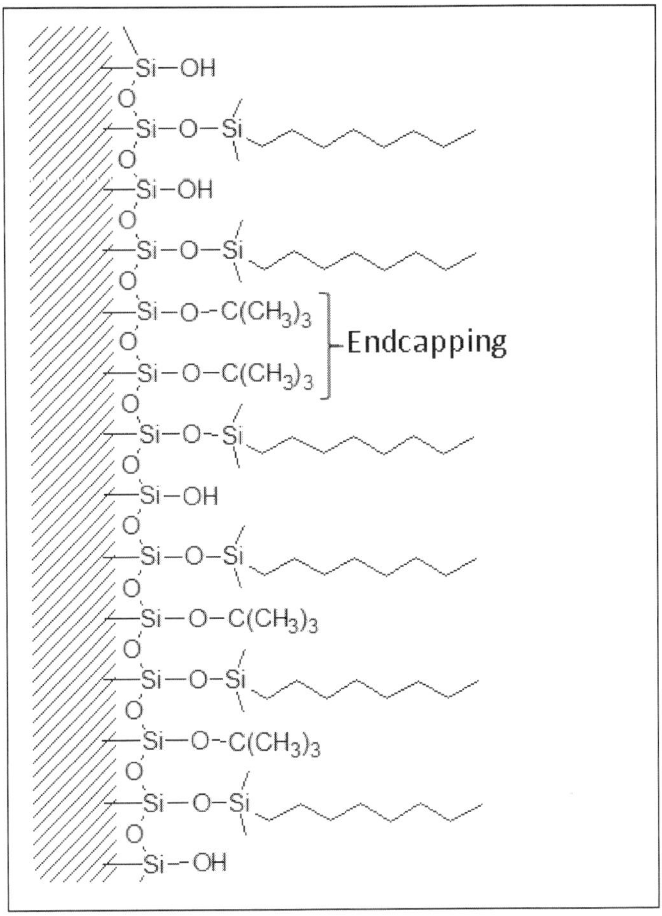

Different columns of the same bonded phase type will differ in silanol exposure and end-capping, resulting in a range of different overall polarities and different separating ability.

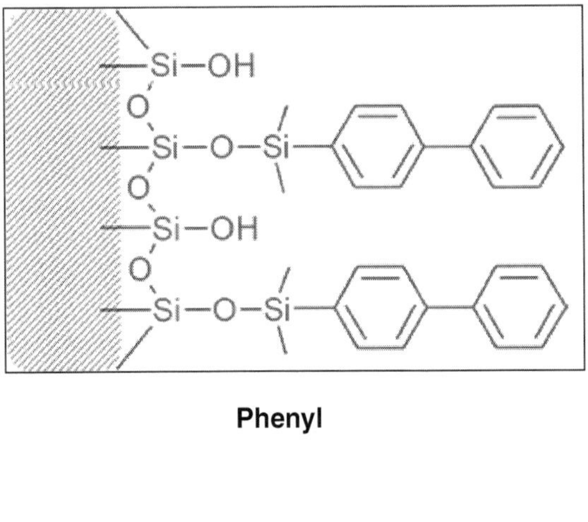

Phenyl

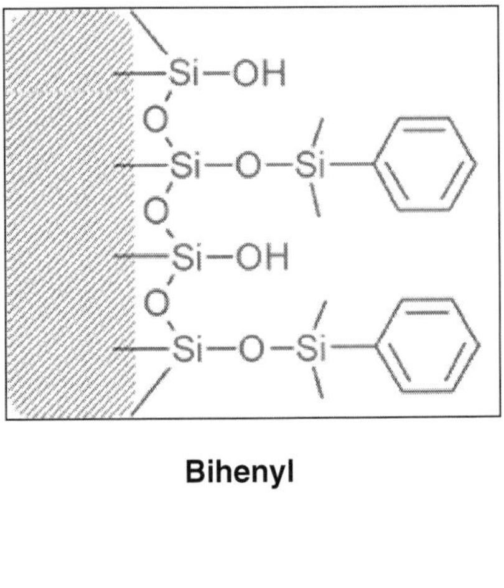

Bihenyl

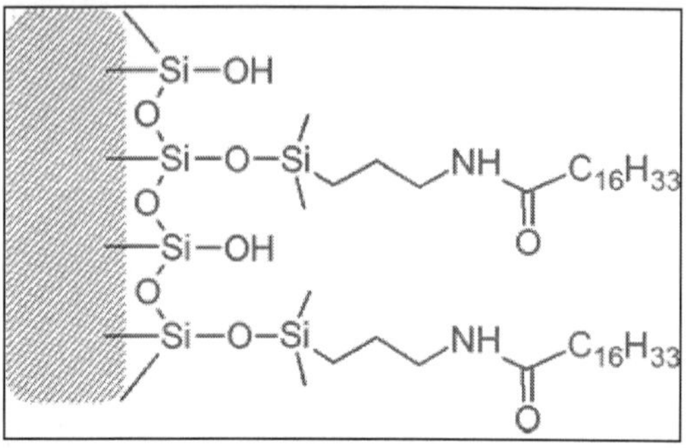

Alkyl Amide

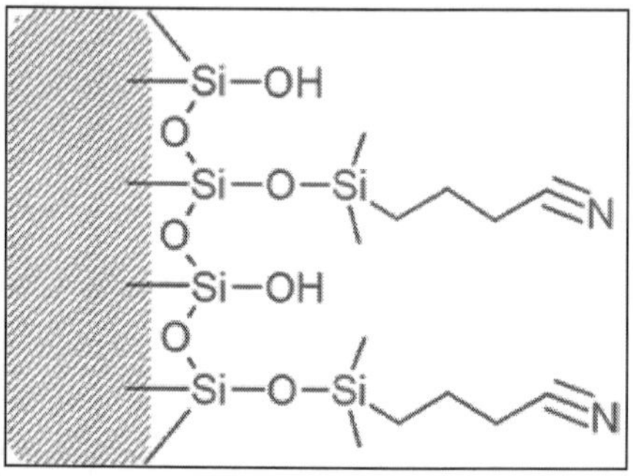

Cyano Propyl

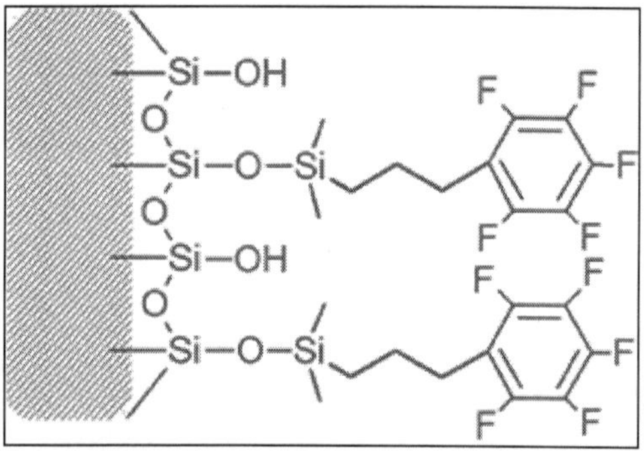

Pentafluorophenyl Propyl

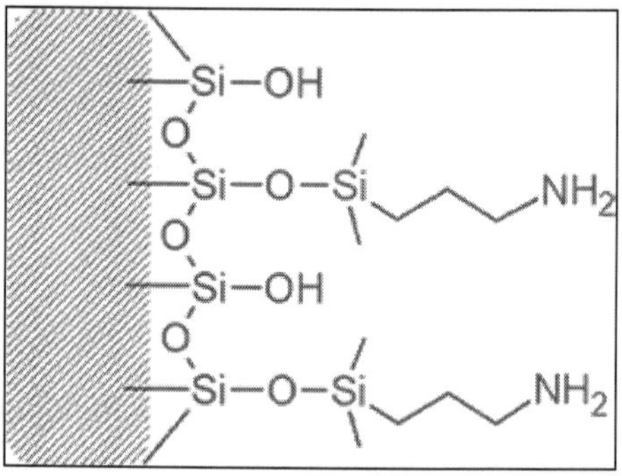

Amino Propyl

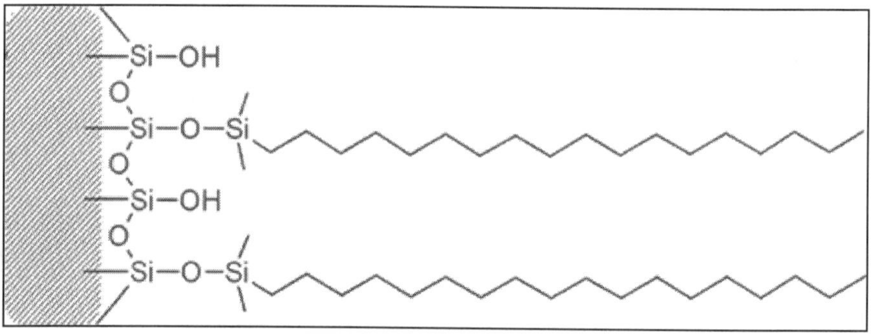

C-18

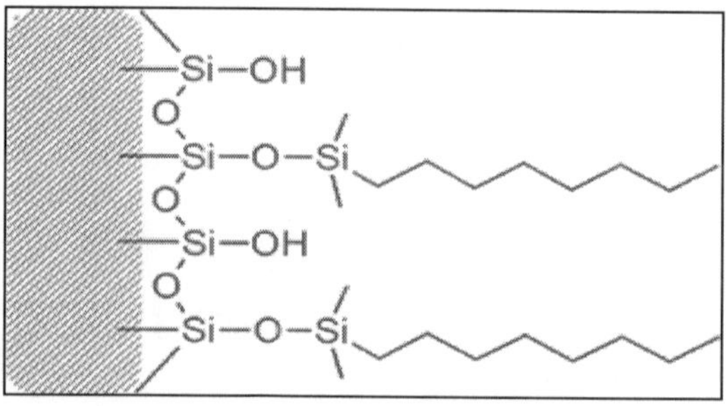

C-8

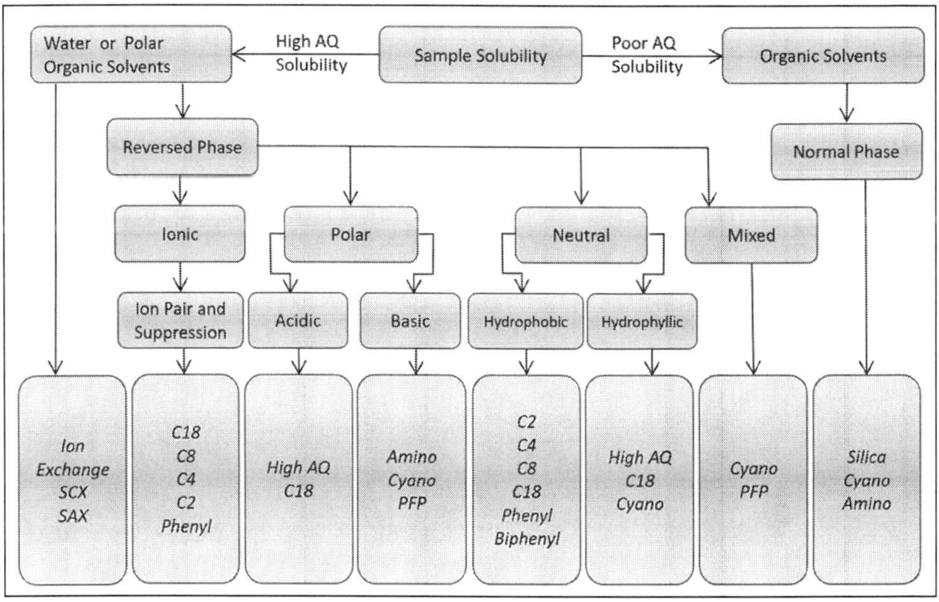

Figure 9.3: Column Chemistry Choices Based on Sample Solubility.

The Phenomenon of Hydrophobic Retention: So-called reversed phase columns have a bonded phase on the surface of the silica particles which is hydrophobic in nature. Examples are C30, C18, C8, C4. In principle, "like dissoves like". Such a column will tend to retain sample constituents which are also hydrophobic, as long as the mobile phase is not stronger in its attraction for that sample constituent. More polar sample constituents will tend to elute from the column faster because they are retained to a lesser degree. For a reversed phase column, use a combination of solvents which give an overall "greater" polarity (less hydrophobic) than the stationary phase. In this way, the sample constituents have a reason to be attracted to the stationary phase (column bonded phase) and resist the tendency to flow with the mobile phase (solvent). Use the "weakest" or most polar condition necessary to differentiate between sample constituents as they migrate through the column. Examples of common mobile phase solvent pairs are Acetonitrile/Water and Methanol/Water. Changing the ratio of organic/water changes its overall polarity.

9.3.2.3 Stationary Phases Used in various Chromatographic Modes

a) Adsorption Chromatography

Silica gel is a popular column packing material for adsorption chromatography because of its silanol groups (Si-OH). Silanol groups are slightly acidic and can interact with polar or unsaturated functional groups of solutes through hydrogen bonding or dipole interactions.There are three types of silanol groups on silica gel: 1. Geminal: Two silanol groups bonded to the same silicon atom; 2. Vicinal: Two silanol groups bonded to adjacent silicon atoms; 3. Isolated: A single silanol group bonded to a silicon atom. The number and arrangement of these silanol groups determine the activity of the adsorbent and the retention of solutes. The number and topographical arrangement of the several types of OH groups, as shown in Figure 8.7, determine the activity of the adsorbent and thereby the retention of the solutes. These OH groups can be divided into three types: • silanol (free OH), • siloxane bond (Si-O-Si) and • hydrogen bond (Si-OH...O).

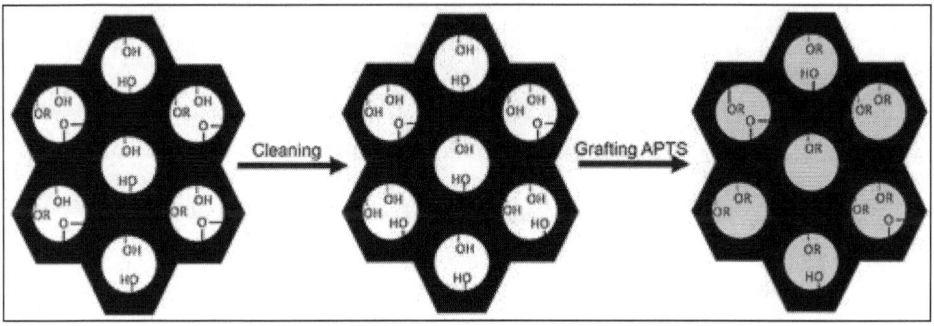

Figure 9.4: Structure of Silica Gel Depicting the Various Types of Hydroxyl Groups that Interact with the Functional Groups of Solute/Solvent Molecules.

Each of these groups has different activity that increases in the following order: Bound < free < H-bond. According to current models of adsorption process, it is assumed that adsorption sites are completely covered by either of solute or solvent molecules that are adsorbed depending on their relative strength in this competitive interaction. The competition between the solute and the mobile phase molecules for an active site provides the driving force and selectivity in separations. Interaction between a solute molecule and the adsorbent surface is best when functional groups overlap adsorption sites. Adsorption chromatography is less influenced by difference in molecular weight but certainly more by functional groups. For compounds of low to moderate polarity, adsorption chromatography often makes possible the separation of complex mixtures into classes of compounds with similar chemical functionality. Typical examples of group separations are polynuclear aromatics from a petroleum sample and the triglycerides from a liquid extract.

b) Partition Chromatography

Liquid chromatography can be divided into two types: liquid-liquid chromatography (LLC) and bonded phase chromatography (BPC). The main difference between the two is how the stationary phase is attached to the support particles. In LLC, the stationary phase is a liquid that is held to the surface of the support particles by physical adsorption. In BPC, the stationary phase is a liquid that is chemically bonded to the surface of the support particles. Bonded phase chromatography has become more popular than liquid-liquid chromatography in recent years because it has several advantages, including:

- ☆ Better reproducibility
- ☆ Faster analysis times
- ☆ More versatility

Bonded phase chromatography columns are typically made from rigid silica or silica-based materials. These materials are formed into uniform, porous, and mechanically sturdy particles that are typically 3, 5, or 10 micrometers in diameter. The surface of fully hydrolyzed silica is made up of chemically active silanol groups. The most common type of bonded phase coating is a siloxane, which is formed by the reaction of the hydrolyzed silica surface with an organochlorosilane. The surface of fully hydrolysed silica is made up of chemically silanol groups. The most useful bonded phase coatings are siloxanes formed by the reaction of hydrolysed surface with an organochlorosilane as shown below:

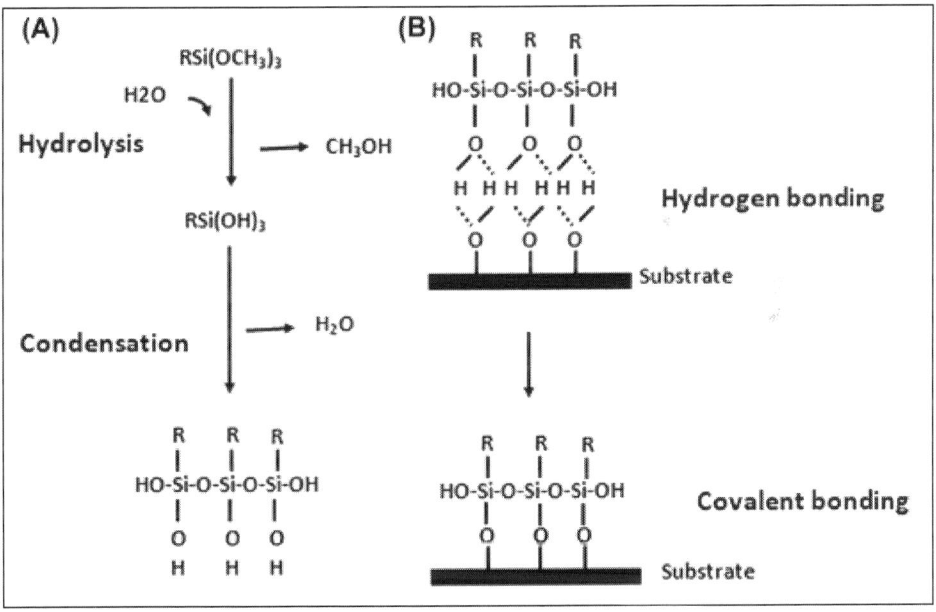

Surface coverage by silanization is limited to 4 $\mu mol/m^2$ or less because of steric effects. This means that the number of silane molecules that can bond to a given surface area is limited by the size of the silane molecules themselves. Unreacted SiOH groups are polar, meaning that they are attracted to water and other polar molecules. This can lead to chromatographic tailing, which is a phenomenon where the peaks on a chromatogram are not sharp and well-defined. To avoid this effect, siloxane packing's are often capped by further reaction with chloromethylsilane, which reacts with many of the unreacted silanol groups and makes the surface less polar. There are two types of partition chromatography, normal phase LC and reversed phase chromatography. In normal phase LC, the stationary phase is more polar than the mobile phase. This means that polar compounds will be attracted to the stationary phase and will elute more slowly than nonpolar compounds. In reversed phase chromatography, the stationary phase is less polar than the mobile phase. This means that nonpolar compounds will be attracted to the stationary phase and will elute more slowly than polar compounds. Reversed phase chromatography is the most common type of partition chromatography, and it is used in about three quarters of all HPLC separations. This is because reversed phase chromatography is very versatile and can be used to separate a wide variety of compounds.

Most commonly, the R group of the siloxane in these coatings is a n-octyl (C-8 chain) or n-octadecyl (C-18 chain). With such preparations, the long chain hydrocarbon groups are aligned parallel to one another and perpendicular to the particle surface, giving a brush or bristle-like structure. The relationship the between polarity of the sample with that of the column packing material and mobile phase is illustrated in Fig. 8.8. Retention increases with the hydrophobic character of the solute samples. Generally, the lower the polarity of the mobile phase, the higher is its eluent strength. The effect of chain length of the alkyl group upon column performance is illustrated in Figure 8.8 where it is observed that longer chains produce packings that are more retentive. For example, maximum sample size for a C18 packing is roughly double that for a C4 preparation under similar experimental conditions.

Table 9.3: Suitability of Column per Packing

Sample	*Column Packing*	*Mobile Phase*
Low/moderate polarity(Soluble in aliphatic hydrocarbon)	Bonded C-18	Methanol/Water
Moderate polarity(Soluble in methyl ketone)	Bonded C-8	Acetonitrile/Water
High polarity(soluble in lower alcohol)	Bonded C-2	1,4 Dioxane/water

In commercial normal-phase bonded packing's, the R in the siloxane structure is a polar functional group such as cyano ($-C_2H_4C_N$), diol ($-C_3H_6OCH_2CHOHCH_2OH$), amino ($-C_3H_6NH_2$), and dimethylamino ($C_3H_6N(CH_3)_2$). The polarities of these packing materials vary over a considerable range with the cyano type being the last polar and the amino types the most. Diol packing's are intermediate in polarity. With normal phase packing's, elution is carried with relatively non-polar solvents such as ethyl ether, chloroform and n-hexane.

c) Ion-exchange Chromatography

In this case, column packings have charge bearing functional groups attached to a polymer matrix. The functional groups are permanently bonded ionic groups associated with counterions of the opposite charge. Some ion-exchange packings bear negatively charged groups and are used for exchanging cationic species whereas others are designed for exchanging anionic species. Similarly, some functional groups such as –COOH or -PO3 2– have weak acidic or basic properties whereas some others have considerable affinity for heavy metal cations. Several structural types of packings, have been used in ion-exchange HPLC. Of these, the pellicular type consists of a resin coating, about 1-2 µm thick, on a glass bead of 30-40 µm diameter. Superficially porous resins are obtained by coating glass beads with a thin layer of silica microspheres on which ion exchanger is bonded. This increases the interface between the resin and mobile phase. Either type of these packings have low exchange capacity, 0.01 – 0.1 meq/g. The exchanger may also be bonded to silica microparticles by means of silylation reactions or polymerized into pores of a superficially porous silica gel.

d) Size Exclusion Chromatography

Column packing's can be made of either semi-rigid, cross-linked macromolecular polymers or rigid, controlled pore size glasses or silicas. Semi-rigid materials swell, so they should only be used at pressures up to 300 psi to avoid bed compressibility. Styrene-divinyl benzene polymers can be used to fractionate samples with molecular weights between 100 and 5000 million. Partially sulfonated polystyrene beads are compatible with aqueous systems, while non-sulfonated beads are compatible with non-aqueous systems. Both types of beads have diameters of around 5 µm. Another class of hydrophilic porous packing is prepared by suspension polymerization of 2-hydroxyethyl methacrylate with ethylene dimethacrylate. These packings can withstand pressures up to 3000 psi and can be used with aqueous systems and a variety of polar organic solvents. Porous glasses and silicas cover a wide range of pore size diameters. For example, a series of particle size diameters and operating ranges of molecular weights are listed in Table 9.3.

Table 9.3: Correlation of Pore Size Diameter and Operating Range of Mol. Wt

Pore-Size Diameter (µm)	Operating Range (Daltons)
4	1000-8000
10	1000-3000
25	2500-125000
55	11000-350000
150	100000-1000000
250	200000-1500000

These packing's are chemically resistant at pH <10 and can be used with aqueous and polar organic solvents. With nonpolar solvents, it is desirable to deactivate the surface by silylation. Porous inorganic packing's have distinct advantages over organic exclusion packing's. The surface of a typical hydrophilic packing has the following structure;

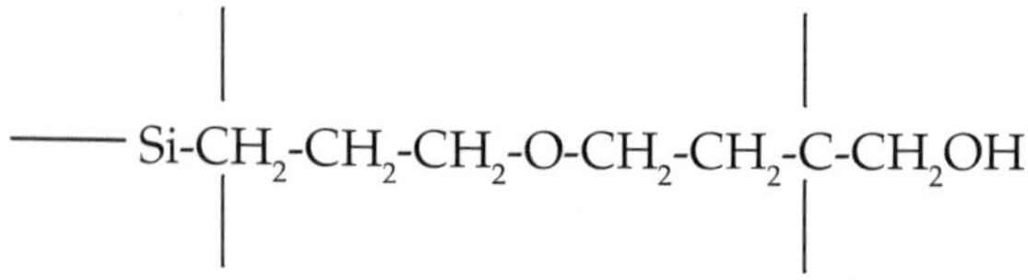

e) *Ion Chromatography*

It differs from ion-exchange chromatography in the nature of exchange resins. The technique involves an ion-exchange column and a means of suppressing (removing) ionic species other than the sample ions in the eluting mobile phase to facilitate detection of the sample by a conductivity monitor as schematically illustrated in Figure 9.5.

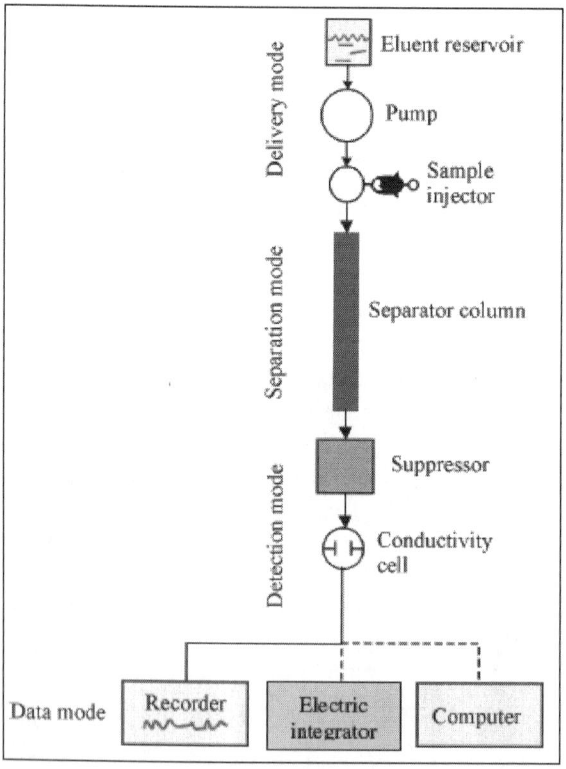

Figure 9.5: Schematic Diagram of Ion Chromatograph with Separation Column.

The column packing consists of a neutral polymer core of ~ 10 μm diameter depending on whether the packing will be used for the separation of cations or anions. Contrary to the conventional ion-exchange chromatography where core is sulphonated or aminated leading to the formation of sulfonic acid or quaternary amine groups, in ion chromatography, a monolayer of aminated or sulphonated polymeric anion exchange beads is used. Similarly, for a cation exchanger, there would be an intermediate layer of aminated groups covered by a thin layer of sulphonated resin beads. Due to the proximity of all the active sites to the eluent-resin interface, this type of exchanger has favorable mass transfer characteristics. It has low exchange capacity, about 0.020 meq/g of copolymer. In most applications, silica based materials are inappropriate due to their degradation in the presence of aqueous eluents and their poor selectivity for some ionic species. The eluent passes through a suppressor column where the eluting or background electrolyte is effectively removed by converting it into water or, water and carbon dioxide *i.e.*, sodium ions are replaced by hydronium ions or methylsulfonate ions with hydroxyl ions. A miniaturized 'self-regenerating' suppressor cartridge incorporating an electrolysis cell is also available where H_3O^+ and O_2 are continually formed by the electrolysis of a stream of deionized water passing through an anode compartment and similarly, OH^- and H_2 are formed in a cathode compartment. Both compartments are separated from the eluent either by cation or anion exchange membranes depending on whether anionic or cationic analytes are to be separated.

f) Chiral Chromatography

Many drugs are only effective in one enantiomeric form, while the other form may be inactive or even harmful. HPLC chiral separations can be used to separate these enantiomers by exploiting their different interactions with a chiral stationary phase (CSP). The CSP is a silica gel support coated with a polymeric material to which an optically active isomer is bonded. In one example, the *l*-form of the amino acid proline is bonded to polystyrene-p-divinylbenzene to create a CSP for separating amino acid enantiomers. Cu^{2+} ions are added to the solution of the analyte enantiomers to form a ternary complex with the stationary phase and the amino acid anion. The formation constant of this complex differs for *d*- and *l*-forms of the analyte amino acid, allowing for their separation. Cyclodextrin-bonded stationary phases are particularly efficient at resolving structural isomers, such as prostaglandin A1, A2, B1, and B2; α- and β-naphthols; o,o2 and p,p2-biphenyls; and the ortho-, meta-, and para- isomers of nitrophenol, nitroaniline, xylene, cresol, and aminobenzoic acid. Recently introduced graphitized carbon and new generation of rigid porous polymeric micro beads based on styrene/divinyl benzene can be used as alternatives to silica-based CSPs. These CSPs can be used over a wide range of pH between 1 and 13. Some examples of column packings used in HPLC and their applications are listed in Table 9.4.

Table 9.4: Some Typical Column Packing's Used in HPLC

Packing	*Mode of HPLC*	*Applications*
Micro-particulate silicas; spherical or irregular particles; mean particle size 3, 5 and 10μm chemically modified versions of the above (bonded-phase packings)	LSC (adsorption)	Non-polar to moderately polar mixtures, *e.g.*, polyaromatics, fats, oils, mixtures of isomers
Octadecyl (ODS or C18)	BP (bonded phase) and Ion Pair Chromatography (IPC)	Wide range of moderately polarmixtures, *e.g.*, pharmaceuticals and drugs, amino acids
Octyl (C8)	BPC, IPC	More polar mixtures, *e.g.*, pesticides, herbicides, peptides, metabolites in body fluids
Short chain (C3 or less) ead of IEC	BPC, IPC	IPC applications of above three packings include bases, dyestuffs and other multiply charged species; used inst
Nitrile	Normal phase and BP C	Alternative to silica, can give better results
Aminoakyl	BPC	Carbohydrates including sugars
Anion and cation exchangers (tertiary amine or sulphonic acid)	IEC (Ion-exchange chromatography)	Ionic and ionizable compounds, *e.g.*, vitamins, water-soluble drugs, amino acids, food and drink additives
Controlled porosity silicas (chemically modified to reduce adsorption effects) modes of HPLC	Size exclusion Chromatograph y	Polymer mixtures, screening of unknown samples. Increasing use for separating mixtures of smaller molecules before other
Chiral amino acids bound to aminopropyl	Chiral Chromato-graphy (CC)	Mixtures of enantiomers especially of drugs
Chiral peptides	CC	
Cyclodextrins CC		

9.3.3 Solvent Supply System

The mobile phase in liquid chromatography (LC) is important because it determines how long it takes for different substances to travel through the column and be separated. In high-performance liquid chromatography (HPLC), high-purity solvents without dissolved gases must be used to avoid affecting the retention time and separation of the constituents. The eluent system consists of reservoirs from which one or more solvents can be selected. Essential features of a modern HPLC system include flow control, an inlet filter, and a degassing facility. The degassing facility is used to remove dissolved gases, which can have a negative impact on the column's performance. The general criteria for the selection of a mobile phase are:

- It should dissolve the sample.
- It should keep the column stable.
- It should be compatible with the detector.
- It should be immiscible with the stationary phase.
- Its viscosity should not be high.
- Active fluorides should be avoided when using glass column.

The eluting power of a solvent determines how well it can remove a solute from an adsorbent. It is affected by the polarity of the solvent, the polarity of the stationary phase, and the nature of the solute. The capacity factor, k', is a measure of how strongly the solute is adsorbed to the stationary phase and is controlled by the strength of the solvent. Snyder defined the solvent strength parameter, ε^o, as the absorptive energy per unit area of absorbent. Solvents are listed in order of increasing solvent strength in Table 9.5. This list is also called the eluotropic series of solvents. The log k' for a given solute varies linearly with εo, which can be used to predict the elution order of solutes in adsorption chromatography. Other properties of solvents that must be considered include boiling point, viscosity, detector compatibility, flammability, and toxicity. Generally, solvents with lower boiling points and viscosities give higher chromatographic efficiency and lower back pressure.

Table 9.5: Solvent Strength Parameter, ε^o and the Physical Properties of Selected Solvents Used in HPLC

Solvent	*ε^o (SiO_2)*	*ε^o (Al_2O_3)*	*Viscosity, 20°C (mN secm%2)*	*Refractive index, 20°C*
Pentane	0.00	0.00	0.23	1.358
Hexane		0.00	0.313	1.375
Cyclohexane	-0.05	0.04	0.980	01.426
Carbon disulphide	0.14	0.15	0.363	1.628
Carbon tetrachloride	0.14	0.18	0.965	1.460
1-Chlorobutane		0.26	0.47	1.402
Di-isopropyl ether		0.28	0.379	1.368
2-Chloropropane		0.29	0.335	1.37
Benzene	0.25	0.32	0.65	1.501
Diethyl ether	0.38	0.38	0.23	1.353
Chloroform	0.26	0.40	0.57	1.443
Acetone	0.47	0.56	0.32	1.359
Acetonitrile	0.50	0.65	0.375	1.34
Methanol		0.95	0.60	1.329
Water		Large	1.00	1.333

As the column flow rate is proportional to the product of the linear velocity and the cross sectional area of the column, the solvent consumption is considerably reduced as illustrated in Table 9.6.

Table: 9.6 Solvent Consumption for Different Diameter HPLC Columns

ID (mm)	*Flow Rate for Linear Velocity of 0.14 cm/sec (mL/min)*	*Volume in a 8 hr day (mL)*
0.51	0.02	6.9
0.71	0.027	13
1.02	0.044	24
1.29	0.079	38
1.59	0.12	57
4.6	1.00	480

When two more solvents with a fixed composition are used, it is called isocratic elution. This, however, is a very cumbersome process and instead gradient elution is used.

9.3.3.1 Gradient Elution

Gradient elution is a liquid chromatography technique that involves gradually changing the composition of the mobile phase from less polar to more polar. This is done by mixing two solvents in a reservoir and pumping the mixture through the column. As the mobile phase becomes more polar, the solutes in the sample elute from the column in order of increasing polarity. Gradient elution is useful for separating complex mixtures of solutes that cannot be separated by isocratic separation, which uses a mobile phase with a constant composition. It is also especially useful for separating components that differ widely in polarity. In modern liquid chromatographs, gradient elution is controlled by a microprocessor, which regulates the flow of solvents from the reservoirs to the mixing chamber. This allows for precise control of the gradient profile, which can be customized to achieve the best separation for a particular sample. Gradient elution can reduce the run time and increase the sensitivity of a liquid chromatography analysis. As the gradient develops, solutes that are more strongly retained by the stationary phase are eluted more quickly. This can reduce the overall time required for the analysis. Gradient elution also has some disadvantages. One disadvantage is that it can cause baseline drift in the detector response. This is because one of the solvents in the mobile phase may give a response at the detector. To minimize baseline drift, a blank gradient run is often performed between samples. Another disadvantage of gradient elution is that it can take longer to set up and run than isocratic separation. This is because the solvents must be mixed in the correct proportions and the gradient profile must be programmed into the instrument. Overall, gradient elution is a powerful liquid chromatography technique that can be used to separate complex mixtures of solutes. It is especially useful for separating components that differ widely in polarity.

Pumps

A variety of pumps are used to maintain flow rate and pressure of the mobile phase. Also a degasser is needed to remove dissolved air and other gases from the solvent. A desirable feature of the delivery system is the capability of generating solvent gradient. A pump should be able to operate up to a pressure of 100 atm (1500 psi) though in some cases 400 atm (6000 psi) is desired. For most analytical columns, only moderate flow rates of 0.5 – 2 mL/min may be required. However, for microbore columns, low flow rates of only a few microlitres/min may be sufficient. Also, a pump should have a small hold up volume. Some typical pumps are described below:

i) Reciprocating Piston Pump

Reciprocating pumps are the most common type of pump because they are inexpensive and can produce a wide range of flow rates. They work by using a small motor-driven piston to move back and forth in a hydraulic chamber. The piston sucks in solvent from a reservoir through check valves. A hydraulic fluid then transmits the pumping action to the solvent through a flexible diaphragm, which helps to minimize solvent contamination and corrosion of the pump parts. Reciprocating pumps can produce a wide range of flow rates by varying either the stroke volume (the amount of solvent pumped per cycle) or the stroke frequency (the number of cycles per minute). They can also deliver solvent continuously without any restrictions on the reservoir or operating time. Reciprocating pumps have several advantages, including:

- ☆ Small internal volume (35 to 400 μL)
- ☆ High output pressures (up to 10,000 psi)
- ☆ Ready adaptability to gradient elution
- ☆ Constant flow rates that are largely independent of column back pressure and solvent viscosity

ii) Syringe Type Displacement Pump

These pumps work by using a piston to move a fixed amount of solvent forward at a constant rate. The piston is driven by a screw feed drive, which is turned by a digital stepping motor. The speed of the motor can be changed to control the flow rate of the solvent. The solvent chamber has a limited capacity of 250-500 mL, but it can be refilled if needed. These pumps can deliver a pulse-free flow of solvent at high pressures of 200-475 atmospheres.

iii) Constant Pressure Pump

Plunger pumps are a type of pump that uses a large piston to generate high liquid pressures. The pressure delivered through the piston is proportional to the ratio of the areas of the two pistons, which is usually between 30:1 and 50:1. This means that a low pressure gas source of 1-10 atm can be used to generate high liquid pressures (1-400 atm). Plunger pumps have a valving arrangement that permits the rapid refill of the solvent chamber, which has a capacity of about 70 mL. This system provides pulse-less and continuous pumping, including high flow rates for preparative applications. Plunger pumps are useful for pumping columns, but they are inconvenient for solvent gradient columns. This is because plunger pumps cannot deliver a smooth and continuous gradient of solvent compositions, which is necessary for solvent gradient columns.

iv) Pneumatic Pump

These types of pumps are simple, inexpensive and pulse free but suffer from limited capacity and pressure output. In this case, mobile phase is contained in a collapsible container housed in a vessel that can be pressurized by a compressed gas. These are not amenable to gradient elution and are limited to pressures less than 2000 psi. Most commercial instruments are equipped with computer controlled devices for measuring the flow rate by determining the pressure drop across a restrictor located at the pump outlet. Any difference in signal from a preset value is then used to decrease or increase the speed of the pump motor. Composition of solvents may be continuously varied or in a stepwise fashion.

9.3.3.2 Detectors

A detector is an essential part of an HPLC instrument and must be carefully selected to achieve selective separation and precise measurement. The most important factor is continuous detection of the separation process of a component, which can be immediately displayed and recorded. However, a good detector must also have the following characteristics:

Linear response to solute concentration in the range of 0.1 µg/mL to 1 ng/mL

- ☆ Only responds to the solute and not to the solvent or changes in the solute-to-solvent ratio
- ☆ Insensitive to changes in temperature, pressure, and flow rate

Although highly sensitive detectors have been developed for HPLC, there is no universal detector that can be used for all types of samples and concentration ranges. The choice of detector depends on the specific application, but sometimes more than one detector may be used. These are of two basic types:

i) Bulk Property Detectors

These types of detectors measure on overall change in a physical property of the mobile phase with and without solute *e.g.* refractive index, dielectric constant, density, electrical and thermal conductivity, vapour pressure *etc.* These types of detectors are somewhat insensitive and require good temperature control.

ii) Solute Property Detectors

These respond to a physical property of the solute that is not exhibited by the pure mobile phase. These are highly sensitive with a detection signal for a few ng or even lesser amount of sample. For example absorbance, fluorescence, diffusion current and electrochemical detectors are considered in this category. Besides low detection limit, a HPLC detector must meet following requirements:

- ☆ Selective response towards one or more classes of solutes.
- ☆ It must be small and compatible with liquid flow.
- ☆ It should have good stability and reproducibility.
- ☆ It must have low dead volume to minimize extra-column band broadening.
- ☆ It must have high reliability and ease of use.
- ☆ It must have small response time, at least 10 times less than the peak width of a solute.
- ☆ It should have a linear response to solute that extends over several orders of magnitude. Some characteristics of commonly used detectors in HPLC systems are listed in Table 9.6.

Table 9.6: Characteristics of Detectors

Detector Property	*Typical LOD* (mass)*	*Linear Range*	*Commercial Availability*
Absorbance	10 pg	3-4	Yes
Fluorescence	10 fg	5	Yes
Electrochemical	100 pg	4-5	Yes
Conductivity	100 pg^{-1} ng	5	Yes
Refractive Index	1 ng	3	Yes
Mass spectrometry	<1pg	5	Yes
FTIR	1μg	3	Yes
Light scattering	1μg	5	Yes
Optical activity	1 ng	4	No
Photo ionization	<1pg	4-5	No

* Actual LOD will depend on the compound.

It is essential to know the band spreading for estimating detection limit of a particular detection system. Connecting tubing must be minimum (not longer than 20 cm). Tubing diameter (< 0.25 mm) is most critical as this would create a 10μL volume and so also dilution factor is important for detection limit. A detector must have a linear dynamic range so that major and trace components can be determined in a single analysis.

a) Optical Detectors

These include uv-visible spectrophotometers and are the most widely used ones in HPLC instruments. Three types of absorbance detectors are available;

- ☆ fixed wavelength (UV) detector,
- ☆ variable wavelength detector using deuterium lamp, and
- ☆ scanning wavelength detector.

These have noise limitations due to thermal instability in the flow cell and in the optical and electronic components. Therefore, thermosetting to 0.01 °C is required as it may put noise limitation of 10–6 absorbance units. A quartz collimating lens focuses the radiation on the sample and reference cell with detector cell volume of 8µL/cm optical path length. A schematic diagram of a typical flow through cell for absorbance measurements in eluents. A rise time of 0.1 sec is needed for fast HPLC measurements. It is essential that the mobile phase solvent must not absorb or may absorb only weakly. Water, methanol, hexane and acetonitrile all permit operation in the for uv to at least 210 nm. Many absorbance detectors are double beam devices where one beam passes through the element cell and the other through a filter to reduce its intensity. Alternatively a chopped beam system in conjunction with a single phototube is used. Detection limit can be estimated if the noise level and the approximate molar absorptivity are known at the operating wavelength. Assuming 1 cm path length and a noise level of 0.00004 absorbance unit, detection limit is

$$2\,(\text{noise})/b\varepsilon = 0.00008/\varepsilon \text{ mol cm}^{-1}\text{litre}^{-1}$$

If ε =10000, the minimum detectable concentration is 8 nM/L or 4 ng/mL for a compound having mol wt = 500. If it is required in terms of sample weight instead of concentration, the sample volume and system dilution factor must be considered. For 5µL sample and a dilution factor of 20 (Mol wt 500), detection limit is

$$2(0.00004)\ 20 \times (5 \times 106\text{ L})\ 500/(10{,}000\text{ Lmol}^{-1}\text{ cm}^{-1})\ 1\text{ cm} = 0.4\text{ ng}$$

b) Fixed Wavelength Detector

The device utilizes a light source that emits light with the highest intensity at specific wavelengths. These wavelengths are isolated using appropriate filters. This results in minimal noise, but the wavelengths cannot be freely chosen. For example, a medium pressure mercury lamp can be used to produce light at wavelengths of 254, 280, 313, 334, and 365 nm by using narrow bandpass interference filters. Visible region wavelengths can be obtained using a quartz iodine lamp and appropriate filters. However, these typically have short-term noise levels.

c) *Variable Wavelength Detector*

A wide band pass UV-visible spectrophotometer is typically coupled to a chromatographic system. It provides a broad selection of wavelengths in the UV and visible regions. A versatile detection system is based on a spectrophotometer equipped with a grating monochromator and a continuum source, such as a deuterium lamp for the UV region and a tungsten-halogen lamp for the visible region (see Figure 9.6). These systems have double beam optics, stable low noise electronics, and are microprocessor controlled. Some can be programmed to select a sequence of optimal monitoring wavelengths during chromatographic runs and to record complete UV spectra.

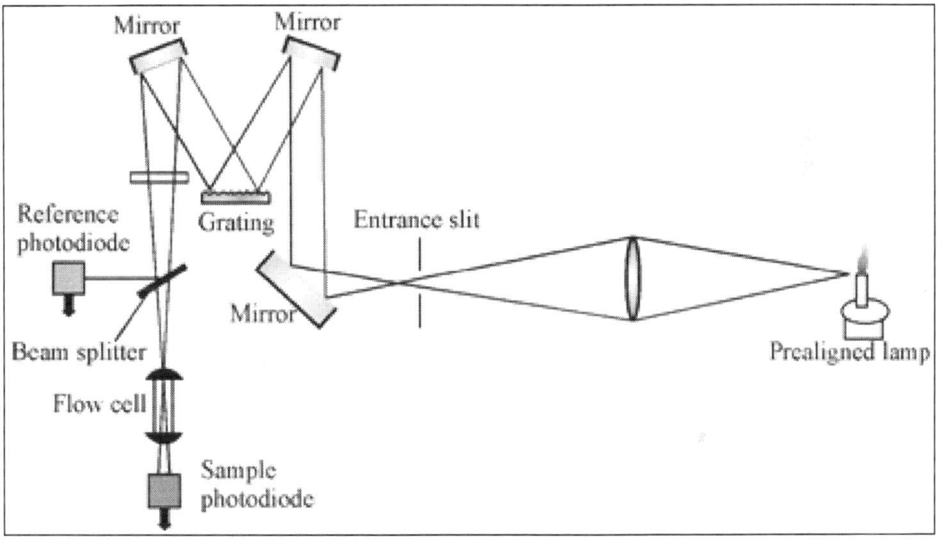

Figure 9.6: A Schematic Diagram of a Variable Wavelength Detector.

iii) *Scanning Wavelength Detector*

Solid-state diode arrays are used to simultaneously record the spectra of all solutes by monitoring all wavelengths at once. This allows for the rapid acquisition of complete spectra, with a range of 190 to 600 nm being obtainable in just 0.01 seconds. Some instruments can be configured to scan the complete spectrum for the monitoring of independent signals. Additionally, all signals can be integrated between two preset wavelengths, enabling the detection of multi-component complex samples.

d) *Infrared Absorbance Detector*

Infrared detector cells share a similar construction to those used with ultraviolet radiation. However, infrared detector cells utilize windows made of sodium chloride or calcium fluoride. Additionally, the cell lengths range from 0.2 to 1.0 millimeters, with volumes ranging from 1.5 to 10 microliters.

Simpler instruments can be operated at one or more single wavelength settings. Alternatively, spectra for peaks can be scanned by stopping the flow at the time of elution. Two types of IR detectors are commonly used. The first type utilizes wavelength scanning provided by three semi-circular wedges with a range of 2.5 to 14.5 micrometres or 4000 to 690 wave numbers per centimetre. The second type of IR detector is based on Fourier transform (FT) instruments, which can be attached to HPLC detectors.

e) Fluorescence Detector

This is similar to the design of fluorometers and spectrofluorometers where a photoelectric detector is located at 90o to the excitation beam. The simplest detectors use a mercury excitation source and one or more filters to isolate a band of emitted radiation. More sophisticated instruments are based upon a xenon source and employ a grating monochromator to isolate the fluorescent radiation. These have the advantage of being highly sensitive, typically greater by more than an order of magnitude than most spectrophotometers and selective (specific) in picking out a specific component of a mixture which fluoresces from a host of other components. These are typically in the range of 10–9 to 10–12 g/mL. Quite often the number of fluorescent species can be enlarged by preliminary treatment of samples with fluorescent derivative forming reagents. Further developments are taking place based on tunable laser sources leading to enhanced sensitivity and selectivity.

B) Differential Refractometer

It monitors the difference in refractive index between the references (mobile phase) and the column eluent. It responds to any solute where refractive index is significantly different from that of the mobile phase. A schematic illustration of a differential refractive index detector is shown in Figure 9.7 where solvent passes through one half of the cell and then it passes through another chamber where eluent flows.

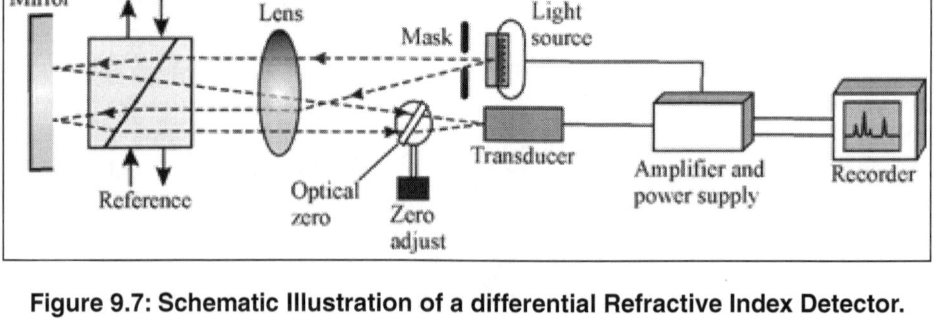

Figure 9.7: Schematic Illustration of a differential Refractive Index Detector.

Two compartments are divided by an angled glass plate. If the refractive indices of the two solutions differ, the incident light beam will bend. The resulting displacement of the beam relative to the detector's photosensitive surface causes a change in the output signal. These detectors, which are based on Fresnel's law of reflection, have the advantage of responding to most solutes and have a wide range of linearity where a single cell can cover the entire refractive index range. The cell volume is 15-25 µL. Additionally, they are highly reliable and unaffected by flow rate. However, they are temperature sensitive and do not produce sensitive results. Reflection type refractometers measure the change in the percentage of reflected light at a glass-liquid interface as the refractive index of the liquid changes. Two collimated beams from the projector light illuminate the reference and sample cells, which are made of a Teflon gasket clamped between the cell prism and a stainless steel reflecting back plate. As the light beam passes through the interface, it travels through the flowing liquid film and strikes the surface of the reflecting back plate. This diffuse reflected light appears as two spots of light that are imaged by lenses onto dual photodetectors. Because the ratio of reflected light to transmitted light is a function of the refractive index of the liquids, the illumination of the cell back plate is a direct measure of the refractive index in each chamber. With mobile phase flowing through both compartments, coarse zero is adjusted by rotating the entire projector assembly. Fine adjustment is made with the optical plate, a glass plate which can be rotated ±30° from normal. Two different prisms must be used to cover the useful range of refractive index.

C) Electrochemical Detectors

Electrochemical detectors can only be used if the solute molecules in aqueous and aqueous-organic phases exhibit voltammetric characteristics. These detectors are classified into several types based on amperometry, polarography, coulometry, and conductometry. Electrochemical detectors have not been as widely used as optical detectors, despite their advantages of simplicity, high sensitivity, convenience, and broad applicability. A variety of commercially available HPLC/electrochemical detectors are available. These detectors have the potential to be universal detectors, meeting a long-standing need. A nanoampere-level current can be measured at a controlled potential as a function of time using a detector that consists of a flow cell sandwiched between two blocks, one plastic and the other stainless steel, with a 50 µm thick polyfluorocarbon gasket. A working electrode made of Pt, Au, glassy carbon, or carbon paste is placed on one side of the channel, and a reference electrode (usually Ag/AgCl) is connected to the working region by tubing. The cell volume is 1-5 µL. A polarographic detector consists of a mechanically controlled dropping mercury electrode where the eluent flows around the mercury droplet. During elution, the potential of the electrode is maintained at a suitable level. A plot of current versus time provides the elution pattern for species that are reduced at the chosen potential. A dual electrode detector offers additional specificity. In one configuration, two working electrodes are placed parallel to the flowing stream, with each electrode held at a different potential, resulting in two simultaneous chromatograms. The current ratio at the two potential settings is calculated and used for peak confirmation. In the second configuration, two electrodes are arranged in series. The upstream working electrode generates an electroactive product from the analyte, which is subsequently detected at the downstream working electrode. The series arrangement limits the number of electroactive compounds that are detected. Only compounds that generate stable electroactive products and reach the second electrode are sensed.

D) Mass Spectrometric Detector

Liquid chromatographs (LCs) and mass spectrometers (MSs) are powerful analytical tools that can be combined to provide detailed information about the composition of complex mixtures. However, there is a fundamental challenge in coupling these two instruments due to the mismatch between the relatively large solvent volume used in LC and the vacuum requirements of MS. Several interfaces have been developed to overcome this challenge. These interfaces work by transferring the analytes from the LC eluent into the MS vacuum chamber without introducing excessive amounts of solvent. Some common interfaces include:

- Direct liquid introduction (DLI): In DLI, a small fraction of the LC eluent is introduced directly into the MS ion source. This approach is simple and can be used for a wide range of analytes. However, it can suffer from problems with solvent overload and analyte discrimination.
- Moving belt/wire interfaces: In these interfaces, the LC eluent is deposited on a continuously moving belt or wire. The solvent is then evaporated as the belt or wire moves through a heated chamber. The resulting analyte residue is then introduced into the MS ion source. Moving belt/wire interfaces are well-suited for the analysis of thermally stable analytes.
- Thermospray (TSP): In TSP, the LC eluent is passed through a heated capillary tube. The resulting aerosol is then introduced into the MS ion source. TSP is a versatile interface that can be used for the analysis of a wide range of analytes. However, it can be sensitive to the composition of the LC mobile phase.

In addition to the interface, a LC-MS system also includes a liquid chromatograph and a mass spectrometer. The liquid chromatograph is used to separate the components of the sample. The mass spectrometer is then used to measure the mass-to-charge ratio of the separated components.

LC-MS is a powerful analytical technique that can be used for a wide range of applications. Some common applications include:

- Drug discovery and development
- Environmental analysis
- Food safety analysis

- ✰ Clinical diagnostics
- ✰ Proteomics

9.3.4 Optimization of Separation

Chromatographers are constantly striving to achieve higher column efficiency. This is driven by the resolution equation, which demonstrates that improvements in column efficiency always lead to enhanced resolution. A plethora of experimental variables influence chromatographic efficiency, and its optimization can be accomplished through various methods, depending on the number of variables that one is willing to modify. These variables include pressure, temperature, particle size, column length, and eluent velocity. In the early stages of high performance liquid chromatography (HPLC), the selection of column formats (particle size, type, and column diameters) was rather limited. As a result, optimization was often achieved by adjusting operational variables such as eluent velocity, column temperature, and operating pressure.Today, the selection of column formats is considerably wider, with a range of particle sizes available between 1.7 µm and 5 µm. Additionally, numerous column lengths can be achieved by coupling columns in series. This makes the optimization of these nominally "discrete" variables (*i.e.*, particle size and column length) possible. The optimization of performance (*i.e.*, efficiency versus time) in HPLC has been a topic of research for decades. Van Deemter was among the first to understand how plate count on a given column could be optimized by varying the eluent velocity. When column length is also allowed to vary, the Poppe plot or kinetic plot techniques can be used to consider pressure and time constraints as part of the optimization process. Ultimately, particle size can also be varied, and this optimization has been studied by several LC pioneers.

As the need for faster separations without sacrificing efficiency continues to grow, so does the demand for more efficient separations. This demand is driven by the ever-present need to analyze more and more samples in the same or less time. The definition of "fast" varies depending on the application. For example, in the pharmaceutical industry, a "fast" separation for organic impurities analysis could be 10 minutes, while a "fast" separation for cleaning analysis in drug manufacturing could be 1 minute. In this column, we focus on the use of fast LC in pharmaceutical drug product dissolution testing. In dissolution testing, the primary goal is to quantify the amount of active pharmaceutical ingredient (API) released from dosage forms. As a result, only moderate plate counts are required, and speed of analysis is more important than high efficiency. Dissolution samples are often analyzed by either direct UV or HPLC-UV, with both techniques being used in roughly equal proportions. While UV-VIS spectroscopy is cheaper and simpler, HPLC offers many advantages, such as better specificity and a greater linear dynamic range. HPLC is also more versatile, especially for early drug development when different formulations and strengths are being screened. The main disadvantage of traditional HPLC is its relatively slower speed (about 3-5 times slower than UV analysis). Clearly, increasing the speed of HPLC would make it more attractive for dissolution testing.

There is a need for theoretical guidance to develop better ultrafast isocratic separations (approximately 30 seconds) for dissolution testing. The goal of this research is to make the speed of HPLC competitive with UV analysis. In addition, ultrafast HPLC is critical to the implementation of automated dissolution testing systems with online HPLC analysis. There are many approaches to speeding up HPLC analysis. One popular approach is to use monolithic columns at very high eluent velocities. Sub-1-minute separations have been achieved using this method. However, a recent review of different technologies for enhancing speed suggests that the most effective way to achieve very high speed is to use a combination of higher temperatures, higher pressures, and smaller particles. In addition, substantially lower solvent consumption can be achieved with narrow-bore columns packed with small particles compared to monolithic columns. Therefore, in this column, we focus on optimizing an ultrafast separation under high temperature and high pressure conditions.

The goal of most separation optimizations is to either maximize the number of plates (a measure of separation efficiency) in a given amount of time, or to achieve a desired number of plates in the shortest possible time. For simplicity, we will only discuss optimization for maximizing the plate count in a given amount of time (*i.e.*, with a fixed column dead time to). We will first present three different optimization schemes in order of increasing complexity, followed by a simple, practical, stepwise optimization procedure.

One-Parameter Optimization: In this case, one has already chosen a particle size and column length. The only variable left to optimize is velocity. Here, the van Deemter equation is used to calculate the eluent velocity that gives the minimal plate height, provided that this velocity (and flow rate) can be reached within the pressure limit and flow limit of the instrument. Equations 9.1 and 9.2 give the van Deemter equation and the optimal linear velocity, respectively, where Dm is the solute diffusion coefficient, dp is particle size, and A, B, and C are van Deemter equation coefficients. It should be noted that throughout this column, it is assumed that the plate height is related to velocity through the van Deemter equation. This is always a rather good approximation as long as one does not work far from the minimum in the plate height curve.

$$h = A + \frac{B}{v} + C * v \tag{9.1}$$

$$U_{c,opt} = \frac{D_m}{d_p}\sqrt{\frac{B}{C}} \tag{9.2}$$

However, the problem of operating at the van Deemter optimum velocity specified by equation 2 is that one frequently cannot achieve the separation in the desired time (using t0 as a proxy for the analysis time) because column length is already chosen. To achieve the target t0, one needs to operate at the velocity given by equation 3, where L is the column length, εt and εe are the total and interstitial porosities of the column, respectively (typically about 0.55 and 0.38), and λ is the ratio of the porosities.

$$U_c = \frac{€_t}{€_o} * \frac{L}{\lambda} * t_o \tag{9.3}$$

The resulting plate count obtained at this velocity can be calculated easily by inserting equation 3 into equation 1. This plate count is always lower than the plate count calculated based upon the van Deemter optimum velocity; however, it is also the more realistic estimate.

Two-Parameter Optimization: In this case, one has chosen a particle size but needs to optimize both column length and velocity. This is precisely the optimization that is dealt with by the Poppe and kinetic plot techniques. For a given analysis time, there is exactly one optimal combination of column length and velocity that will produce the maximum plate count. Equations 9.4 and 9.5 give the optimal column length and velocity; P_{max} is the maximum operating pressure, η is mobile phase viscosity, and t0 is the column dead time. It is important to note that the optimal velocity as described here is almost never the same as the van Deemter optimal velocity.

$$\mathbf{L} *= \frac{\sqrt{\mathbf{Pmax}\ \lambda\mathbf{t0}}}{\phi\eta\lambda\mathbf{t0}} \tag{9.4}$$

$$Uc *= \frac{\sqrt{\mathbf{Pmax}}}{\phi\eta\lambda\mathbf{t0}}\ d_p \tag{9.5}$$

Three-Parameter Optimization: In this case, one simultaneously optimizes the particle size, column length, and eluent velocity. The solution to this optimization is that one must work at the van Deemter optimum but then must also use the particle size, column length, and velocity that are dictated by equations 9.6–9.8. This scenario often is referred to as the Knox–Saleem limit, which is the absolute best separation one can do assuming the optimal particle size and column length are available and the optimal velocity can be delivered by the pumping system.

$$dp * = \frac{[[e\phi\eta\mathbf{B}]]^{\frac{1}{4}}}{P_{\max C}} * ([e\lambda\mathbf{t0}])^{\frac{1}{4}} * D_m^{\frac{1}{2}} \tag{9.6}$$

$$L * = \frac{[P_{\max}\ \square]^{\frac{1}{4}}}{\phi\eta\mathbf{C}} * (\lambda\mathbf{t0})^{\frac{1}{4}} * (\lambda\mathbf{t0})^{\frac{1}{4}} \tag{9.7}$$

$$U_c * = \frac{[P_{\max\ \mathrm{xB}}]^{-\frac{1}{4}}}{\Phi\eta\lambda\mathbf{t0}} * D_m^{\frac{1}{2}} \tag{9.8}$$

It is easy to show that at a given analysis time the three-parameter optimization gives the highest maximum plates followed by the two-parameter and then the one-parameter optimization.

9.4 Application of HPLC

HPLC is a powerful and versatile technique that can be used to separate and analyze a wide variety of compounds, from small ions to large macromolecules. It is well-suited for separating organic, inorganic, and biological compounds, including chiral compounds and labile materials. The most appropriate type of HPLC for a given separation problem depends on the properties of the compounds to be separated, such as their molecular mass, solubility, and polarity. HPLC is widely used in many different fields, including pharmaceutical research, quality control, food and beverage analysis, environmental monitoring, and forensic science. It is used to separate and analyze a wide range of compounds, including amino acids, nucleic acids, proteins, drugs, steroids, carbohydrates, vitamins, dyestuffs, pesticides, and polymers. Some specific examples of how HPLC is used include:

- ☆ Assay of pharmaceutical products
- ☆ Monitoring drugs and metabolites in body fluids
- ☆ Detection of drugs of abuse
- ☆ Determination of additives in food and beverages
- ☆ Quality control of polymers, plastics, and resins

HPLC is a valuable tool for many different industries and research fields, and its applications are still growing. A summary of typical applications in various fields are listed in Table 9.7.

Table 9.7: Typical Applications of HPLC Methods to Various Fields

Sl.No.	Field	Typical mixtures
1	Pharmaceuticals	Antibiotics, sedatives, steroids, analgesics
2	Biochemicals	Amino acids, proteins, carbohydrates, lipids
3	Food products	Artificial sweetners, antioxidants, aflatoxins, additives
4	Industrial chemicals	Condensed aromatics, dyes, surfactants, propellants.
5	Clinical medicines extracts,	Bile acids, drug metabolites, urine estrogens
6	Polymers	Molecular weight determination and distribution
7	Forensic	Chemistry Drugs, poisons, blood alcohol, narcotics
8	Pollutants	Pesticides, herbicides, PCBs, phenols in environmental samples
9	Quality control	Purification from mixture

9.4.1 Polyaromatic Hydrocarbons

Silica is well suited for the separation and analysis of non-ionizing, water insoluble, and relatively simple molecules which are very closely related such as polyaromatic hydrocarbons and fats and oils with different functional groups The order of adsorption follows the general polarity scale for various classes of compounds. It is less influenced by molecular weight differences and more by specific functional groups. Therefore, the separation of compounds differing in the degree or type of alkyl substitution, such as members of a homologous series is usually by adsorption. However, adsorption chromatography has been used to isolate a number of polynuclear aromatics from a petroleum sample, from a totally porous silica column having dimensions of 25 × 0.46 cm and acetonitrile/water (70:30) as mobile phase.

9.4.2 Isomeric Compounds

Adsorption chromatography is uniquely suited to separate isomeric compounds because it can differentiate them based on the polarity of their functional groups and the geometry of their molecules. For example, in Figure 9.8, the positional isomers of syn- and anti-pyrazolines (also known as cis- and trans-pyrazolines) are separated on a silica column using a methylene chloride/ isooctane (50:50) mobile phase. The syn-isomer is eluted first because it is more polar and has a better geometric fit to the silica surface. Therefore, adsorption chromatography can be a powerful tool for separating isomeric compounds that are difficult to separate using other methods. A 100 × 0.3 cm pellicular silica along with methylene chloride/isooctane (50:50) mobile phase was used at a flow rate of 0.225 mL/min. A fixed wavelength of 254 nm UV detector was used.

9.4.3 Sugars in Popular Drinks

A typical application of HPLC may be illustrated by the identification of glucose, fructose and sucrose in a popular drink such as Pepsi using a 4 mm × 30 cm column of Macropak-NH_4 and acetonitrile-water system at a flow rate of 2.0 mL/min as shown in Figure 9.9. It is observed that fructose is eluted first followed by glucose and fructose and the whole separation takes about 15 min.

However, size exclusion chromatography has been used for qualitative identification and quantitative determination of these sugars in four types of canned fruit juices of apple, orange, pineapple and cranberry. Also shown is chromatogram for the standards which can be used for quantitative determination of the constituent sugars. The packing which had an exclusion limit of 1000, was a cross linked-polystyrene polymer made hydrophilic by sulphonation. A 25 cm long column with this packing contained 7600 plates at 80°C. The size exclusion chromatography or gel permeation chromatography is the most suited technique for the solutes with molecular weights 2000 or more and is also useful for preliminary investigation of unknown samples. It can be used for the determination of relative mass distribution for components of

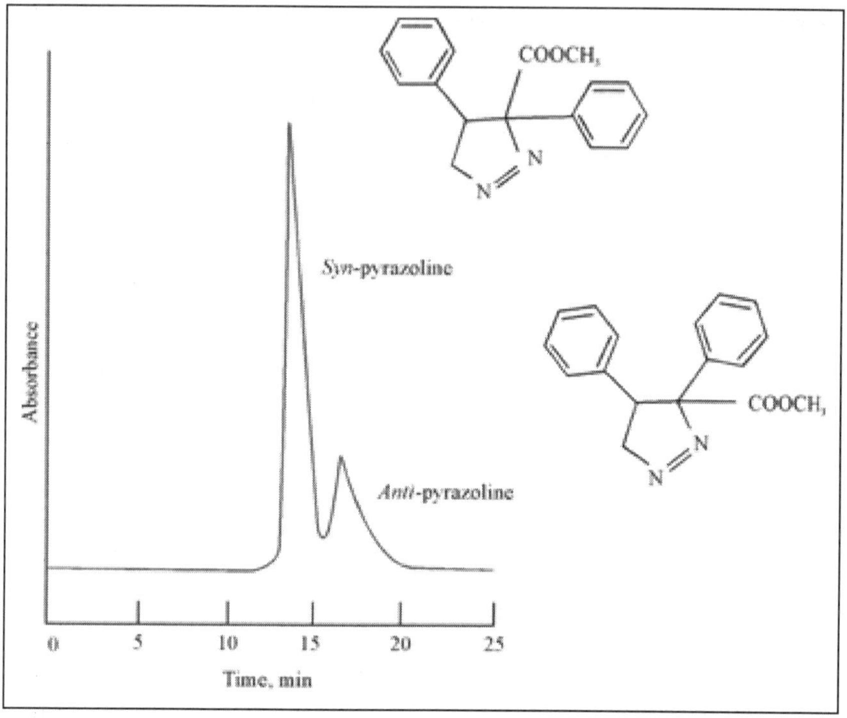

Figure 9.8: Separation of Syn- and Anti-pyrazoline Isomers.

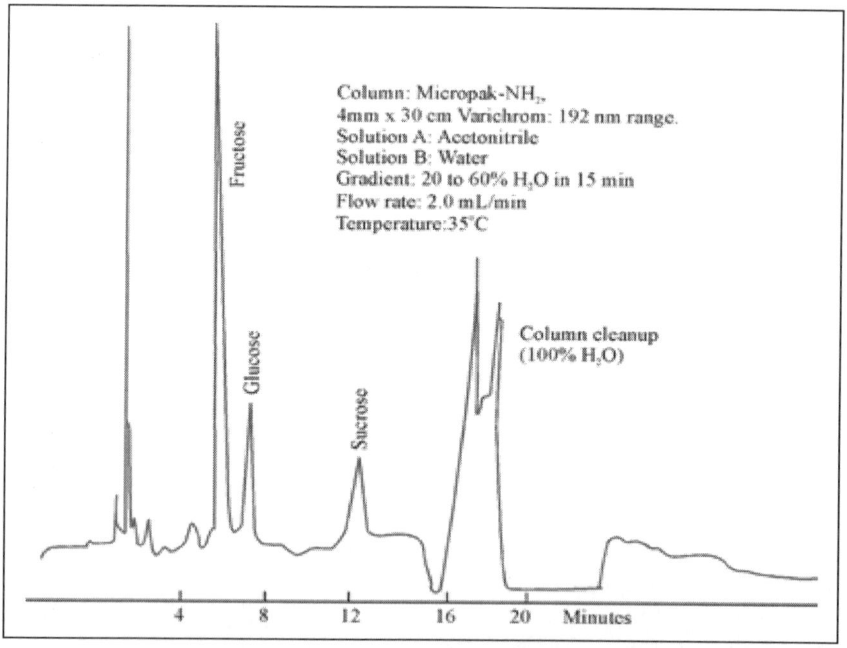

Figure 9.9: The HPLC Separation of Glucose, Fructose and Sucrose in Pepsi Shown as Three Distinct Peaks.

biochemical and polymer systems with an accuracy of 10 per cent. Desalting is commonly employed to isolate the macromolecules from biochemical materials, where both simple and macromolecules may be present in an electrolyte solution. Dilute solutions of macromolecules can be concentrated and isolated by adding dry gel beads to absorb the solvent relative molecular mass solutes.

9.4.4 Drug Abuse

The HPLC is being widely used for the detection of drug abuse especially for the steroids by the athletes at the International games. A typical case of drug abuse is illustrated in Figure 9.10 where oxphenbutazone, phenylbutazone and furosemide were detected in horse plasma.

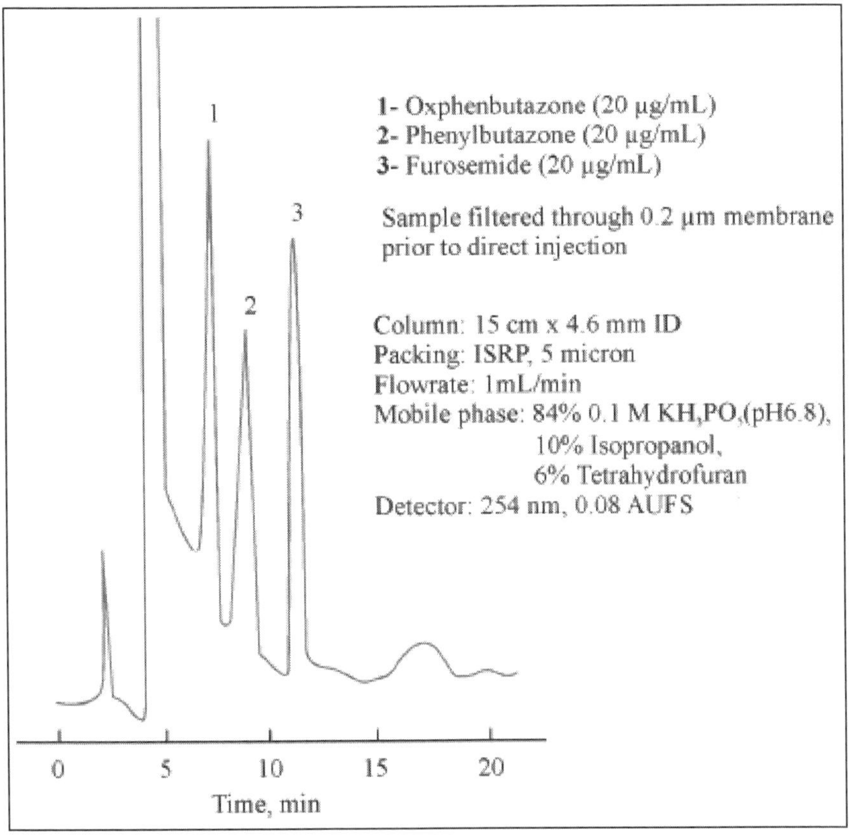

Figure 9.10: Reversed Phase HPLC Separation of Drug Abuse in Horse Plasma.

The sample is filtered through 2 µm membrane prior to direct injection into a 15 cm × 4.6 mm with 5 micron ISRP packing. A ternary mobile phase of isopropanol, tetrahydrofuran along with potassium hydrogen phosphate buffer was used at a flow rate of 1 mL/min.

9.4.5 Separation of Nucleic Acids

The HPLC is a rapid separation method as illustrated by a comparison of separation of nucleic acids by HPLC and conventional ion-exchange chromatography in Figure 9.11 whereas conventional ion-exchange method takes more than 2 hours for the separation of cytosine, uracil, guanine and adenine, it takes only 5 minutes by HPLC method requiring only microgram quantities.

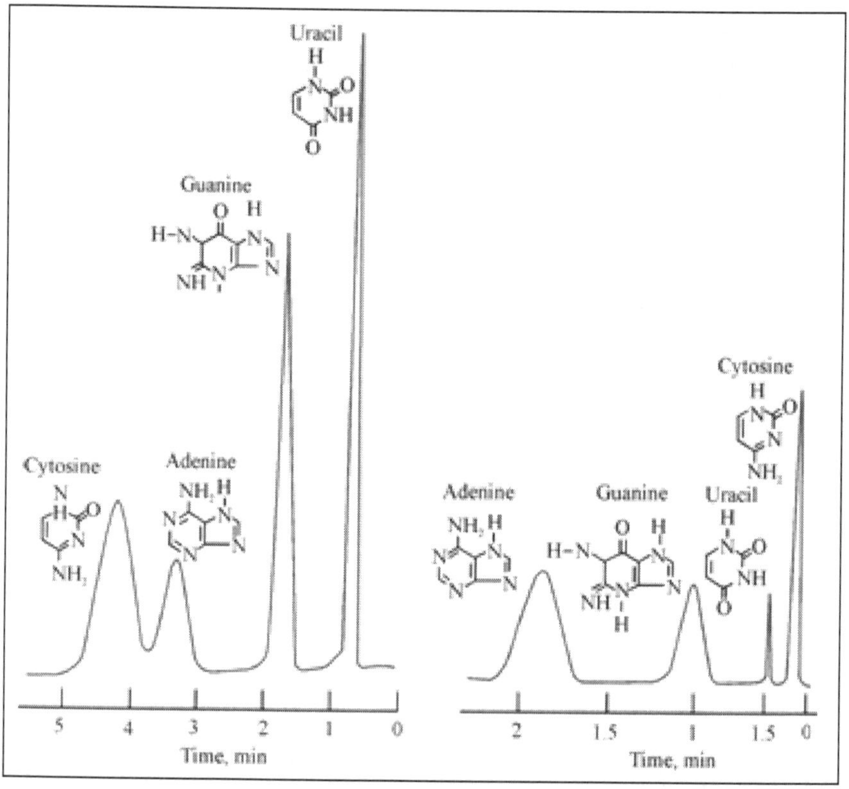

Figure 9.11: HPLC Separation of Nucleic Acids: (a) on Zipax cation exchange packing using HPLC; and (b) using conventional ion-exchange chromatography.

9.4.6 Analysis of Amino Acids

Commercial amino acid analyzers have been available for many years. They work on the principle of HPLC and are widely used for the separation analysis of a large number of amino acids from a mixture as they form cations by adding protons in the pH range below their isoelectric point. They are separated on a cation exchange column where more acidic component emerges first and the most basic in the last.

9.4.7 Partition Chromatography

As already pointed out it is being used in two forms of bonded-phase (BPC) and reversed phase chromatography (RPC). Both of these forms are suitable for most HPLC separations ranging from mixtures of weakly polar to highly polar and ionic compounds. Reverse phase chromatography using octadecyl (C18) columns and methanol/aqueous buffers or acetonitrile/water mobile phases is by far the most widely used. Ion pair chromatography offers the advantages of greater efficiency and column stability and more selectivity in the separation of ionic compounds compared to bonded phase or conventional ion exchangers. The reverse phase technique comprises nearly half of all the LC methods described in literature. It provides optimum retention and selectivity when compounds have no hydrogen bonding groups or have a predominant aliphatic or aromatic character as typically illustrated in Figure 9.12 showing separation of solutes based on the size and structure of alkyl groups in a series of phthalate esters.

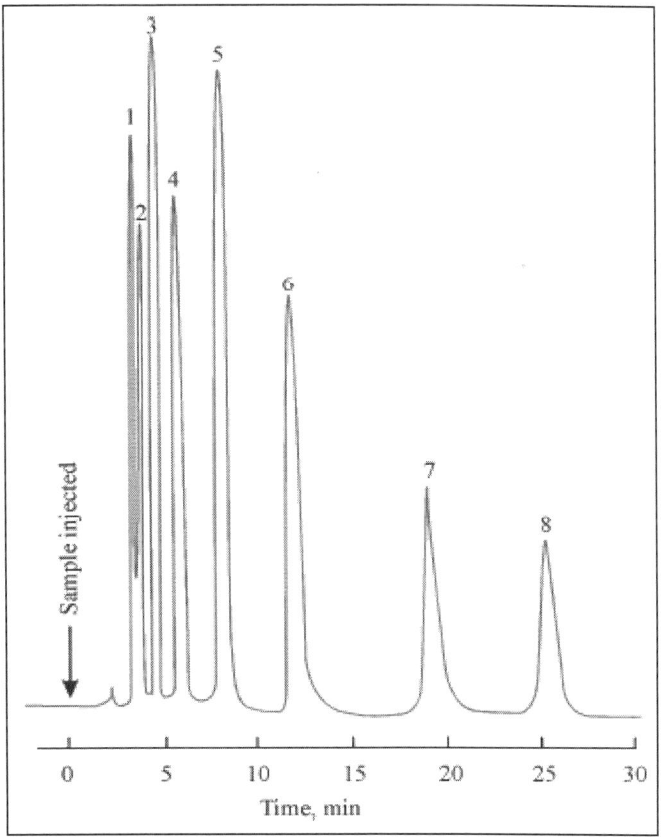

Figure 9.12: Chromatogram of O-phthalate Esters Run on a Column with Octadecyl Packing and Methanol/Water (90:10) as eluent. Peaks; 1. Dimethyl, 2. Diethyl, 3. Dipropyl, 4. Dibutyl, 5. Dipentyl, 6. Dihexyl, 7. Diheptyl, 8. Dioctyl.

9.4.7 Ion Chromatography

It is being routinely used for the separation of inorganic anions such as Cl^-, Br^-, F^-, SO_4^{2-}, NO_3^- NO_2^-, and PO_4^{3-} at ppm levels in surface waters, industrial effluents, food products, pharmaceuticals and clinical samples. Typical ion chromatograms in Figures 9.13 and 9.14 shows the analysis of various anions and cations using two column flow method. Thus, inorganic cations such as Cu^{2+}, Mg^{2+}, Sr^{2+}, Ca^{2+} and Cd^{2+}can be monitored in foods such as dietetic foods low in sodium, and urine samples where the efficiency of the separation is markedly influenced by the use of complexing agent in the eluent. However, separation of organic acids and bases, alkali, alkaline earth and transition metal cations are also being achieved. Ion-exchange resins having a proportion of the ionic sites replaced with hydrophobic reversed phase groups commonly used in HPLC columns, typically octadecylsilane, enable separation of both non-ionic and ionic species in a mixture. It has become a widely used technique in a pathological and environmental analysis laboratory where commercial instruments are used.

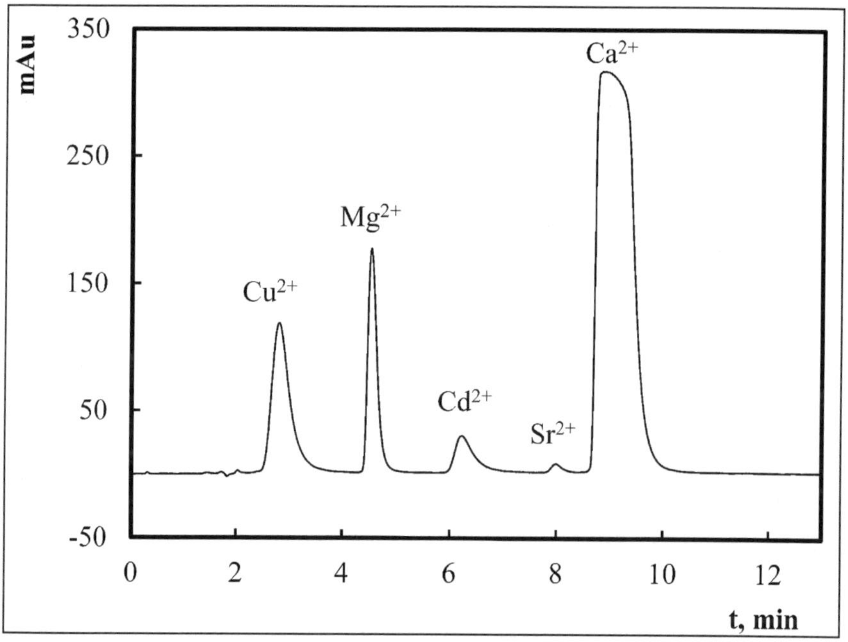

Figure 9.13: IC Separation of Cu^{2+}, Mg^{2+}, Sr^{2+}, Ca^{2+} and Cd^{2+}.

Another version of reverse-phase chromatography is ion-pair chromatography which deals with separation of ionized or ionizable species on a reverse phase column. The method can handle samples that are very polar, multiply ionized and/or strongly basic. In ordinary reverse phase HPLC, organic ions show poor peak shapes and inadequate retention. Ion-suppression method is limited to the pH range 2.0 to 7.5 by the instability of stationary bonded phases

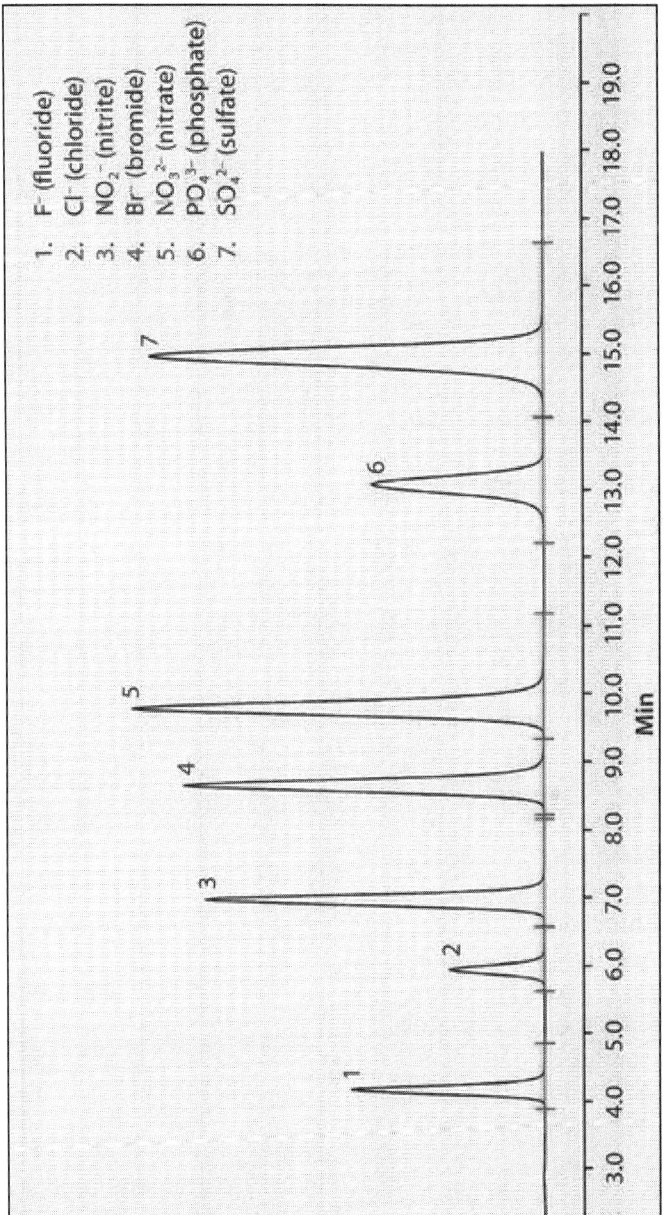

Figure 9.14: IC Separation of Chloride, Fluoride, Nitrite, Bromide, Nitrate, Phosphate, and Sulfate Ions.

outside this pH range. In case of ion-pair chromatography, an ion pair reagent, a large organic counter-ion which is ionized, is added at low concentration to the mobile phase. One ion of the reagent is retained and separate organic solute ions of opposite charge by forming a reversible ion-pair complex with the ionized sample as represented by the equilibrium

$$RCOO^- + R_4N^+ \leftrightarrow [R_4N^+, {}^-OOCR]^o \text{ ion-pair}$$

Thus, electrically neutral compounds are partitioned between the mobile and non-polar stationary phases. Unlike conventional ion-exchange, ion-pair partition can separate non-ionic and ionic compounds in the same sample. First of all separation of the nonionic solutes is optimized and then counterion is added to the mobile phase whereby ionic solutes are retained. Let us consider the separation of water soluble vitamins, strongly ionic thiamine, nonionic riboflavin and less ionic pyridoxine and niacinamide by a two step process. In the first step, water/methanol ratio is adjusted to obtain good retention of non-ionic riboflavin and then the organic counterion is added to the eluent to separate three ionic compounds which is affected by the alkyl chain length of the counterion. Thus, thiamine, a quaternary amine shows greatest sensitivity to change in counterion. The optimum separation is achieved with a 50/50 mixture of C-5/C-7 alkyl sulphonic acids as shown in Figure 9.15 where a mixture of counterions is added to the mobile phase producing a retention proportional to the concentration of each counterion.

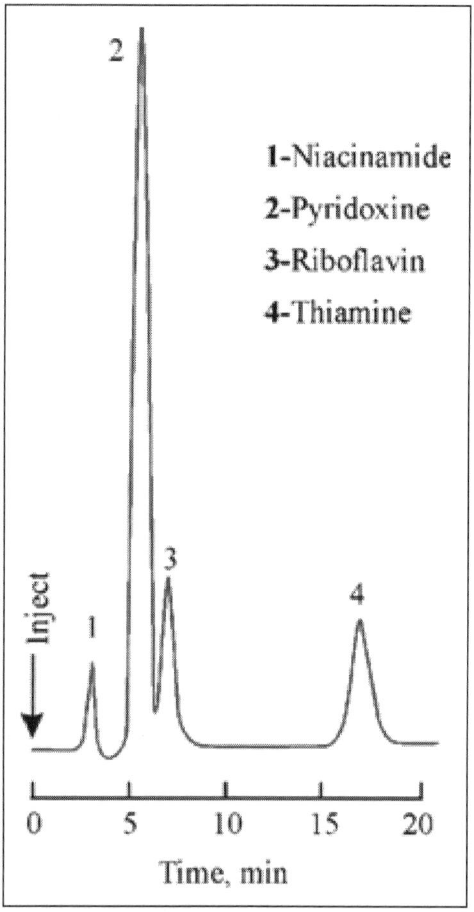

Figure 9.15: Separation of Ionic and Non-ionic Compounds in a Mixture by Ion-Pair Chromatography.

9.4.8 Chiral Separation of Enantiomers

A typical example of the separation of enantiomers of benzodiazepine Temazepam drug is illustrated in Figure 9.16 where immobilized human serum albumin (HSA) bonded to silica was used as a reverse phase material and phosphate buffer-acetonitrile as mobile phase. A number of manufacturers now supply these columns specifically tailored for particular application.

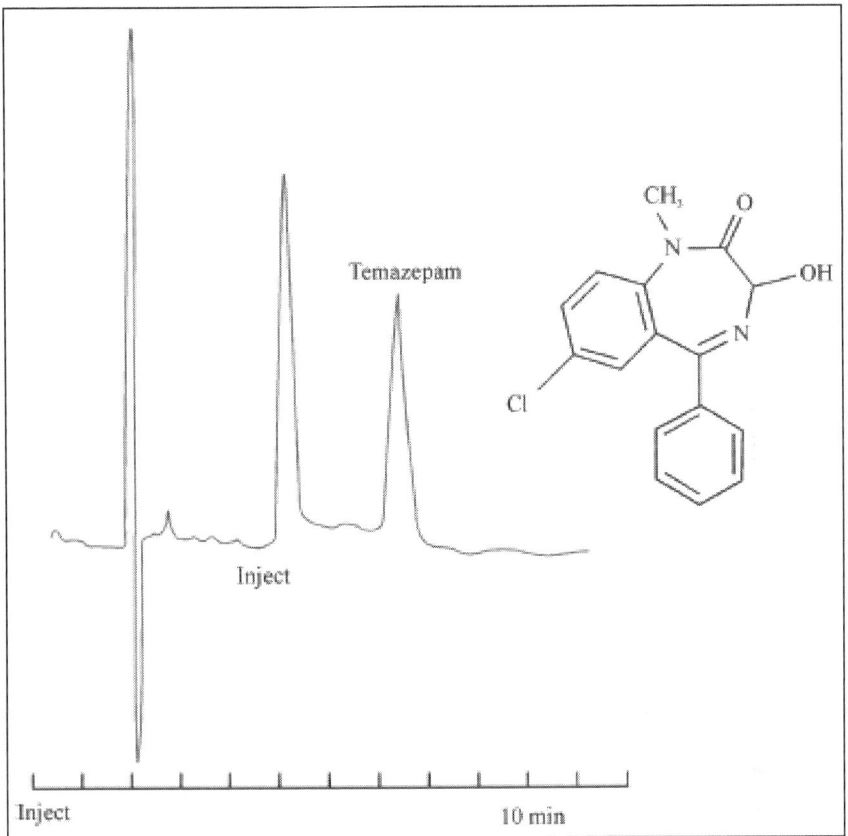

Figure 9.16: Chiral Separation of Enantiomers of the Benzodisazepine Temazepam Drug Showing Separate Peaks Corresponding to D and L Forms.

9.4.9 Ion-Exclusion Chromatography

Similar to ion chromatography, it also employs ion-exchange columns to achieve separations. However, it differs from ion chromatography in that it is used for the separation of neutral species rather than ions as typically illustrated by the separation of simple carboxylic acids in Figure 9.17. A cation exchange resin in acidic form was used and elution was accomplished with dil HCl. The analytical column was followed by a suppressor column packed with a cation exchange resin in silver form where H^+ were exchanged for Ag^+ which then precipitated Cl^- thus removing the ions contributed by the eluent. The undissociated analyte acids were distributed between the mobile phase in the column and the immobilized liquid held in the ores of the ion packing. Ion-exclusion chromatography has found numerous applications for the identification and determination of acidic species in milk, coffee, wine and other commercial products. Similarly, weak bases and their salts can also be separated by using anion exchange column in OH^- form.

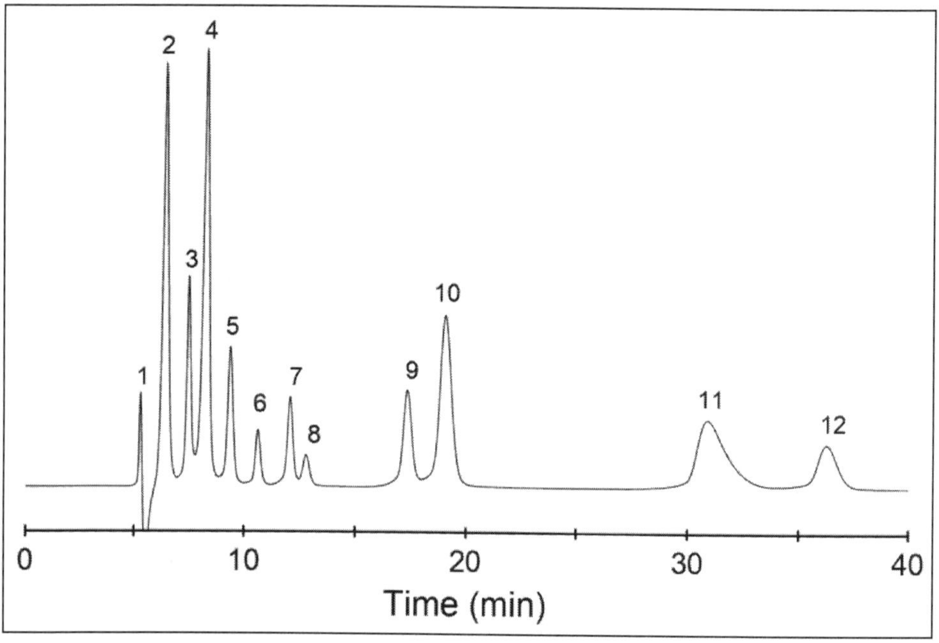

Figure 9.17: Organic Acids by Ion Exclusion Chromatography (1), maleic acid (2), glyxoxylic acid (3), malonic acid (4), malic acid (5), glycolic acid (6), formic acid (7), lactic acid (8), acetic acid (9), succinic acid (10), fumaric acid (11), acrylic acid (12).

Chapter 10

Gas Chromatography

10.1 Introduction

Gas chromatography (GC), a technique developed by Nobel Laureate Martin and colleagues in 1952, has become an indispensable tool in various industries over the past six decades. Its ability to both quantify and qualify compounds makes it suitable for a wide range of applications, including residual solvent analysis in pharmaceuticals, pesticide residue analysis in food safety, and trace level analysis in environmental, petrochemical, and other chemical industries. In GC, the mobile phase, typically an inert gas like helium or nitrogen, carries the sample molecules through a stationary phase, which is either a thin layer of non-volatile liquid bound to a solid support (gas-liquid chromatography, GLC) or a solid adsorbent (gas-solid chromatography, GSC). In GLC, separation occurs due to partitioning between the gas and liquid phases, while in GSC, separation is based on selective adsorption onto the solid surface. GSC, developed earlier than GLC, has limited applications compared to GLC, which is the more versatile and widely used technique. GLC's versatility stems from the ability to tailor the stationary phase to selectively interact with the target compounds, enabling the separation of a vast range of analytes. Both GSC and GLC share similarities with column chromatography, but the use of gas as the mobile phase distinguishes them. The choice between GSC and GLC depends on the specific properties of the analytes and the desired separation mechanism. Martin, who had theoretically predicted the feasibility of GLC a decade earlier, along with A.T. James, played a pivotal role in its development. Their work marked the beginning of a transformative era in analytical chemistry, making GC an essential tool for scientists and researchers worldwide. Gas solid chromatography was developed by G.D. Kohler and K.Thile in 1943 followed by E. Cremer, T. Janak, H.W. Patton, J.S. Lewis and W.I Kaye. C.S.G. Philips contributed to frontal analysis whereas P.Schuftan, N.C. Turner, S. Claesson and N.M. Turkel gave displacement development of elution technique. The gas-liquid chromatography is more popular than gas-solid chromatography and has many more applications. Gas-solid chromatography was originated by A.J.P Martin in 1951 together with A.T James. It is noteworthy that Martin had theoretically predicted the feasibility of GLC about a decade earlier while presenting his work on liquid-liquid chromatography (LLC). In 1952 Martin and Synge were awarded the Nobel Prize in chemistry for their work on the development of partition chromatography.

The main advantages of gas chromatography are given below:

i) The technique has strong application power and even complex mixture can be resolved into constituents
ii) The sensitivity of the method is quite high. It is micro-method and only few mg of the sample is sufficient for analysis.
iii) It gives good precision and accuracy.
iv) The analysis is completed in short time
v) The cost of instrument is relatively low and is life is generally long
vi) The technique is suitable for routine analysis because the operation of a gas chromatograph and related calculations do not require highly skilled person.

Gas chromatography is at present one of the most widely used and powerful tools available for separations. The major factors for this are the speed, resolving power and exteme sensitivity of the technique. For example, using this technique it is possible to separate thye ten isomers of heptane in less than ten seconds. Detectors are available with detection limit as low as 10^{-12} to 10^{-14} g.

10.2 Principle of Gas Chromatographic Separations

When a gas or vapour in contact with an adsorbent, certain amount of it get adsorbed on solid surface. The phenomenon takes place according to the well-known law of Freaundlich ($x/m = Kc^{1/n}$) or Langmuir ($x/m = K_1c + K_2c$)where x is mass of gas or vapour sorbed on the mass m and c is vapour concentration in gas phase and K1 and K2 are constants. Similarly if vapour or gas comes in contact with liquid, a fixed amount of the gas dissolved in the liquid. The phenomenon takes place according to Henry's law of partition ($x/m = Kc$). Now both the phenomenon is selective and there are different K-values for different vapour –sorbent pairs.

The principle of gas chromatography can be explained in terms of following experiments. A gas that is flowing smoothly at the rate of 3ft/min down in an empty tube that is 6ft long and gas takes 6/3= 2 min to flow from one end of the tube to the other. If such tube were fitted filled with sand the gas would flow through it more slowly. If the rate at which gas flows in the sand –filled tube is 2 ft/min, it will take the gas 6/2 = 3 min to traverse the tube. The sand filled tube in this example has same properties of a gas chromatography column. The gas is the moving mobile phase and the sand is the stationary phase. The gas that emerges after it has passed through the column is called effluent. In practice the mobile phase should be relatively insoluble in the stationary phase, otherwise stationary phase becomes overloaded.

To further our analogy, consider the following property of soluble gases. In a vessel containing non-volatile liquid A we place a soluble gas B. A short time after the vessel has been sealed, the gas B in the gaseous phase comes to equilibrium with the gas dissolved in the liquid phase, and the distribution of gas between the two phases remain constant. Although the total number of gas molecules above the liquid and in the liquid stays the same, a rapid interchange take place between the molecules in the two states that is molecules from the gas pass into the liquid at the same rate that dissolved gas molecules leaves the liquid and becomes gaseous. The molecules are said to be in dynamic equilibrium. This is constant if the temperature is kept constant. It can be shown that on, the average each molecule of B spends a constant fraction of the time in the liquid phase and the remainder of the time in the gas phase. For the purpose of our discussion, we hypothesize that gas B spends 40 per cent of its time in liquid A and 60 per cent of its time as a gas. Also, the equilibrium will be established more quickly if the thickness of the liquid is small otherwise dissolved molecules W wander from the interface and the equilibrium is disturbed.

To proceed with our illustration we remove the sand from the tube coat it with a thin film of liquid A and replace the sand in tube. Liquid A is known known as stationary phase, or substrate and gas B is known as carrier gas. The sand is support. Again we allow gas B to flow the packed tube. We noted earlier that it took 3 min for gas B to flow through the tube filled with uncoated sand, moreover we have hypothesized that when this gas is in equilibrium with liquid A, it spends 40 per cent of its time in the liquid. Therefore for 40 per cent of the time gas B spends in the tube, it does not travel down the tube. During the remaining 60 per cent of the time (during the time in which it is in gaseous state), gas B moves down the tube at the normal rate of 2 ft/min. On the average the gas flow at (60/100) x 2ft/min = 1.2 ft/min. The time taken to pass through the column at this rate is

$$6ft/1.2ft/min = 5\ min$$

Alternatively we can say that if a gas were flowing all the time, it would take 3 min to flow through the tube. Therefore it must spend 3 min flowing to reach the end of the tube. But it spends only 60 per cent in the gas phase and 3 min is 60 per cent of the total time in the column. Therefore the total time in the tube is 3 x (100/60) = 5 min.

The distribution coefficient for the gas is given by

K = concentration in liquid/concentration in gas = 40/60

And this is reflected in the time it takes for the gas to elute from the chromatography column.

Based on this relationship, we can state that the time spent in the gas phase or mobile phase is the same as the time it takes for the carrier gas to pass through the column that is t_0. The time spent in the solvent phase or stationary phase is the extra time it takes for the sample to pass through the column that is t_t-t_0 where t_t is the total time for the sample to pass through the column or retention time. We can therefore present K in terms of the chromatographic experiments as

$$K = \frac{t_t - t_o}{t_o} \tag{10.1}$$

$$K = \frac{t_r}{t_o} \tag{10.2}$$

If we substitute a second gas C for gas B and it is found that gas C spends only 20 per cent of its time in liquid A and 80 per cent of its time as a gas then when C is permitted to flow through the tube packed with coated sand, it moves at the rate of 2ft/min for 80 per cent of the time and is at rest for 20 per cent of the time. The average flow rate should be 2 x (80/100) = 1.6 ft/min. The total time needed for the gas C to flow through the tube is 6/1.6 = 3.75 min. If we were to pass a mixture of gases B and C down the tube, gas C emerges after 3.75 min and gas B after 5 min. Passage of a mixture of the two gases down the tube result in separation of the mixture.

Inert gases such as Helium flow through chromatographic column at a constant velocity that is such gases do not interact with and thus spends no time in the stationary phase. For this reason inert gases are often used as carrier gases. In practice the carrier gas flow steadily through the column. The sample is injected into the carrier gas and is swept onto the chromatographic column. The sample then shares its time between stationary phase and moving carrier gas.

We have seen with liquid substrate, the process can be used to separate from each other (gas-liquid chromatography). By using solid substrates, we can separate liquids from each other (liquid-solid chromatography). Different species of molecules spend different periods of time in the stationary phase and therefore a different percentage of time in the mobile phase. The variation in the time spent in the mobile phase, in turn affects the length of time a gas takes to pass through and emerges from a chromatographic column. A record of the emergence of various compounds plotted against time is called chromatogram.

This process of separation is the basis of all forms of chromatography. The factors that affect the time distribution between the mobile phase and the stationary phase differ from one branch of chromatography to another. The most important branch is gas-liquid chromatography, which has number of applications. The next three most important branches are gas-slid, liquid-liquid and liquid-solid chromatography, where the first phase is the mobile phase and second is the stationary phase.

In the example described previously, a compound B distributed itself between the liquid Phase A and gas phase in the ratio of 40 per cent (liquid) to 60 per cent (gas). If we add a second phase such as helium that can be used as a carrier gas, the distribution of B species between the gas phase and the liquid phase will not be distributed. The distribution can then be described by

$$K_B = \frac{\text{Concentration of B in liquid (stationary phase)}}{\text{Concentration of B in gas phase (mobile phase)}} = \frac{X_s}{X_m} = \frac{40}{60} \quad (10.3)$$

In this case K is the partition coefficient. Frequently it is expressed in terms of moles per litre in which case

$$K_B' = K_B \frac{V_s}{V_m} \quad (10.4)$$

Where V_m and V_s are the volumes of the mobile (gas) and stationary (liquid) phases, respectively and K_B' is the solute partition ratio. Also, more correctly,

$$k' = \frac{(t_t - t_o)}{t_o} \quad (10.5)$$

Suppose that we have a second gas C which spends 80 per cent of its time in the gas phase and 20 per cent in the liquid phase. The partition coefficient for the gas C is given by

$k'_{(c)}$ = moles of C in stationary phase/moles of C in mobile phase = 20/80

When the partition coefficients are widely different, we can expect separation. On the other hand if the partition coefficients are similar, it is likely that separation will be achieved. The separation factor or selectivity factor α can be given as

$$\frac{K_B}{K_c} \text{ or } \frac{k'_B}{k'_C} \quad (10.6)$$

Unfortunately, selectivity factors are only a guide to the actual ability of a column to separate components. Other factors influence the final result. For example, the two gases are in dynamic equilibrium between the liquid phase and the mobile phase. If the mobile phase is flowing continuously, then this dynamic equilibrium is approached at all times but never quite achieved. This results in spreading of the component B and C along the column. Sometimes this result in a lack of separation through the selectivity factors, indicate that the two components should be separable. Here the capacity fraction K is deduced in terms of the retention time of the sample and flow rate of the solvent.

10.3 Instrumentation

Up to this point we have refreshed some of the basic concepts of gas chromatography and understood the significance of different terms affecting the efficiency of separations. But the separations of the desired level are achieved by the optimum performance of each component of the instrument. Figure 10.1 shows schematically, the essential components, of a chromatograph

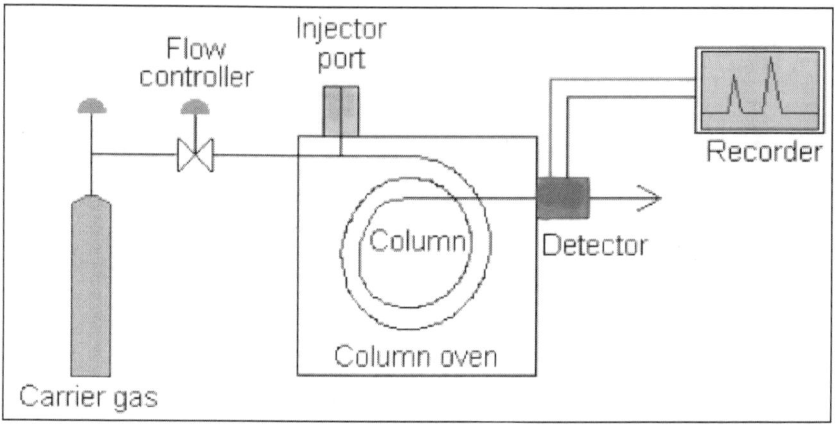

Figure 10.1: Block Diagram of a Typical Gas Chromatograph.

Let us now discuss in brief, the construction and working of each component of a gas chromatograph.

10.3.1 Mobile Phase (Carrier Gas)

The carrier gas, typically stored in a pressurized cylinder, serves as the medium for transporting vaporized analytes from the injection port through the chromatographic column to the detector. The carrier gas must meet stringent criteria:

1. Inertness: It must not undergo chemical reactions with any component of the instrument, column, or analytes.
2. Dryness: Moisture can react with the column, leading to resolution loss, asymmetric or "tailing" peaks, unwanted compounds or ghost peaks, and reduced sensitivity.
3. Purity: Impurities, such as hydrocarbons, oxygen, and water, can contribute to noise, baseline drift, and additional peaks in the chromatogram. Ultra-pure gases (99.9999 per cent) are ideal, but 99.995 per cent purity is generally sufficient. Commercial-grade gases may be usable if impurities are reduced to acceptable levels using specialized driers and scrubbers. Water and trace hydrocarbons can be removed with a 5A° molecular sieve filter, while oxygen requires a BTS (copper

aluminum zinc oxide) catalyst from BASF (Germany). After two gas cylinders have been used, the sieve should be regenerated by heating to 300°C for three hours with a slow flow of nitrogen. Common carrier gases in GC include hydrogen, helium, nitrogen, and argon. Argon and nitrogen are widely used due to their inertness and availability at the required purity levels. Certain detectors may have specific gas requirements for optimal performance.

Detectors	Carrier Gas
Thermal conductivity	Helium
Flame Ionization	Helium or nitrogen
Electron capture	Very dry nitrogen or argon, 5 per cent methane

The solutes of interest in the sample may also dictate the choice of a particular carrier gas. For example, in the analysis of hydrogen in air, nitrogen would be the best choice of carrier. Moreover, there is a need of flow controller.

Flow controller and pressure regulator: Maintaining a constant flow rate enhances the quality and precision of analytical results. The flow rate of carrier gas, along with temperature, determines the retention time, a key qualitative aspect of each sample solute. To achieve optimal results, the flow rate must exhibit certain characteristics: stability, reproducibility, and ease of adjustment. Stability ensures accurate results, reproducibility guarantees precision, and convenient adjustment allows for fine-tuning of parameters to optimize outcomes. Therefore, specialized flow controllers or pressure controllers are employed to maintain precise and accurate flow rates for both column carrier gas and auxiliary gases required for detector operation. A rotometer typically serves as the indicator of column flow rate. Its operation relies on the gas flow lifting a ball within a tapered tube. The position of the ball, influenced by the gas flow, indicates the flow rate. Most modern gas chromatographs incorporate rotometers for flow rate monitoring. Figure 10.2 illustrates the design of a typical rotometer.

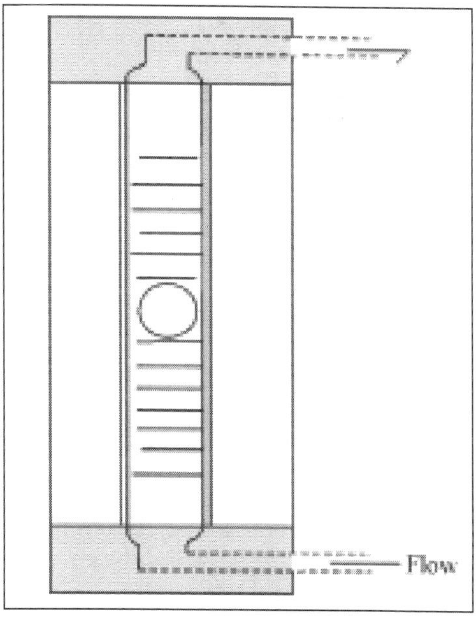

Figure 10.2: Rotometer.

10.3.2 Columns

The column is the pivotal component of a gas chromatograph, often referred to as the instrument's heart, where the separation process unfolds. Typically, most columns are constructed using glass or metal tubing that is filled with a carefully selected granular solid support. These support grains are typically coated with a relatively high boiling liquid phase, also known as the "stationary phase," which is the active portion of the column. The extent of separation achieved is determined by the interaction of sample solutes with this liquid phase. Some columns utilize porous polymers or solid adsorbents instead of a liquid phase to achieve separation. Each type of column described has its own unique set of applications. The most widely used type of column is the packed column, where the tubing is filled with a coated solid support, polymer, or adsorbent. Glasswool, which is commonly used to retain the packing within the column, is placed on both sides of the column. The end of the column that is connected to the sample introduction system typically has 4 to 6 cm of glasswool, while the opposite end, or the detector end, has about 6 mm of glasswool. There are two broad classes of chromatographic columns: packed columns and capillary columns. In general, a packed column can handle larger samples and a capillary column can separate more complex mixtures.

Packed Columns: Packed columns are constructed from glass, stainless steel, copper, or aluminum, and typically are 2–6 m in length with internal diameters of 2–4 mm. The column is filled with a particulate solid support, with particle diameters ranging from 37–44 μm to 250–354 μm. Figure 10.3 shows a typical example of a packed column.

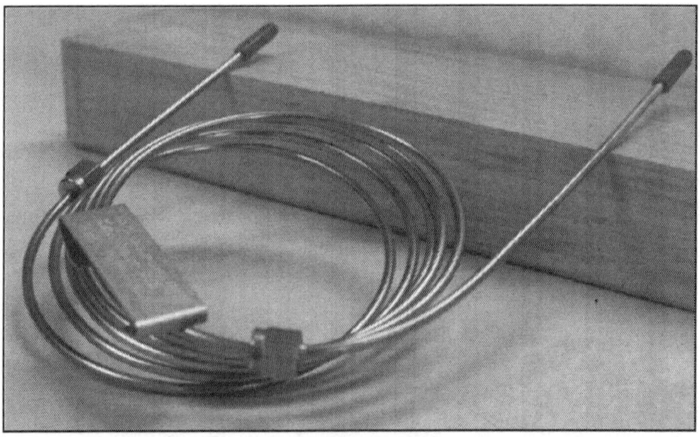

Figure 10.3: Typical Example of a Packed Column for Gas Chromatography.

This column is made from stainless steel and is 2 m long with an internal diameter of 3.2 mm. The packing material in this column has a particle diameter of 149–177 μm. To put this in perspective, beach sand has a typical diameter of 700 μm and the diameter of fine grained sand is 250 μm.

The most widely used particulate support is diatomaceous earth, which is composed of the silica skeletons of diatoms. These particles are very porous, with surface areas ranging from 0.5–7.5 m^2/g, which provides ample contact between the mobile phase and the stationary phase. Two high resolution images of diatomaceous earth are shown in Figure 10.4. When hydrolyzed, the surface of a diatomaceous earth contains silanol groups (–SiOH), that serve as active sites for absorbing solute molecules in gas-solid chromatography (GSC).

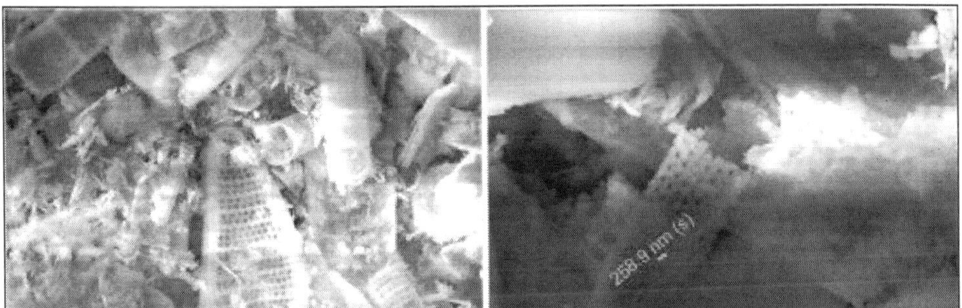

Figure 10.4: Two Scanning Electron Microcopy Images of Diatomaceous Earth Revealing the Porous, High Surface Area Nature of this Widely Used Packing Material. "File:1-s2.0-S0272884217313470-gr4 lrg.jpg" by Zirconia1980 is licensed under CC BY-SA 4.0.

In *gas-liquid chromatography* (GLC), we coat the packing material with a liquid mobile phase. To prevent uncoated packing material from adsorbing solutes, which degrades the quality of the separation, surface silanols are deactivated by reacting them with dimethyldichlorosilane and rinsing with an alcohol–typically methanol–before coating the particles with stationary phase.

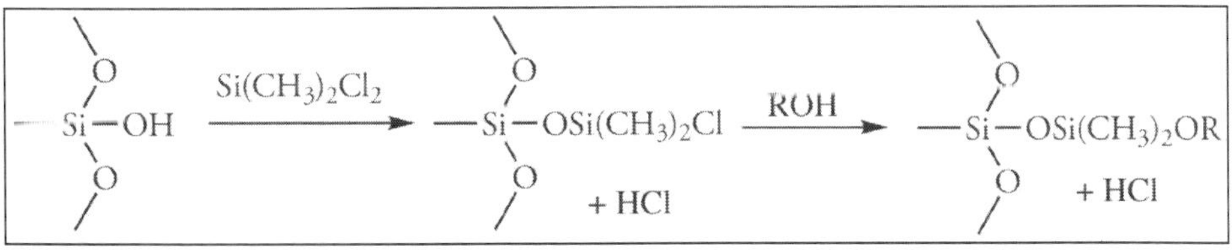

Capillary Columns

A capillary or ***open tubular column*** is constructed from fused silica and is coated with a protective polymer coating. Columns range from 15–100 m in length with an internal diameter of approximately 150–300 µm. Figure 10.5 shows an example of a typical capillary column.

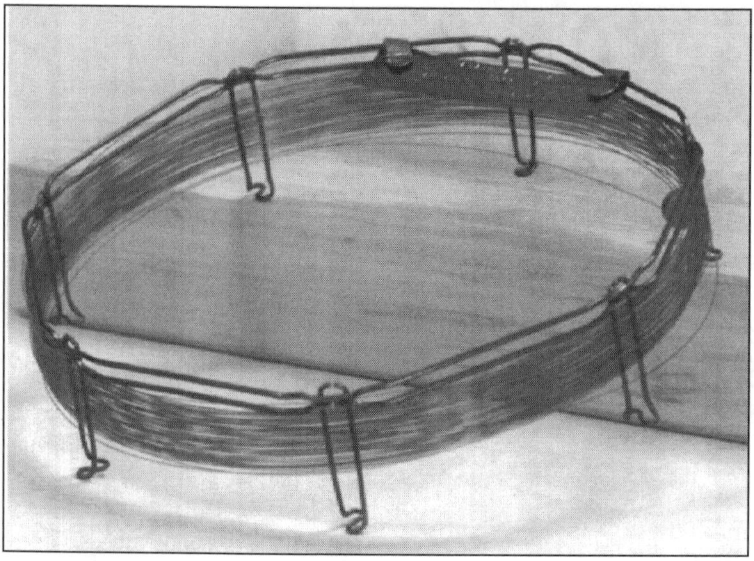

Figure 10.6: Typical Example of a Capillary Column for Gas Chromatography. This column is 30 m long with an internal diameter of 247 µm. The interior surface of the capillary has a 0.25 µm coating of the liquid phase.

Capillary columns are of three principal types. In a ***wall-coated open tubular column*** (WCOT) a thin layer of stationary phase, typically 0.25 nm thick, is coated on the capillary's inner wall. In a ***porous-layer open tubular column*** (PLOT), a porous solid support–alumina, silica gel, and molecular sieves are typical examples–is attached to the capillary's inner wall. A ***support-coated open tubular column*** (SCOT) is a PLOT column that includes a liquid stationary phase. Figure 10.7 shows the differences between these types of capillary columns.

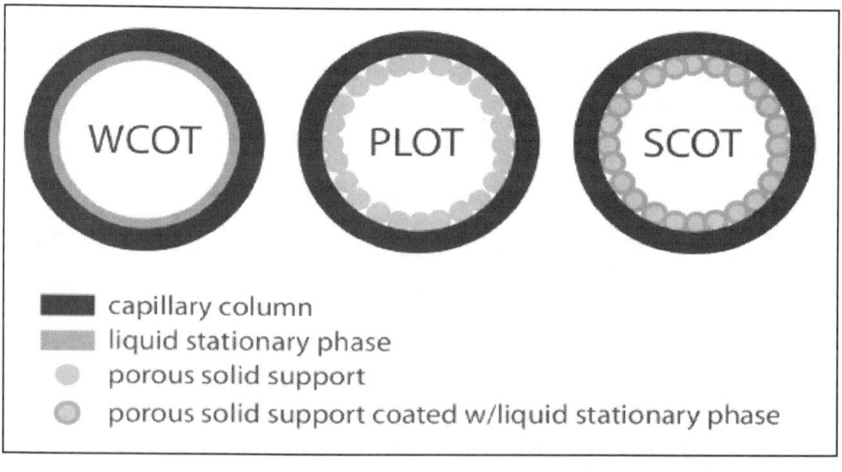

Figure 10.7: Cross Sections through the Three Types of Capillary Columns.

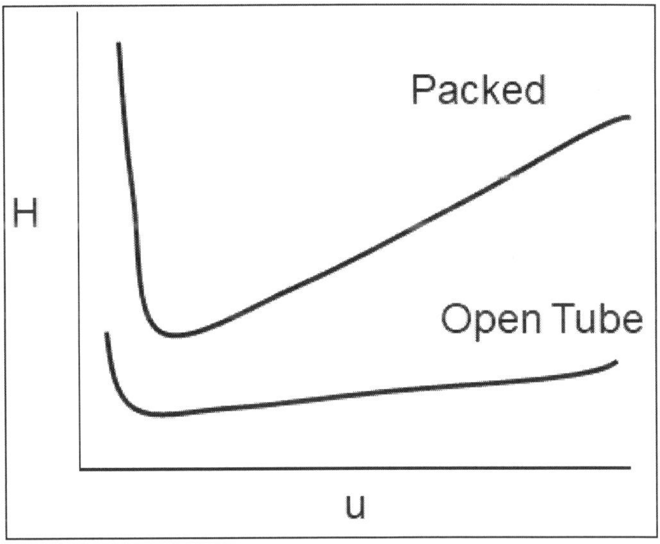

Figure 10.8: A Sketch of the Improved Separation Efficiency for an Open Tube Column versus a Packed Column as Revealed in a van Deemter Plot. In this figure H is the height equivalent of the theoretical plate and u is the linear flow velocity.

10.3.3 Stationary Phases for Gas-Liquid Chromatography

In gas-liquid chromatography (GLC), the order in which solutes elute from the column depends on two main factors: their boiling points and their interactions with the stationary phase. When the boiling points of the solutes in a mixture are very different, the choice of stationary phase is not as critical. However, if two solutes have similar boiling points, selective interaction with the stationary phase is essential for achieving separation. Generally, nonpolar solutes are better separated using nonpolar stationary phases, while polar solutes are better separated using polar stationary phases.

Several important criteria should be considered when selecting a stationary phase:

1. Chemical inertness: The stationary phase should not react with the solutes of interest.
2. Thermal stability: The stationary phase should be able to withstand the elevated temperatures used in GLC without degrading.
3. Low volatility: The stationary phase should not vaporize at the operating conditions used in GLC, as this could lead to contamination of the sample and the detector.

4. Appropriate polarity: The polarity of the stationary phase should match the polarity of the solutes in the sample to achieve optimal separation. Table 10.1 summarizes the properties of several common stationary phases.

Table 10.1: Selected Examples of Stationary Phases for Gas-Liquid Chromatography

Stationary Phase	*Polarity*	*Trade Name*	*Temperature Limit (°C)*	*Representative Applications*
squalane	nonpolar	Squalane	150	low-boiling aliphatics hydrocarbons
Apezion L	nonpolar	Apezion L	300	amides, fatty acid methyl esters, terpenoids
polydimethyl siloxane	slightly polar	SE-30	300–350	alkaloids, amino acid derivatives, drugs, pesticides, phenols, steroids
phenylmethyl polysiloxane (50 per cent phenyl, 50 per cent methyl)	moderately polar	OV-17	375	alkaloids, drugs, pesticides, polyaromatic hydrocarbons, polychlorinated biphenyls
trifluoro-propylmethyl polysiloxane(50 per cent trifluoropropyl, 50 per cent methyl)	moderately polar	OV-210	275	alkaloids, amino acid derivatives, drugs, halogenated compounds, ketones
cyanopropyl-phenylmethyl polysiloxane (50 per cent cyanopropyl, 50 per cent phenylmethyl)	polar	OV-225	275	nitriles, pesticides, steroids
polyethylene glycol	polar	Carbowax 20M	225	aldehydes, esters, ethers, phenols

Many stationary phases have the general structure shown in Figure 10.9a. A stationary phase of polydimethyl siloxane, in which all the –R groups are methyl groups, $-CH_3$, is nonpolar and often makes a good first choice for a new separation. The order of elution when using polydimethyl siloxane usually follows the boiling points of the solutes, with lower boiling solutes eluting first. Replacing some of the methyl groups with other substituents increases the stationary phase's polarity and provides greater selectivity. For example, replacing 50 per cent of the $–CH_3$ groups with phenyl groups, $–C_6H_5$, produces a slightly polar stationary phase. Increasing polarity is provided by substituting trifluoropropyl, $–C_3H_6CF$, and cyanopropyl, $–C_3H_6CN$, functional groups, or by using a stationary phase of polyethylene glycol (Figure 10.9 b).

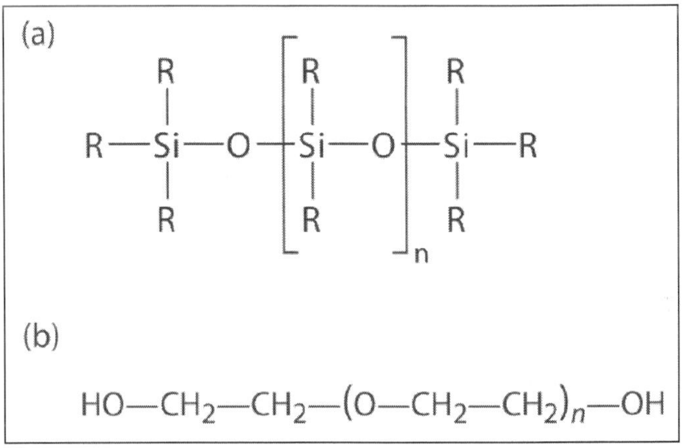

Figure 10.9: General Structures of Common Stationary Phases. (a) Substituted polysiloxane; (b) Polyethylene glycol.

Liquid stationary phases are prone to bleeding, which is the loss of the stationary phase from the column due to heating. To minimize bleeding, temperature limits should be adhered to as specified in Table 10.1. Capillary columns with bonded or cross-linked stationary phases offer superior stability compared to conventional liquid stationary phases. In bonded stationary phases, the stationary phase is chemically attached to the capillary's silica surface, while in cross-linked stationary phases, separate polymer chains are linked together after the stationary phase is introduced into the capillary column.

The thickness of the stationary phase is another critical factor. Thinner films of stationary phase enhance separation efficiency. The most common thickness is 0.25 μm, although thicker films are advantageous for highly volatile solutes like gases due to their increased capacity for solute retention. Thinner films are preferred for separating low volatility solutes like steroids. Chemical selectivity can be exploited by certain stationary phases. Chiral functional groups incorporated into stationary phases are particularly noteworthy for enantiomer separation.

10.3.4 Columns in Gas Chromatography

The column is the heart of a gas chromatograph, where the separation of analytes occurs. Most columns are constructed from glass or metal tubing packed with a carefully sized granular solid support. The support grains are typically coated with a high boiling liquid phase, referred to as the "stationary phase." The interaction of sample solutes with this stationary phase determines the extent of separation achieved. Some columns employ porous polymers or solid adsorbents instead of a liquid phase to achieve separation. Each type of column has its own set of applications.

Packed columns, where the tubing is filled with a coated solid support, polymer, or absorbent, are the most commonly used type. Glasswool is placed on both ends of the column to retain the packing material within the column. The end connected to the sample introduction system typically has 4 to 6 cm of glasswool, while the detector end has about 6 mm of glasswool.

Stainless steel columns are generally used for packed columns. They are packed while straight to ensure uniform packing and then coiled to accommodate longer lengths. The spiral diameter should be at least ten times the column diameter to minimize diffusion and racetrack effects. Packed column lengths range from a few centimeters to several meters. Typical analytical columns are 1 to 3 meters in length. Column diameters vary from 0.25 mm to 50 mm I.D. and larger. Smaller column diameters lead to higher column efficiency. Standard analytical columns have an I.D. of 3 mm to 6 mm.

Open tubular columns, another type of column, typically have an I.D. of 0.25 mm. The liquid phase is directly coated on the inside wall of the tubing, eliminating the need for additional packing. Due to the limited surface area and solute capacity, these columns are necessarily long (20 to 50 m) and require specialized sample inlets. Since capillary columns are not packed with particles, band broadening due to eddy diffusion is not a concern.

Table 10.2: Comparison of Capillary and Packed Columns

<table>
<tr><th></th><th>Capillary</th><th colspan="4">Packed</th></tr>
<tr><td rowspan="12">Inside diam., inches Maximum plates/ft Practical length Max. total plates Amt. liquid phase, per cent Liq. Film, thickness, μ Mesh range Permeability, × 107 cm2. Avg. linear vel, cm/sec/ Avg. flow rate, ml/ min Max. sample size, μl.</td><td></td><td>1/16-in. O. D.</td><td>1/8-in O. D</td><td>1/4 -in. O. D</td><td>3/8-in. O. D.</td></tr>
<tr><td>0.01-0.02</td><td>0.047</td><td>0.065</td><td>0.155</td><td>0.315</td></tr>
<tr><td>1000</td><td>1000</td><td>800</td><td>500</td><td>300</td></tr>
<tr><td>300</td><td>60</td><td>60</td><td>60</td><td>100</td></tr>
<tr><td>300,000</td><td>60,000</td><td>48,000</td><td>30,000</td><td>15,000</td></tr>
<tr><td>–</td><td>3</td><td>5</td><td>10</td><td>20</td></tr>
<tr><td>1</td><td>5</td><td>5</td><td>10</td><td>20</td></tr>
<tr><td>–</td><td>100/20</td><td>80/100</td><td>60/80</td><td>20/40</td></tr>
<tr><td>200-800</td><td>1.5</td><td>2</td><td>3.5</td><td>4.0</td></tr>
<tr><td>25</td><td>10</td><td>10</td><td>7</td><td>7</td></tr>
<tr><td>1-3</td><td>10</td><td>20</td><td>70</td><td>120</td></tr>
<tr><td>0.010</td><td>1.0</td><td>2.0</td><td>20</td><td>1000</td></tr>
</table>

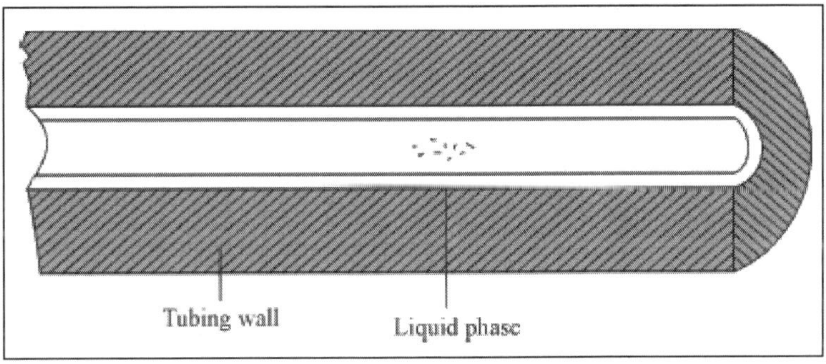

Figure 10.10: Open Tubular Column.

A comparison of packed and capillary columns given in Table 10.2 shows some interesting differences. A major advantage of capillary columns is the total plates obtainable. Plates per metre of column length are comparable with packed columns but much longer columns are usable because capillary columns have high permeabilities, *i.e.* they are "open tubes" with small resistance. The process of separation can be considered as a series of partitions where the vaporized solutes from the injection port go into solution with the liquid stationary phase and are then revaporized. The affinity of the liquid phase for a sample component determines the length of time for which each individual component will remain in solution. After a number of partitions, various solutes will separate into discrete bands and travel at different velocities through the column. The most volatile solute usually passes through the column first and the least volatile or highest boiling solute emerges last. Figure 10.11 demonstrates these stages of separation.

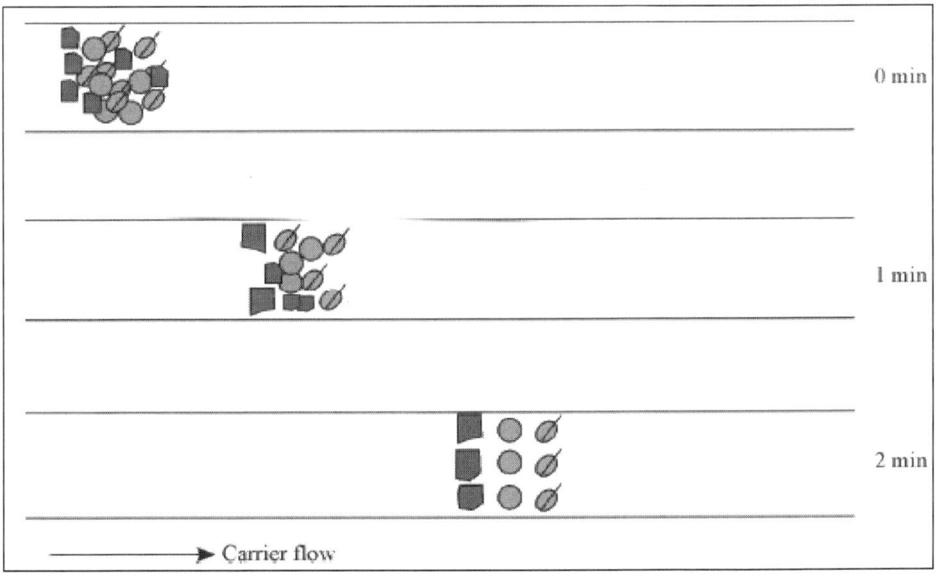

Figure 10.11: Component Separation within the Column.

10.4 Stationary Phase Support and Liquid Phases

After studying different types of columns, you would like to know about the actual packing material inside them particularly their characteristics. This sub- section is devoted to this particular aspect. The discussion will be taken up under the following heads.

A. Stationary phase support. B. Liquid phases. C. Super selective liquid phases D. Liquid phase percentage E. Effect of column temperature F. Porapak as packing material.

A. Stationary Phase Support

The purpose of the solid phase is to support a thin uniform film of liquid phase. An optimum support should have certain characteristics:

- ☆ A large specific surface area - from 1 to 20 sq. metres per gram
- ☆ A pore structure with uniform pore diameter in the range of 10 μ or less
- ☆ Inertness - a minimum of chemical and adsorptive interaction with the solutes
- ☆ Regularly shaped particles, uniform in size for efficient packing.
- ☆ Mechanical strength – it should not crush on handling

Currently, no single material exists that perfectly meets all the requirements for a gas chromatography support. However, several commercially available options offer good performance in different aspects. Typically, the choice between inertness and efficiency (high surface area) is a trade-off. The primary raw material for most gas chromatographic supports is diatomite, also known as diatomaceous silica, diatomaceous earth, and kieselguhr. Diatomite is composed of the skeletons of diatoms, microscopic unicellular algae. These skeletons consist primarily of micro-amorphous hydrous silica, with minor impurities such as metallic oxides. Individual diatom skeletons are extremely small, highly porous, and possess a large surface area. Chromosorb is the registered trademark for a series of gas chromatography support materials produced by Johns Manville. Five forms or types of chromosorb exist: A, G, P, W, and T. Each is available untreated or treated and in various mesh ranges. Gas Chromatography Supports are typically sized by passing them through standard ASTM screens. Mesh numbers correspond to the number of openings per linear inch. For instance, particles that pass through a 60-mesh screen but are retained by an 80-mesh screen are referred to as 60-80 mesh. The uniformity of packing directly influences the extent of Eddy diffusion and, consequently, the column efficiency. For 3 mm I.D. columns, 100/120 or 80/100 mesh support is preferred. For 6 mm I.D. columns, 40/60 or 60/80 mesh support is suitable. The chemical composition of some of these supports is given in Table 10.3.

Table 10.3: Chemical Analysis of Supports

	Firebrick C22	Celite 545	Chromosorb P	Chromosorb W
SiO_2	89.7	99.9	89.2	91.2
Al_2O_3	5.1	3.6	5.1	4.1
Fe_2O_3	1.55	1.65	1.56	1.15
TiO_2	0.30	0.30	0.30	0.25
CaO	1.30	1.75	0.90	0.40
MgO	0.90	0.70	1.00	0.65

B. Liquid Phases

There is no fool-proof method for selecting the best liquid phase for a particular separation. The right selection is based mainly on experience and/ or trial and error. Following are the basic requirements of a liquid phase.

- It should be a good absolute solvent for sample solutes. If solutes are sparingly soluble in the liquid phase, they will elute rapidly and separation will be poor.
- It should exhibit differences in solubilities for different solutes in the sample.
- It should have low vapour pressure (0.01 to 0.1 mm) at operating temperatures for a reasonable column life.
- It should be thermally stable.
- It should be chemically inert toward solutes of interest at the column temperature.

The choice of liquid phase depends upon several considerations. The liquid phase chosen depends on the composition of the sample. It is expected that the type of solutes likely to be present in the sample is known before the analysis is started. The more one knows about a sample, easier it is to select the proper column and operating temperatures. For an efficient normal separation, the liquid phase should be similar in chemical structure to the solutes of the sample. For example, hydrocarbon compounds are best separated with a hydrocarbon solvent. Paraffins are best separated on squalane (a long chain hydrocarbon). Polar solutes like alcohols are best separated on a polar - Hailcomid (an amide) solvent. Table 10.4 gives the names of common liquid phases in GC separations.

Table 10.4: Recommended Liquid Phases by Sample Types

Classification of Compounds	Stationary Phase
Hydroxy	2,4-Xylenyl Phosphate
Olefinas	AgNO3/Benzyl cyanide
C1-C5	Dimethylsulfolane

Classification of Compounds	Stationary Phase
	Propylene Carbonate
C5-up	Carbowax 20M
Polynuclear	SE-30 on DMCS treated support FFAP on DMCS treated support PMPE (5 ring)
Ketones	Lexan FFAP
Nitrogen Compounds	
Amines	Dowfax 9N9/KOH
Amides	Versamid 900
Ammonia	Ethofat or Carbowax 600 cc
	Chromosorb T
Nitriles	Tetracyanoethvlated Pentaerythritol
	FFAP XF-1150
Organo-metallic	FFAP SE-30
Alcohols	
C1-C5	Halleomid M-18 OL Carbowax 600 or 1540
C1-C18	FFAP Carbowax 20M
Di-Poly	FFAPQF-1
Aldehydes	
C1-C5	Ethofat C5-C18 Carbowax 20M
Alkaloids	QF-1 SE-30
Amino Acid Derivatives	
N-Butyltrifluoroacetyl esters	DEGS/EGSS-X
Boranes	Apeizon L
Essential Oils FFAP(General)	Carbowax 20 M
Esters(Mixed)	Dinonyl phthalate Porapak Q
Ethers	Carbwax 20 M
Glycols	Porapak Q

Classification of Compounds	*Stationary Phase*
Halogen Compounds	Carbowax 20 M QF-1 (FS-1265) FFAP
Freons	Dibutyl Tetrachlorophthalate UCON Polar 2000
Hydrocarbons	
C1-C5	Propylene Carbonate Carbowax 100 Tributyl phosphate
C5-C10	Didecylphthalate SE-30
Aromatics	Tetracyanoethylated Pentaerythrtiol Dibutyl tetrachlorophthalate
Acids	
C1-C18	(free) FFAP
Bile and Urinary	
Fatty Acid Methyl Esters	DEGS FFAP Apleron L TCEPE EGSS-X

C. Super Selective Liquid Phases

Certain liquid phases may be said to be super selective. Some of these liquids depend on the formation of loose chemical adducts with particular solutes to effect separation. Silver ions form loose adducts with olefins of the type shown below

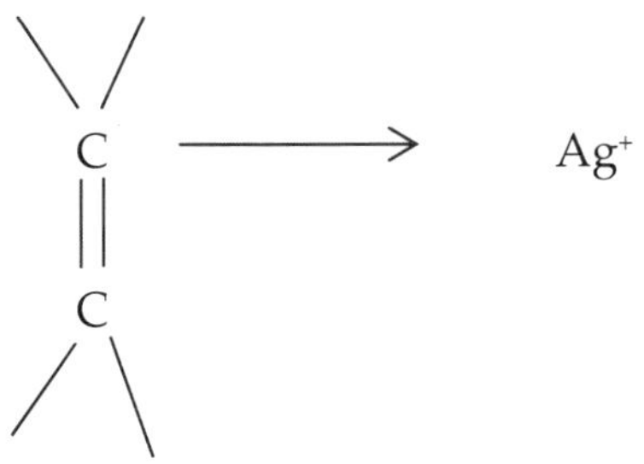

Solvents like silver nitrate in ethylene glycol, glycerol, triethylene glycol, or benzyl cyanide can be used to selectively separate olefins. Benzyl cyanide is preferred over glycols due to its non-hygroscopic nature and lack of requirement for a moisture-free carrier gas. Paraffins pass through the column quickly without being retained, while alkynes react with silver nitrate to form silver acetylides and remain on the column. The temperature should be kept below 40°C, as higher temperatures prevent the formation of $AgNO_3$ adducts and destabilize the solutions. Mercuric perchlorate can be used to remove unsaturated compounds by forming π-complexes with olefins, alkynes, and aromatics. Tetracyanoethylated pentaerythritol (TCEPE) is highly selective for aromatics. Liquid crystals are an intermediate state between crystalline solids and isotropic liquids. They can exist in two liquid crystal phases: smectic and nematic. As the temperature is raised, the solid transitions into the smectic liquid phase, then the nematic phase, and finally into the 'normal' liquid phase. A typical formula for a liquid crystal is:

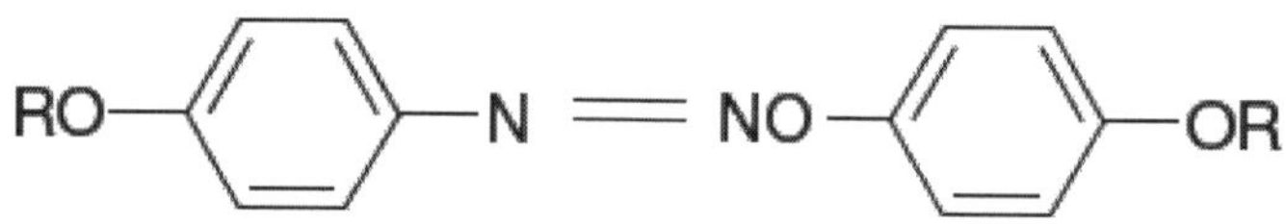

Where R is CH_2 or n-C_6H_{13} or n –C7H_{15}

The liquid crystals show a selective affinity for linear molecules. They will retain p-disubstituted benzenes relative to o-and m-isomers.

D. Liquid Phase Percentage

The quantity of liquid phase employed should be sufficient to create a thin, uniform coating over the particles. Excessive liquid phase accumulation between particles can diminish column effectiveness. For diatomaceous earth supports, liquid loadings beyond 30 per cent result in a significant decrease in efficiency. The retention time is directly proportional to the amount of liquid phase present; therefore, low liquid loading facilitates rapid analysis. However, an excessively low liquid load can leave adsorbing sites on the support exposed, potentially leading to irreversible adsorption or decomposition of the sample. To prevent this, highly inert supports should be utilized with minimal liquid loadings. Additionally, consider the volatility of solutes when determining the appropriate amount of liquid phase. Solutes with low volatility are best analyzed using low-loaded columns (~ 3 per cent or less) for materials like steroids. Conversely, highly volatile solutes, such as light hydrocarbons, require higher liquid loadings (~ 20 per cent to 30 per cent) due to their limited solubility in the liquid phase. The greater the liquid phase volume, the longer the time spent in the liquid phase, enhancing partitioning efficiency.

E

The value of the distribution constant (Kc) is highly sensitive to changes in temperature. Typically, a 30°C temperature increase leads to a 50 per cent reduction in Kc, effectively doubling the rate at which solutes migrate. As illustrated in Figure 10.12, the most immediate consequence of a temperature rise is a decrease in analysis time. Generally, lowering the temperature can improve resolutions. In most cases, the optimal temperature is chosen as a compromise between achieving sufficient resolution and avoiding excessively long retention times. When selecting the column temperature, it is crucial to consider the operating temperature range of the liquid phase being used. To ensure effective operation, the operating temperature should exceed the melting point of the liquid phase.

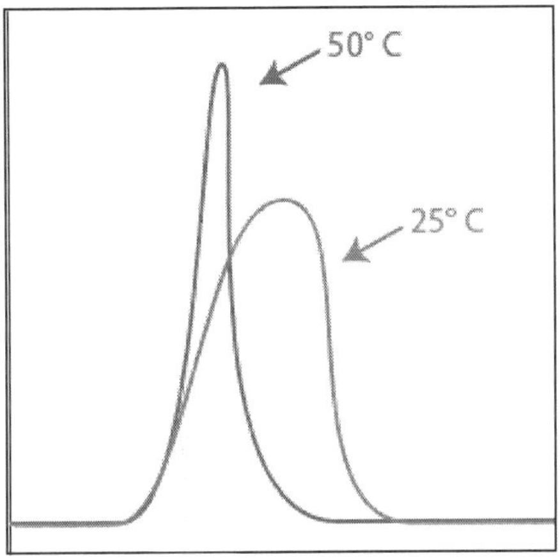

Figure 10.12: Retention Time vs Temperature.

F

Porapak is a porous polymer composed of ethylvinylbenzene cross-linked with divinylbenzene to form a uniform structure of a discrete pore size. Porapak column contains only the porous polymer beads screened to a definite mesh range and no liquid phase or solid support is used. The porous polymer beads serve the function of both the liquid phase and the solid support. Apparently, the solute molecule positions directly from the gas phase into the amorphous polymer. There are six Porapak materials available like type P,Q,R,S,T, and N. Type P and Q are non-polar phases type, and R,S,T and N are moderately beads. Porapak N has been developed for separation of formaldehyde mixtures Gas Chromatography and for the separation of acetophenone and ethylene. All beads are available in various particle sizes ranging from 50/80 mesh to 150/200 mesh. All the polymers are stable to 250°C except Porapak T which is stable upto 200°C. Porapak is used primarily for efficient separations of a wide variety of relatively low molecular weight solutes. Porapak columns should be conditioned prior to use to remove residual solvents from the polymerization process. The utility of the material is very well demonstrated by a variety of a separations shown in Figure 10.13. The conditions are mentioned.

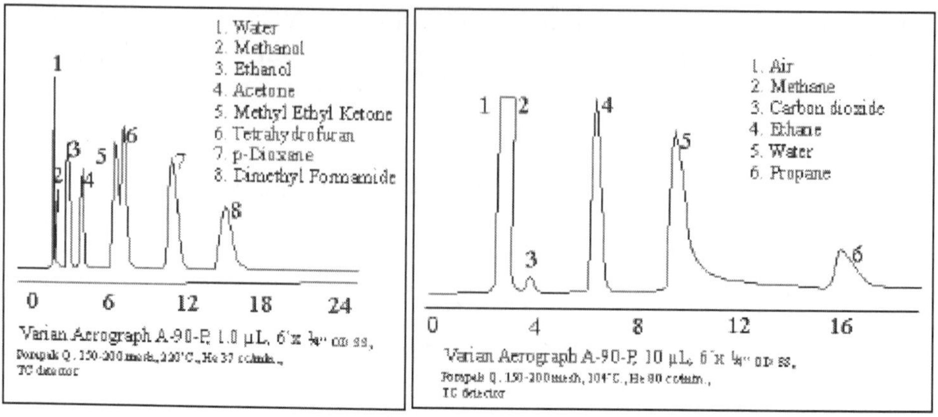

Figure 10.13: Applications of Porapak Column.

10.5 Detectors

After the sample is introduced and the dissolved compounds have been separated and are eluting from the column, a detection method is necessary. The selection of a detector will hinge on the molecular structure of the sample and the concentration ranges of the components of interest. Choosing the appropriate detector along with the column can optimize the results for the investigator. Since chromatographic detectors operate on vastly different principles, it is challenging to compare them directly. However, certain characteristics are indicative of a detector's usefulness. Some of these are listed below:

- ☆ Sufficient selectivity
- ☆ Sensitivity and detectability
- ☆ A linear response to solutes that spans several orders of magnitude
- ☆ Operational in a temperature range from ambient temperature to at least 400°C
- ☆ High reliability and ease of use
- ☆ A rapid response time that is independent of flow rate
- ☆ Consistent response to all solutes
- ☆ Non-destructive for solutes
- ☆ Low noise and minimum detectable level

Needless to say, no detector exhibits all of these characteristics and it seems unlikely that such a detector will ever be designed. Detectors may be classified as "integrating" or "differentiating" type. An integrating detector gives a response proportional to the total mass of solute in the eluted zone. A differentiating detector gives a response proportional to the concentration or mass flow rate of the eluted component. The most familiar example of a detector responding to concentration is the thermal conductivity detector (TCD). The most familiar example of a detector responding to mass flow rate is flame ionization detector (FED). The chromatogram produced by a differentiating detector consists of a series of peaks, each of which corresponds to a different solute. The area under each peak is proportional to the total mass of that solute (Figure 10.14). Differentiating detectors are most commonly used because of their convenience and accuracy.

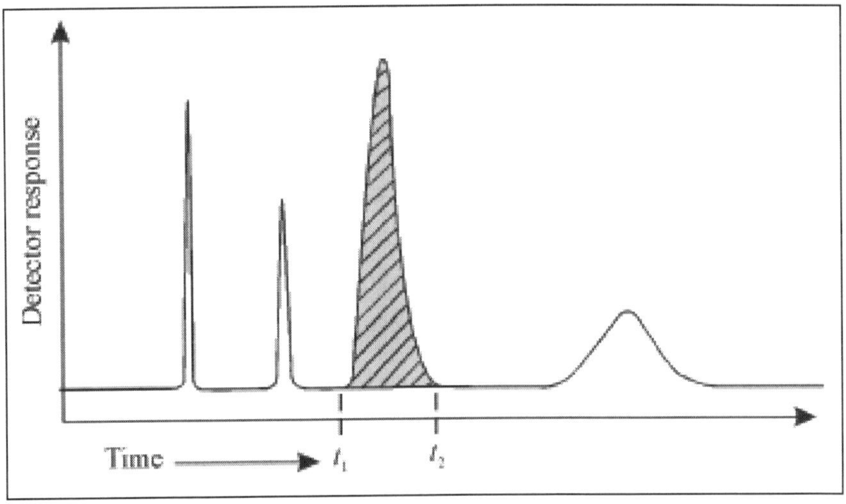

Figure 10.14: Differential Chromatogram.

Let us first consider a detector responding to concentration. Ideally, we have the following relationship

$$R = K_1C \quad (10.1)$$

where, R is detector response in millivolts, K_1 is a constant of proportionality, and C is the concentration of the solute passing through the detector

If the detector response is plotted against time t, we get a curve shown in Figure 10.15.

If the shaded area under curve is A, substituting the value of Ras given in Eq. 10.2, we get

$$A = \oint_{t_1}^{t_2} k_1 \, Cdt = k_1 \oint_{t_1}^{t_2} Cdt \quad (10.7)$$

Now let us consider a solute zone in the form of a plug in which the component concentration is constant (Figure 10.16) and equal to *M/V* where, Mis the total mass of a solute in the plug and Vis the plug volume

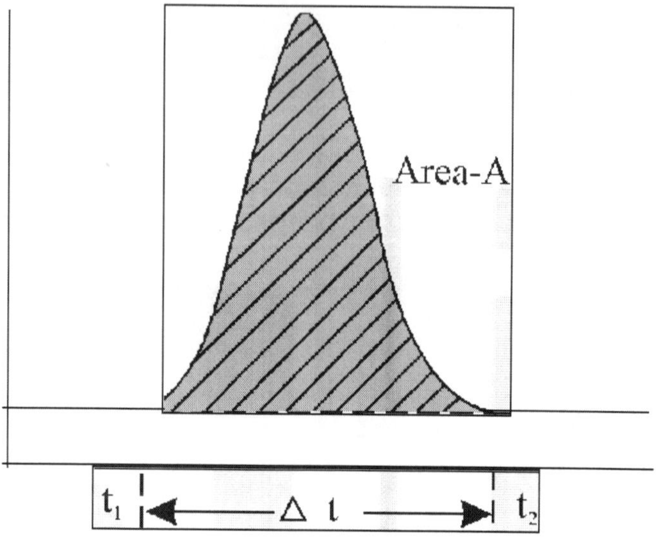

Figure 10.15: Detector Response vs. Time.

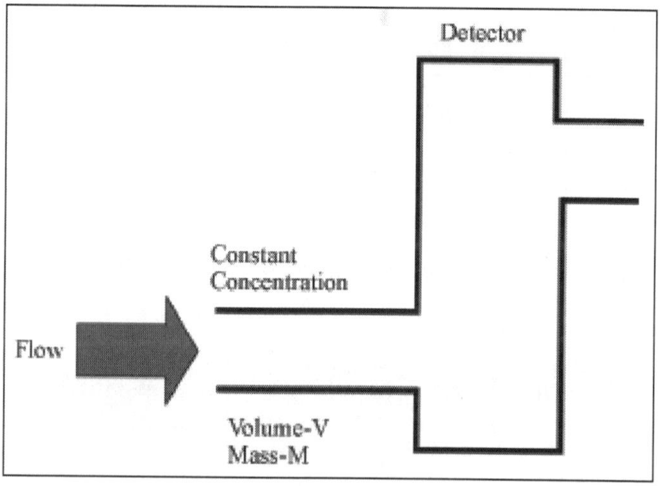

Figure 10.16: "Plug" Flow.

If C is constant, then

$$A = k_1 C \oint_{t_1}^{t_2} dt = k_2 C(t_2 - t_1) = k_1 C\Delta t \tag{10.8}$$

But C = M/V, therefore

$$A = \left(\frac{k_1 M}{V} \Delta t\right) \tag{10.9}$$

However, V= F, t, where F is carrier gas flow rate. Thus, Δt e= V/F

$$A = \left(\frac{k_1\, M}{V}\right)\left(\frac{V}{F}\right) = k_1\frac{M}{F} \quad (10.10)$$

Thus, finally we obtain

$$A = k_1\frac{M}{F} \quad (10.11)$$

Eq. 10.6 shows that peak area is directly proportional to the total mass of component. The above equation also shows that for a detector responding to concentration, the peak area is inversely proportional to the carrier gas flow rate. Thus, for accurate quantitative analysis with TCD, the flow rate must be kept constant. For detectors responding to mass flow rate (dm/dt), such as the FID, we have

$$R = k_2\left(\frac{d_m}{d_\tau}\right) \quad (10.12)$$

where, K_2 is a constant of proportionality, m is the instantaneous mass of component within the detector, and R and t have their earlier meaning. By similar reason:

$$A = \oint_{\tau_1}^{\tau_2} [Rdt = \oint_{\tau_1}^{\tau_2} k_2\,(\square]\,\frac{d_m}{d_\tau}\Big)\,dt = k_2 1\oint_{\tau_1}^{\tau_2}\left(\frac{d_m}{d_\tau}\right)dt \quad (10.13)$$

Cancelling dt within the integral sign, and integrating, we obtain

$$A = k_2 m \quad (10.14)$$

Eq. 10.9 shows that for a detector responding to mass flow rate, the peak area is proportional to the total mass of the eluted solute. However, unlike the concentration detector, the peak area for a mass flow detector is independent of carrier gas flow rate. Constant flow rate is not as critical for FID as for a TCD.

10.5.1 Detector Sensitivity

The sensitivity of a detector depends upon its operating principle. Sensitivity or detectability has been defined as the ratio of the smallest detectable quantity measured at twice the noise level to the peak width at base in seconds. For example, a 0.1 ng (10–10 g) sample with a 60 sec peak width at base would show a sensitivity of 1.6×10^{-12g}/sec If the column temperature is lowered and the same sample now has a 180 sec peak width at base (due to longer retention time) the resulting sensitivity would be 0.55×10^{-12} g/sec.

10.5.2 Detector Response

The response of a detector is the quantity of a signal generated by a given amount of the solute. Noise and minimum detectable quantity Detector sensitivity can be made as large as desired when the electrical output of a detector is increased to almost any desired value by electronic amplification. However, electrical noise inherent in the detector and electronics is also amplified and a point is reached where the noise level is high enough to obscure the detector's response. Therefore, the noise level limits the concentrations or mass flow rate of solutes that can be detected. The minimum detectable quantity is that amount which gives a detector response equal to twice the noise level.

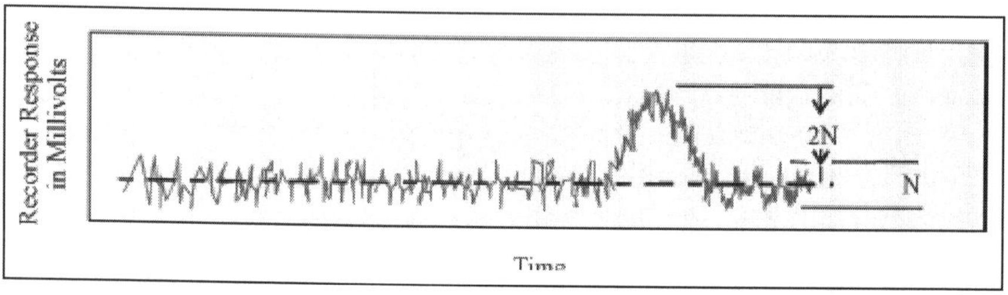

N1 = Noise level; N2 = Minimum Detectable level

Figure 10.17: Minimum Detectable Quantity.

10.5.3 Linear Detector Range

Accurate quantitative analysis depends upon a linear relation between concentration and detector response. The linear range of a detector may be defined as the ratio of the largest to the smallest concentration within which the detector is linear. There is a variety of detectors used in a gas chromatograph. Here, it may not be possible to go into the details of the different detectors. The design, working and utility of some of the important detectors may be essential to know to make a proper choice of the detector. Keeping this in mind, some of the detectors that are being briefly discussed here are

1. Thermal Conductivity Detector (TCD)
2. Flame Ionization Detector (FID)
3. Electron Capture Detector (ECD)
4. Helium Ionisation Detector (HID)
5. Flame Emission Detector (FED)
6. Flame ionization detector (FID)
7. Sulfur Chemiluminescence Detectors (SCD)
8. Flame Thermionic Detectors (FTD)
9. Barrier Discharge Ionization Detectors (BID)-:

10.5.3.1. Thermal Conductivity Detector (Katharometer)

TCD The thermal conductivity detector is based on the principle that a hot body will lose heat at a rate which depends upon the composition of the surrounding gas. Thus, the rate of heat loss can be used as a measure of the gas composition. The TCD operates by comparing the thermal conductivity of a sample component as reference against the carrier gas. Figure 10.18 shows a typical TC cell consisting of a spiral metal filament supported inside a cavity within a metal block.

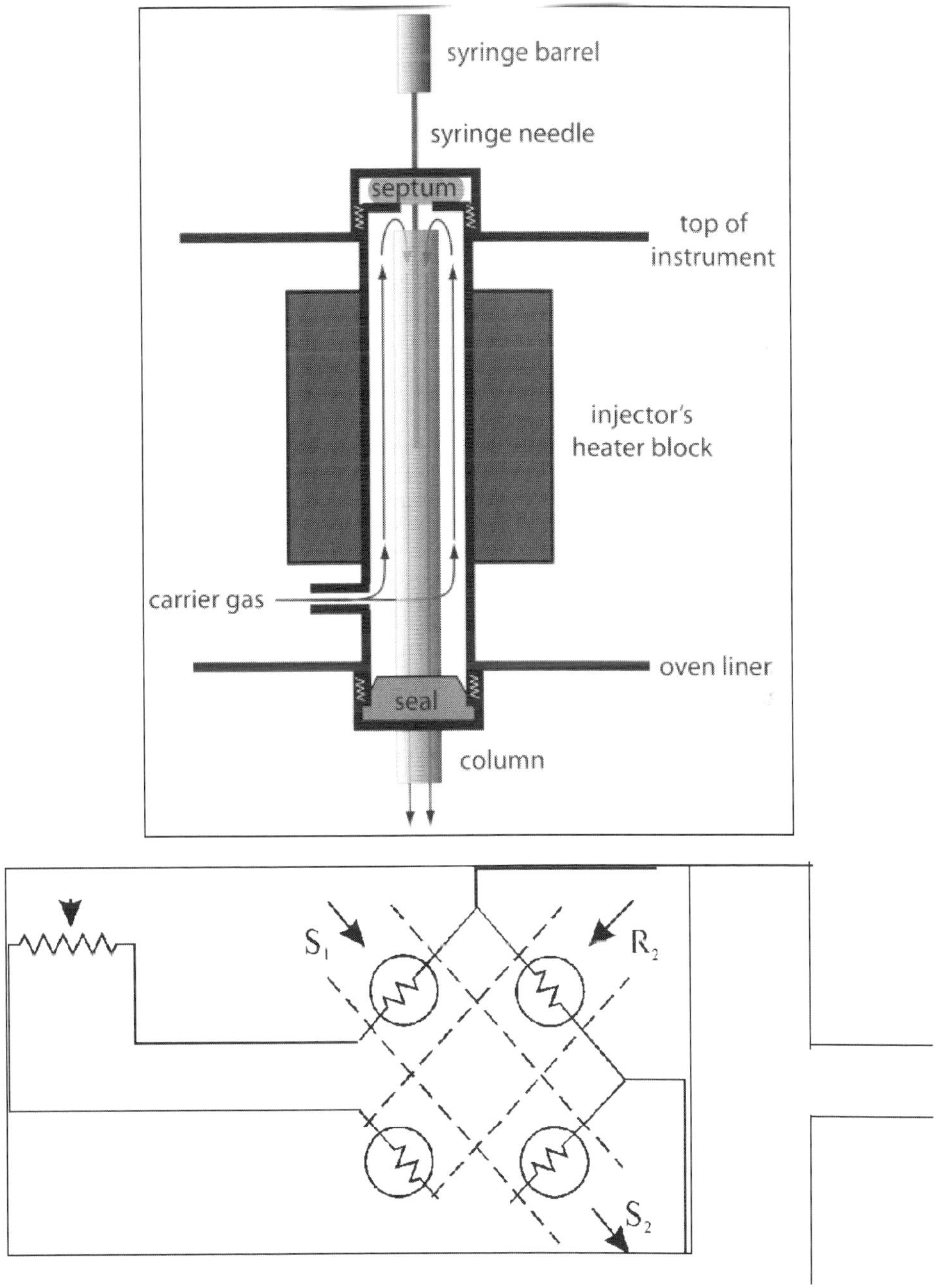

Figure 10.18: Schematic of TC Cell.

The filament is made of a material whose electrical resistance varies greatly with temperature. It has a high temperature coefficient of resistance. A constant current is passed through the filament causing its temperature to rise with pure carrier gas. With pure carrier gas flowing the heat loss is constant and thus, the filament temperature is also constant. If the gas composition changes, *e.g.* when solute peak emerges, the filament temperature changes, causing a corresponding change in electrical resistance. It is this resistance change which is measured by a Wheatstone bridge circuit (Figure 10.19).

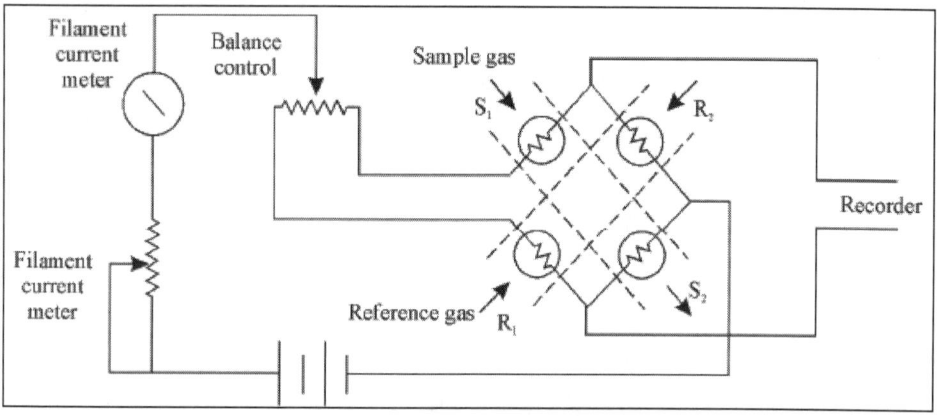

Figure 10.19: T C Wheatstone Bridge Circuit.

The TCD is universal *i.e.*, it will respond to any solute that has a thermal conductivity different than the carrier gas. It is mainly used for analyzing inorganic gases such as the components of air and its impurities. The sensitivity ranges from large percentages of components down to a few parts per million.

10.5.3.2. Flame Ionisation Detector (FID)

In operation as the solute elutes from the column, it is mixed with hydrogen and then combusted. In the combustion process, transitional ions are created that can be collected by a metal conductor.

These collected ions cause an electronic signal which can be amplified and displayed Gas Chromatography on a recorder. Ideally, the final combustion products are carbon dioxide and water so that they can be safely vented to the atmosphere. The number of these transitional ions formed is proportional to the total number of solute molecules passing through the detector. The FID will respond to organic compounds only *i.e.*, those solutes that have a carbon-hydrogen bond and are combustible. Table 10.5 lists the solutes for which FID does not respond. The lack of response to air and water makes the FID especially suitable for the analysis of air pollutants or aqueous samples such as alcoholic beverages, biological materials *etc.* Similarly, the absence of a "solvent peak" makes carbon disulphide a convenient solvent for use with the FID.

Table 10.5: Compounds Giving Little or No Response in the Flame Ionization Detector

He	CS_2	CO
Ar	COS	CO_2
Kr	H_2S	H_2O
Ne	SO_2	$SiCl_4$
Xe	NO	$SiHCl_3$
O_2	NO_2	SiF_4
N_2	NH_3	N_2O

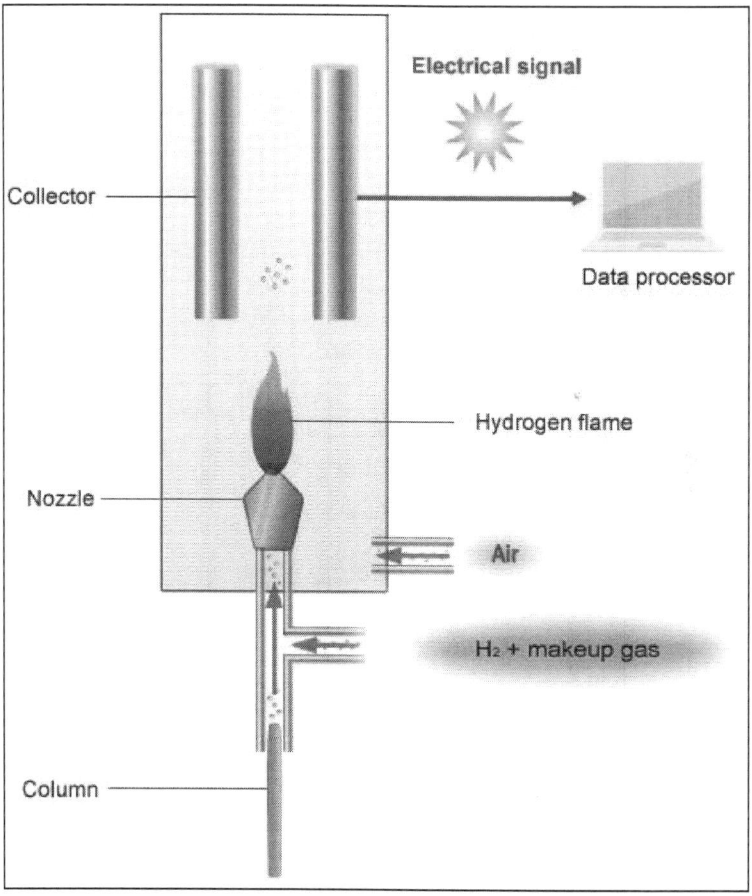

Figure 10.20: Flame Ionization Detector.

In addition to normal carrier gas, auxiliary gases of hydrogen and air are needed to fuel the flame. The hydrogen flame, being perhaps the most popular detector in gas chromatography, is extremely sensitive and under proper conditions, the detector is capable for responding to a few parts per billion of the solute component. In general, good sensitivity and stability are obtained with a carrier gas flow of 30 cm^3/min, hydrogen flow of 30cm^3/min and air flow at 300 cm^3/min.

The FID has the widest linear range of any detectors in common use (10^6 to 10^7). The combination of high sensitivity and wide linear range makes the FID the choice in trace analysis.

10.5.3.3. Electron Capture Detector (ECD)

The characteristic features of an ECD are as follows:

i) Specific
ii) Selective to halogens
iii) Sensitive to a few parts per trillion

The electron capture detector measures the loss of signal rather than a positively produced electrical current. As the nitrogen carrier gas flows through the detector, a tritium or Ni^{63} source (characteristics given in Table 10.6) ionizes the nitrogen molecules and slow electrons are formed. The slow electrons migrate to the anode under the fixed voltage which is formed as "cell voltage". The slow electrons produce a steady current. If a sample containing electron absorbing solutes is introduced, this current will be reduced. The loss of current is a measure of the amount and electron affinity of the solute.

Characteristics	H^3	Ni^{63}
Half Life	12.5	125 years
Type of Decay	β	β
Particle Energy	18 kev	67 kev
Specific Activity	9,800 curie	5 curie/gram
Physical Properties	Gas	Solid
Fall Temp. Limitation	220° C	200° C
Radio Toxicity	Very low	Moderate
Cost Per Curie	$0.024	$17.5

The ECD is extremely sensitive to certain solutes, such as alkyl halides, conjugated carbonyls, nitriles, nitrates and organometals, but is virtually insensitive to hydrocarbons, alcohols, ketones *etc.* Selective sensitivity to halides makes ECD especially valuable for the analysis of pesticides, Both H^3 (tritium) and Ni^{63}have been employed in the ECD.

N_2, which is used as the carrier gas, is ionized by β waves emitted from the 63Ni radiation source.

$$N_2 \rightarrow N_2+ + e^-$$

A current flows when the ions gather in the collector.

When an electrophilic compound is placed in this equation,

$$PCB + e^- \rightarrow PCB^-$$

PCB^- is much larger and heavier than e^- and so takes more time to reach the collector.

-A higher voltage is needed for a constant ion current to flow.

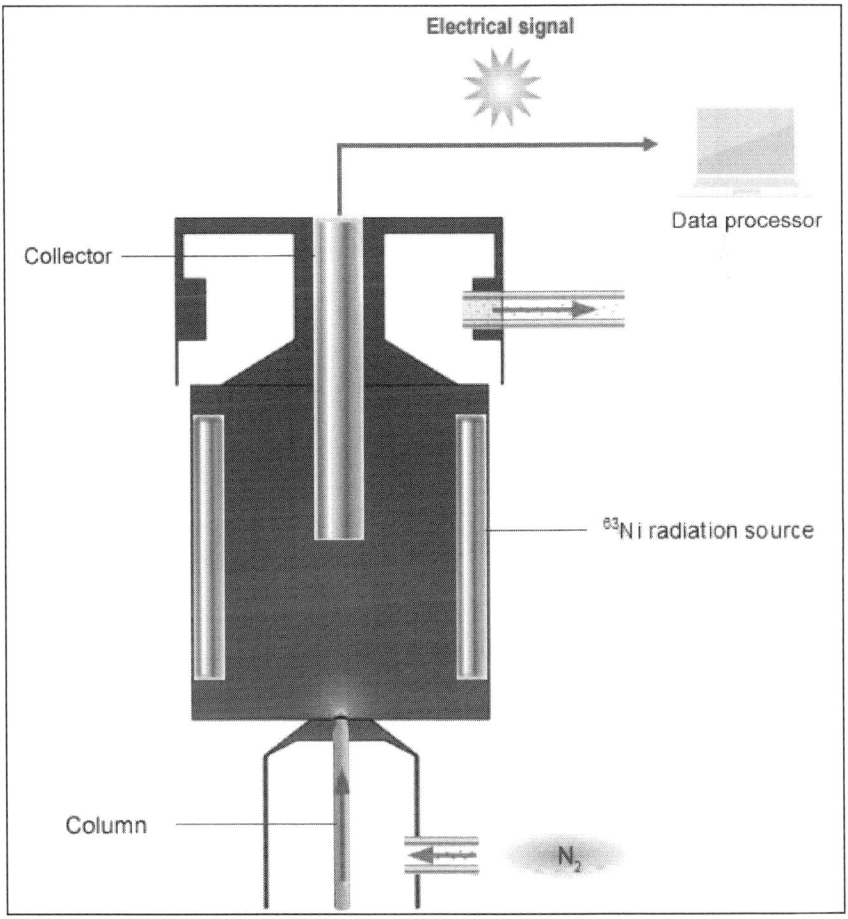

Figure 10.21: Electron Capture Detector.

105.3.4 Helium Ionisation Detector (HID)

Some of the characteristics of this detector are given below. i) Universal ii) Photo-ionization iii) Sensitive to fractional parts per million. The helium ionisation detector is a universal detector except of course, it will not sense helium. The combination of the tritium beta radiation and high field gradient

(4000 volts per cm) raise the helium to a metastable state with an ionisation potential of 19.8 eV. All solutes having a lower ionisation potential will be ionized giving rise to a positive signal. The HID was developed for the ultratrace analysis of permanent gases.

10.5.3.5. Flame Emission Detector (Alkali Flame Detector)

The flame emission detector consists of a normal flame detector plus the addition of a small alkali salt pellet placed at the burner set with precise control of the hydrogen and air flow rates. This detector can be made to give an enhanced response to phosphorus containing compounds. The three major differences between the standard FID and alkali flame detector are the alkali salt tip, the air flow rate and the electrode configuration. Both CsBr and Rb_2SO_4 have been used with good results. These salts give a selectivity of 5000 L for phosphorus and 50/L for nitrogen over normal hydrocarbons. Alkali flame detectors operate at air flow rate of approximately 130 cm^3/min which is significantly below that of a normal FID (300-400 cm^3/min). Since constant air and hydrogen flow rates are important for the stability of this detector, flow controllers are required. The alkali flame detector has been used extensively in the analysis of organo-phosphorus pesticides and for nitrogen containing solutes.

10.5.3.6. Flame Ionization Detector

Sulfur compounds, phosphorus compounds, and organic tin compounds each emit light at unique wavelengths when burned. By passing light through a filter, only light of these unique wavelengths reaches a photomultiplier tube. The photomultiplier tube then converts the detected light intensity into an electrical signal.

10.5.3.7. Sulfur Chemiluminescence

Detectors (SCD): The SCD is a selective, high-sensitivity detector for sulfur (S) compounds. The SCD is highly sensitive and capable of detecting infinitesimal amounts of sulfur compounds. Compared to the FPD, which is similarly capable of selective detection of sulfur compounds, the SCD is around one order of magnitude more sensitive and exhibits a proportional linear relationship between the SCD sensitivity and the sample concentration. (This relationship is quadratic for the FPD.) The SCD also exhibits equimolar sensitivity and measures sulfur compounds with the same relative sensitivity regardless of compound structure. This characteristic of the SCD allows the use of calibration curves for other compounds to determine an approximate concentration of a target compound, even when no standard sample is available. The SCD also differs substantially from other detectors in that a low-pressure environment is maintained inside the SCD.

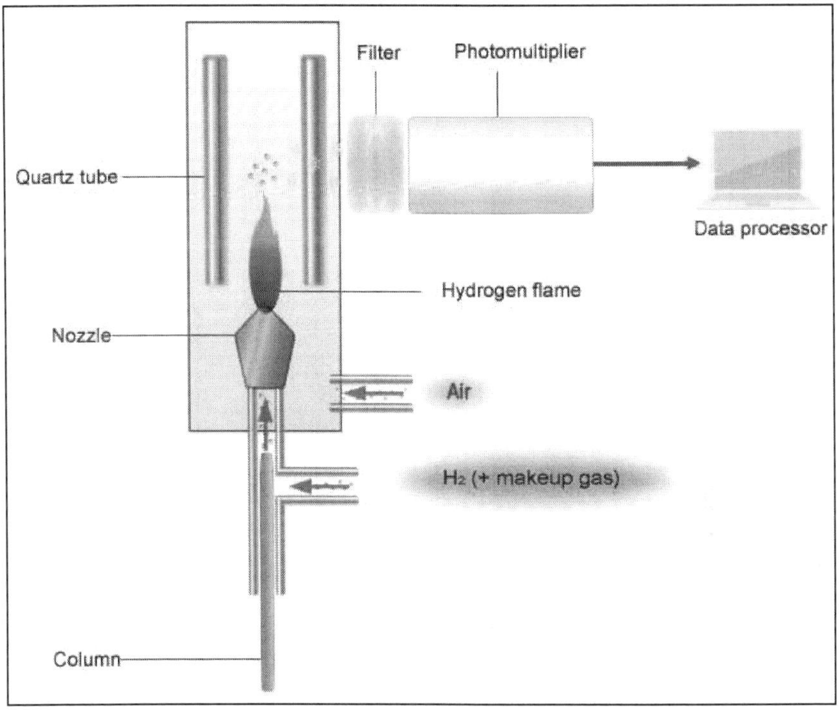

Figure 10.22: Flame Ionization Detector.

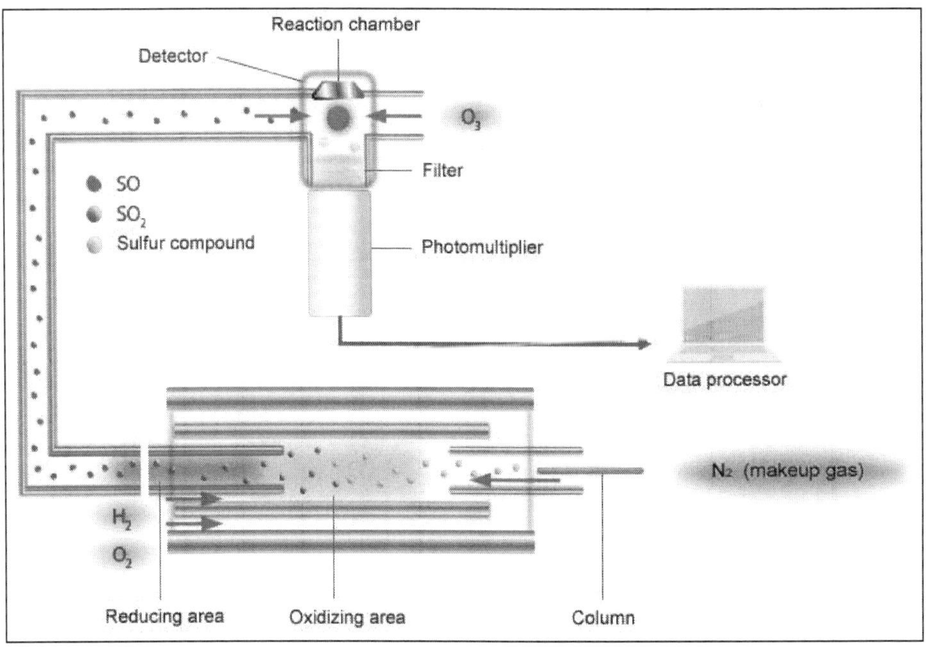

Figure 10.23: Sulfur Chemiluminescence Detectors.

10.5.3.8. Flame Thermionic Detectors (FTD)

The FTD is a selective, high-sensitivity detector for organic nitrogen compounds and inorganic and organic phosphorus compounds. (The selectivity of the FTD for phosphorus compounds is not as good as the FPD.) The FTD does not react to inorganic nitrogen compounds. When a current is passed through the platinum coil with an alkali source attached to the coil (rubidium salt), the coil increases in temperature, which creates plasma around the alkali source.

Rubidium radicals (Rb*) are generated within this plasma.

-Capable of oxidizing CN and organic phosphorus compounds

-PO_2 reacts with Rb* as shown below, creating ions.

$CN + Rb^* \rightarrow CN^- + Rb^+$

$PO_2 + Rb^* \rightarrow PO_2^- + Rb^+$

A current flows when ions gather in the collector.

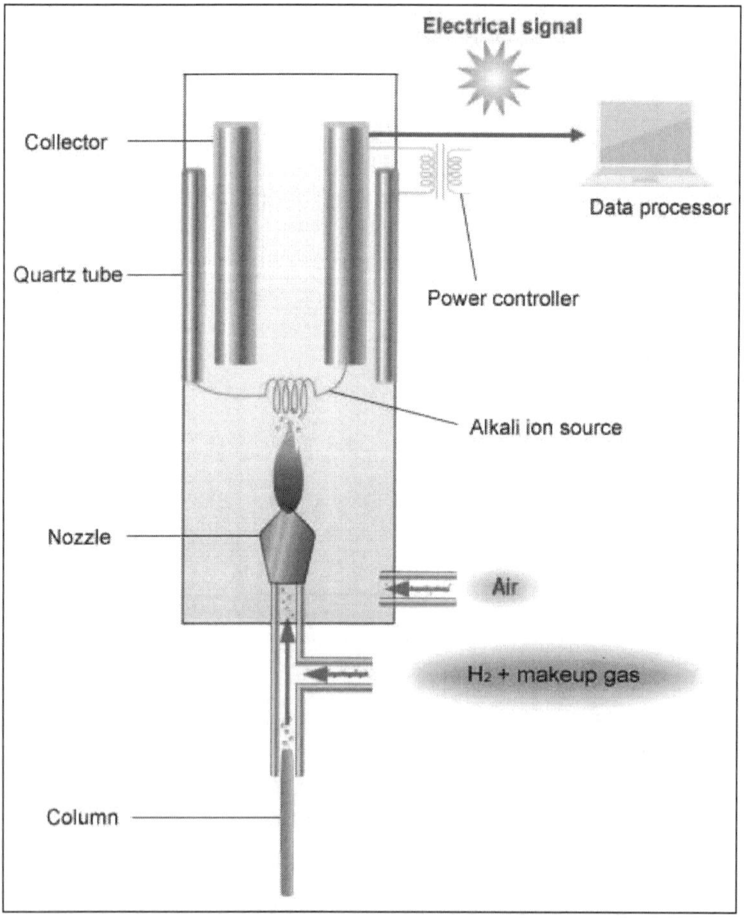

Figure 10.24: Flame Thermionic Detectors.

10.5.3.9. Barrier Discharge Ionization Detectors (BID)

The BID that can detect all inorganic and organic compounds other than He and Ne. The BID is also capable of detecting trace amounts of impurities at the ppm level that the TCD failed to detect during an inorganic gas analysis. The BID generates a stable He plasma, uses the energy emitted by the excited He to ionize compounds, then attracts these ions to a collector. The He plasma energy emitted is extremely high and capable of ionizing all compounds other than He, which is used to create the plasma, and Ne, which has extremely high ionization energy. As a result, the BID can detect any compound, in principle, other than He and Ne.

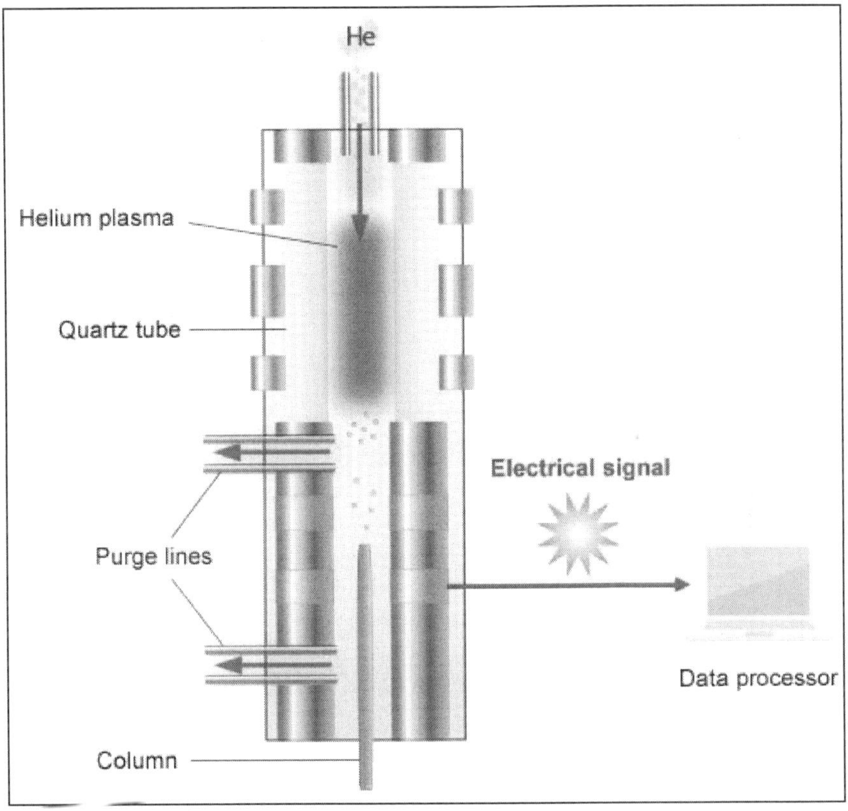

Figure 10.25: Barrier Discharge Ionization Detectors.

10.6 Sampling and Introduction of Sample into GC Unit

For analytical results to have any validity or meaning, adequate sampling procedures should be adopted. Sampling is the process of extracting from large quantity of material a small portion which is truly the representative of the composition of the whole material under scrutiny. Proper sampling in GC assumes a particular significance because here we deal with the analysis of gaseous substances or components which are volatile. The sampling process involves several steps:

1. Removal of an aliquot from a bulk quantity which is representative of that bulk.
2. If necessary, alteration of the physical state of the sample (*e.g.* that of gas) so that it can be manipulated.
3. Transportation to the laboratory for analysis.
4. Storage prior to analysis.
5. Introduction of the sample into the GC unit for analysis

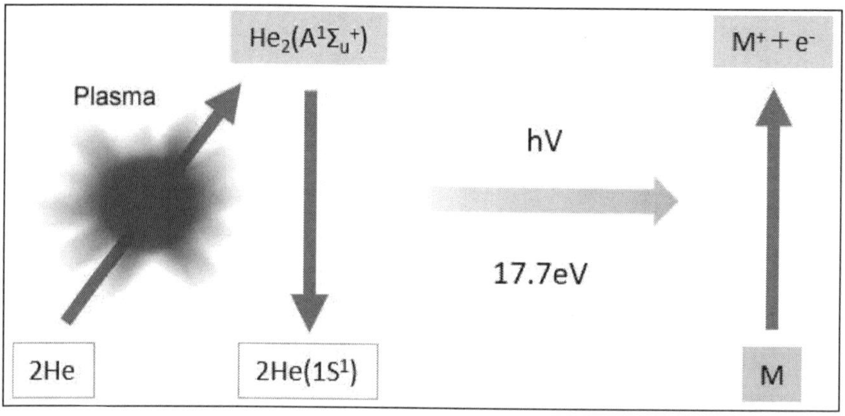

Figure 10.26: Principle of BID Detector.

10.6.1 Hazards in Sampling

In the sampling, handling and storage of the gaseous and volatile materials, there are some hazards involved and due precautions should be taken. The analyst must, as a general rule, try to obtain a prior knowledge of the material which is expected to be analysed. The analyst must wear adequate protective clothing during sampling. Adequate antidotes *etc.* must be kept ready before proceeding for sampling work. No naked flame should be allowed near a sampling area, if the sample is known to contain inflammable components. Many gases/vapours may be toxic. Pressure changes may release excessive quantities of gas. Spontaneous ignition of inflammable gases may occur. Sudden vaporisation of liquefied gases may take place. Volatile and inflammable liquids may pose sufficient dangers to a careless analyst. Using a facemask/goggles during sampling is highly recommended.

10.6.2 Introduction

Introduction of the Sample into the GC Unit The introduction of the sample into the GC unit is the first step in the chromatographic process. Its efficiency, therefore, gets reflected in the overall efficiency of the separation procedure and the accuracy and precision of the qualitative and quantitative results. In this regard, some aspects given below need to be considered.

A. Sample size.

B. Injection systems.

C. Sample injection port.

A. Sample Size

The sample volume may vary depending upon the amount and rate of evaporation of the sample in the sample port in the case of liquids/solutions of solids or on the pressure of gas/vapour sample contained in the sample container. In the case of gases, it is a common practice to vary the amount of sample by introducing different volumes at a constant pressure. The sample size will obviously be related to the kind of solute-stationary phase interaction (linear or non-linear) and consequently to the shape of the final chromatographic peak obtained.

B. Injection Systems

Various kinds of injection systems have been incorporated in commercial GC units. Almost as important as column choice is the choice of injection method. The various approaches are mentioned below:

i) Direct injection: Injection into the packed column.

ii) Split injection and splitless injection: These are adopted for capillary GC.

iii) Solid injection (moving needle): In this case, the solvent is evaporated before injection, resulting in improved separation of fast eluting components.

iv) On-column injection: The sample is introduced jnto the column as a liquid, reducing sample decomposition and giving reliable quantitative analytical results. This is considered as being an ideal solution for quantitative capillary GC.

v) Valve injection: Reproducibility of the valve injection volume is superior to any other injection technique. Valve injection can easily be automated which makes it very appropriate for routine on - line analysis.

vi) Thermal desorption cold trap injection: Gaseous samples are concentrated on an adsorbent tube. The components are concentrated on a cold trap and then injected into the GC unit by rapid heating.

vii) Purge and trap injection: Aqueous samples are purged with inert gas and the components are concentrated on a cold trap. Entrained water is frozen out in the condenser. The components are injected by rapid injection of the cold trap for multidimensional analysis, multiple switching intelligent controllers are available. The sample is pre-separated on a pre-column. Coeluting components are trapped on a cold trap and reinjected into a column for further separation and analysis.

C. Sample Injection Port

The general requirements for an injection port are controlled temperature, low volume and inertness in construction. The temperature must be maintained within the proper range to vaporise all the solutes instantly but not too hot to cause thermal decomposition of labile solutes. The entire injection port must also be swept with carrier gas to ensure that all of the vaporized solutes are moved quickly into the column. This means that its internal portion must be kept to a minimum volume. The materials of construction of the injection port should be such that no reaction with the solutes occurs. It must be as inert as much as possible to minimize any possibility of catalytic decomposition because of contact with exposed surface area. The syringes used to inject samples are usually of small volumes. The normal ranges are from zero to 1 microlitre, zero to 5 microlitres and zero to 10 microlitres. In the one microlitre syringe, the sample is totally confined to the needle and can never be observed by the operator. For the other two syringes, the sample enters and its level can be directly measured in the barrel of the syringe. The barrel is calibrated in microlitres and fractions of microlitres.

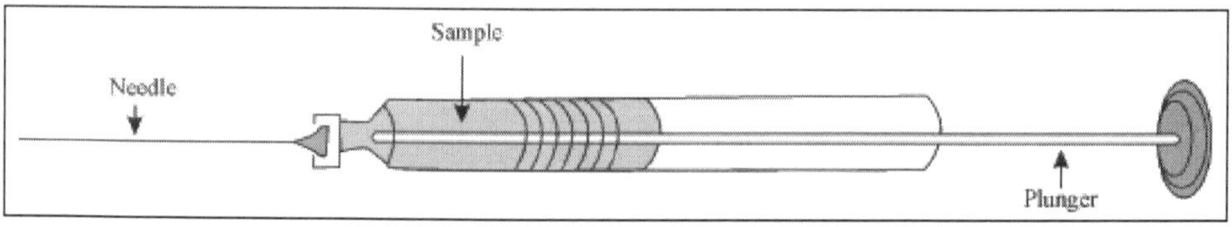

Figure 10.27: Sampling Syringe.

Sample introduction for liquids on packed columns is often accomplished with a microsyringe through a self-sealing silicon septum as shown in Figure 10.28.

In this figure, the column is lined up colinearly with the syringe needle providing for two possible injection modes: on-column or flash vaporization. For on-column operation, the column is positioned as shown in Figure 10.28 with the column packing beginning at a position just reached by the needle. When the syringe is pushed as far as it will go into the port, its contact will be delivered into the first part of the column packing, ideally on a small glass wool plug used to hold the packing in the column. In the second configuration, the column is placed so that its front end and its packing barely extends into the injection port and cannot be reached by the syringe needle. Efficient sampling for this configuration requires that the solutes evaporate quickly (flash vaporization) when injected into the port.

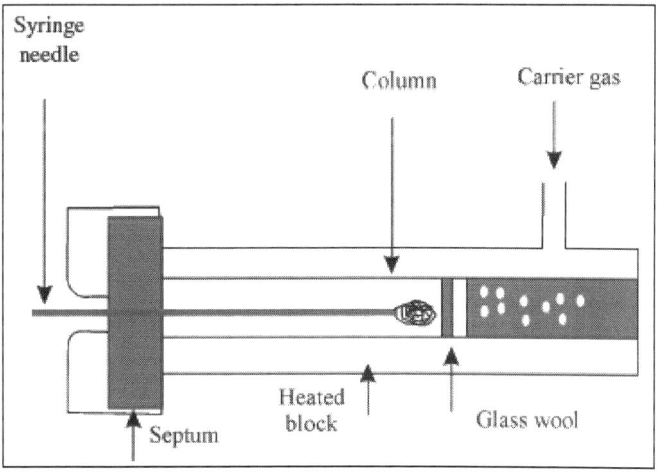

Figure 10.28: Simplified Injection Port for On-column Injection.

As carrier gas passes through the valve on the way to the column, a sample can be filling the sample loop. When the actuator is turned the carrier gas stream is redirected through the sample loop and sweeps the sample into the column. The loop itself may have a volume as low as one quarter of milliliter or as large as 10 milliliters. The choice of loop size is made to be compatible with the analyst's experimental needs.

10.7 Application of Gas Chromatography

Gas chromatography is a versatile analytical technique employed for both qualitative and quantitative analysis. It's crucial to recognize that the column in a gas chromatograph serves as the primary separation mechanism. However, the complete gas chromatograph system, comprising a detector and a recorder, provides detailed information about the eluting bands. Consequently, the instrument furnishes both qualitative and quantitative insights into the components of a sample. This section will delve into three key aspects: 1. Identification of compounds (Qualitative Analysis) 2. Quantitative analysis

10.7.1 Identification of Compounds

A) Qualitative Analysis

In gas chromatography, the retention time or retention volume of a solute is a characteristic property of that substance, just like its boiling point or refractive index. This means that the way a solute interacts with the stationary phase in a chromatography column can be used to identify it. To identify unknown compounds in a mixture, standards are run under identical conditions as the sample, and the chromatograms are compared. If the retention times of the unknown compounds match those of the standards, then the unknown compounds are identified. However, gas chromatography is not the most powerful technique for identification. Instruments like mass spectrometers can provide more detailed information about the structure of molecules, making them better suited for identifying unknown compounds. Gas chromatography is primarily a separation technique, but when coupled with instruments like mass spectrometers or infrared spectrometers, it becomes a powerful tool for both separation and identification of compounds in mixtures. The mass and IR spectra provide valuable information about the eluants from the chromatography column. There are various other approaches that can be used in conjunction with GC to identify separated components, but these are not discussed here.

B) Quantitative Analysis

The chromatogram obtained on a recorder chart can be used to measure quantitatively the concentration of components in a mixture. In general, three methods are used for quantitative evaluation:

Area normalization method: This method assumes that the entire sample is eluted from the column. The area of each peak is measured, and the per cent composition is calculated by dividing the area of each peak by the total area of all peaks and multiplying by 100. This method is only accurate if the detector response is the same for all components of the mixture. If the detector response is not the same, the detector response factor for each component must be determined, and the measured areas must be corrected accordingly.

C) Internal Standardization Method

In this method, known amounts of sample and standard are mixed and chromatographed. The peak areas for sample component and for standard are measured and ratios of both peak areas are determined. Either area ratios are plotted against weight ratios to obtain a graph or area ratios for unknown are compared directly with those for the known amounts. Thus, accurately known quantity of the internal standard is added to the unknown sample and this mixture is chromatographed and area ratios are measured. The requirements for a suitable internal standard are summarized as follows:

- ☆ The standard must be a compound, which is well separated from all components of the mixture being analyzed.
- ☆ The standard must not react with any component of the sample, nor should it influence the physical properties of the other components, *e.g.* volatility.
- ☆ It should yield a symmetrical chromatographic peak.
- ☆ It must be chemically similar to the unknown.
- ☆ It should have concentration comparable to the components of interest.

Comparison method: In this method, a synthetic mixture containing known quantities of the components of interest in the range of concentration expected in the unknown sample is prepared and analyzed. The values for the peak areas for different known volumes of synthetic blends are estimated and a calibration curve is plotted An exact quantity of the unknown sample is then injected and the peak areas so calculated are used to read from the calibration curve the concentration of the component in the unknown mixture.

10.7.2 Some Examples of Applications

There are too many applications of GC to be listed here. It has been estimated that perhaps 20 per cent of the known chemical compounds and almost 80 per cent of the organics can be analysed by GC. Some of the important applications from different areas are being discussed to illustrate the versatility of GC.

1. Bacterial Identifications

Long chain fatty acids found in the bacterial cell can be used to distinguish between various microorganisms. Fatty acids with chain length from Cl0 to C20 can be separated and estimated on a 3 m glass column of 2 mm internal diameter packed with 3 per cent SP–2100 DOH at oven temperature of 150°C to 225°C with nitrogen gas at a flow rate of 20 rnL/min.

2. Environmental Analysis

a) *Water Analysis*

The organic pollutants in water are concentrated from water samples by solvent extraction or by purge and trap technique. The volatile pollutants are analysed on 80/100-mesh carbopack C/0.2 per cent carbowax 1500 column.

b) *Air Analysis*

The analysis of organic vapours in the industrial air environment for the assessment of exposure to workers is done as follows:

i) Organic vapours are collected on a charcoal adsorbent with a portable pump.

ii) Desorption from charcoal is done in a closed vial with carbon disulphide.

iii) Analysis of the desorbed sample is done on a GC using a 6 m, 3 mm internal diameter S.S column packed with 10 per cent SP–1000 by temperature programming from 100°C to 200°C. By this method, pollutants such as vinyl chloride, xylenes and aromatic hydrocarbons can be estimated.

Clinical and Toxicological Analysis

Toxicologists have recognised the usefulness of GC for the analysis of toxic substances. The analysis of lidocaine and diphenhydramine has been done using flame ionisation detection. A 15m x 0.25mm i.d. 5 per cent methylsilicone (DB-5) column has been used, temperature programmed from 180°C to 230°C at 5°C min–1 and helium used as a carrier gas. d) Forensic toxicology: It is highly specialised branch of analytical chemistry concerned primarily with the analysis of specimen from different organs of the human body for toxic substances. A simple GC system utilising four columns and three liquid phases (SE–30, Hallcomid M-18, and Carbowax 6000), complemented by a direct solvent extraction scheme designed to detect common poisons, drugs, and human metabolites to a sensitivity limit of 2 μg/ml in blood, urine and tissue is developed.

c) Food Analysis

Gas chromatography plays a crucial role in the food industry, serving a multitude of purposes. It is employed for quantitative and qualitative analysis of food components, identification and quantification of food additives, and detection and analysis of contaminants such as environmental pollutants, pesticides, fumigants, and naturally occurring toxins. Gas chromatography is instrumental in ensuring the safety of food products by preventing contaminated items from reaching consumers. Moreover, it contributes to maintaining food quality by ensuring consistent flavor, taste, texture, and aroma. While other techniques are utilized in the food industry, gas chromatography remains a preferred method due to its ease of use and cost-efficiency.

d) Quality Control

Gas chromatography plays a crucial role in quality control across a wide range of industries, with particular emphasis in the manufacturing of automobiles, chemicals, and pharmaceuticals. Within the pharmaceutical domain, gas chromatography is indispensable for ensuring the purity and consistency of drug products. It enables the detection and elimination of trace contaminants that could compromise the quality and effectiveness of medications. Additionally, gas chromatography has emerged as a valuable tool for separating chiral compounds, which are molecules with distinct properties due to their spatial arrangement. In the automotive industry, gas chromatography is employed to identify and quantify volatile organic compounds (VOCs) emitted from car interiors. These VOCs can originate from various sources, including carpets, door linings, pedals, seat covers, and other interior materials. By measuring VOC levels, scientists can assess the potential health risks associated with exposure to these chemicals and develop strategies to reduce their emission. Producers of chemicals also extensively utilize gas chromatography to maintain consistent product quality, particularly in the manufacturing of emulsifiers, solvents, and co-solvents. As production scales up, gas chromatography ensures that these essential chemicals adhere to strict quality specifications.

e) Research

Gas chromatography is fundamental to many research areas, in particular, for the analysis of meteorites and natural products. Scientists use gas chromatography to analyze the composition of meteorites that fall to the earth. This provides vitally important information regarding the nature of life outside of earth, as well as revealing details about primitive life on earth. Specifically, many studies have been conducted with gas chromatography to determine the presence of ribose in meteorites, the building block of RNA.

Chapter 11

Calibration of HPLC and GC

11.1 Introduction

Calibration of an instrument is the demonstration that the instrument or device produces results within specified limits by comparison with those produced by reference or traceable standard over an appropriate range of measurement. HPLC Calibration is the most critical activity in the laboratory. Calibration of HPLC is performed quarterly, half yearly and yearly basis depending upon the quality management system procedure. The parameters for calibration are listed as below:

Pressure Test

Drift and Noise

Column oven and sample cooler

Pump by flow rate accuracy measurement.

Pump by gradient flow measurement.

UV-Vis/PDA detector by reference energy check.

Recommended frequency for these tests are done quarterly

Half Yearly Calibration Parameters are enlisted as-:

UV-VIS/PDA detector by linearity measurement.

Auto sampler by carry over check.

Auto sampler by linearity check.

UV-VIS/PDA detector by wavelength accuracy measurement.

Auto-sampler for RI detector by linearity measurement.

Yearly Calibration Parameters

RI Detector by linearity measurement.

Fluorescence detector by linearity measurement.

Fluorescence detector by wavelength accuracy measurement.

Before start, the Calibration activity, Ensure your system is neat and clean,

Clear the surroundings of your HPLC system,

Check the system Power connections, Reservoir Bottles, tray, *etc.*

Ensure the required consumables like Ferrules, Peak nut, Tubing are available.

Make sure your drain collection container is placed and properly attached to drain tubing.

Pressure Test

The pressure test shall be performed before pump calibration as per the procedure as given below.

Purge the system for each flow line with the mobile phase and use any of the flow lines for the pressure test.

Ensure that the tubing from the reservoir to the column inlet shall be removed.

Plug the outlet of the pump using a dead nut.

Set the flow rate of the pump to 0.1 ml/min.

Observe the pressure until it reaches the maximum psi/kgf.

Repeat the same procedure for further two times and observe the same.

In case the pressure does not reach the maximum and fluctuate at some psi/kgf/cm^3, it implies poor check valve performance or leaks within the pumping system.

After satisfactory completion of the pressure test, proceed for further calibration of other parameters.

Drift and Noise.

Mobile Phase Preparation- Mix 700 ml of HPLC grade water and 300 ml of methanol and filter through 0.45-micron nylon filter, degas for 10 min by sonication.

Flush the system with hot water for 15 minutes.

Ensure that, the column is conditioned before injecting the sample.

After conditioning, the column equilibrates the system for 15 min and then run the blank injection as per test condition for 15 min.

After blank injection run completion, process the data using the processing method.

Record the results in the observation table

System Requirement

Column: C18 (25 cm x 4.6 mm), 5 micron or equivalent.

Flow rate: 1.0 ml/min

Wavelength For UV: 254 nm

Wavelength For PDA: 254 nm (2D data collection at 3.6 nm bandwidth)

Injection volume: 0.0 ml

Run time: 15 min

Acceptance Criteria

*Noise should be less than or equal to 60 micro AU for a UV detector and less than or equal to 80 micro AU for PDA detector.

*Drift should be less than 10 milli AU/Hr for UV detector and less than 10 million AU/Hr for PDA detector.

Column oven and sample cooler.

Procedure

Carry out the calibration for the temperature point at 30°C, 60°C and 15°C using digital temperature indicator.

Set the temperature of the column oven at 30°C, keep the temperature probe in the centre of the column oven and allow stabilizing the set temperature for at least 15 to 20 min or more requirements.

Check the displayed temperature of the column oven and temperature indicator device and record the reading in the observation table.

Similarly, carry out the calibration for 60°C temperature calibration in the same manner and note all the reading in the observation table.

If column oven is available with cooler, set the temperature of column oven at 15°C, keep the temperature probe in the centre of column oven and allow stabilizing the set temperature for at least 15 to 20 min and noting the reading in observation Table 11.1.

Similarly set the sample cooler temperature at 4.0°C and 25.0°C and repeat the procedure as described in sec 5.2 to 5.4 and record the results in the observation table.

Note: The location shall decide the calibration range for the column oven and sample cooler temperature based on the working range. Sample cooler temperature calibration shall be applicable for only that system which has a sample cooler facility.

Acceptance Criteria

Observed temperature reading should be within ± 2.0°C of set temp.

11.2 Pump by Flow Rate Accuracy Measurement

Requirements

- ✰ Where four inlet lines (A, B, C and D) are mixed into GPV (Gradient proportioning valve) assembly, before the pump (system operated with a single pump). The mixing accuracy of lines is decided by GPV functionality, this is normally measured by GPV test in such case only one-time flow rate accuracy shall be performed. In the case of a binary system where two pumps exist for each line then individual pump flow rate calibration shall be done.
- ✰ Mobile Phase: HPLC grade water

System Requirements

Testing Condition: Flow Rate = 0.5 ml/min, 1.0 ml/min, and 3.0 ml/min.

Sample/Standard preparation: None.

Procedure

Connect the stainless steel tubing, or other restriction devices, in place of column connections with the instrument.

Set the vacuum degasser (if installed) to run continuously, or set the purge rate to 100 per cent.

Keep the mobile phase in the respective reservoir.

Purge the system for each flow line with filtered HPLC grade water and allow it to run the water for about 15 min before checking the flow.

Ensure that the tubing from the reservoir to column inlet shall be free from the air bubble and system pressure should be constant.

Take 10 ml clean and dried volumetric flask and keep at the restricted capillary's outlet in such a way that when the pump is on, the water drops fall into the volumetric flask

Set the pump for 0.5 ml/minute flow-rate, start the pump and allow to run about 10 min. Then keep the flask at restricted capillary's outlet and then simultaneously start a stopwatch.

Stop the Stopwatch when the volume of water reached the mark in the volumetric flask. Derive the time taken and record the observation in the observation table.

Repeat the same procedure for the flow rate of 1.0 ml/min and 3.0 ml/min in 10 ml volumetric flask and record the observations in below Table 11.1.

Table 11.1

Sl. No.	Set Flow Rate (ml/min)	The Volume Delivered by the Pump (ml)	Theoretical Time (In seconds)	Limits (± 2 per cent of the theoretical limit, seconds)	Actual Time (Seconds)
1	0.5	10	1200	1176-1224	
2	1.0	10	600	588-612	
3	3	10	200	196-204	

Acceptance Criteria

The actual time observed should be within ± 2.0 per cent of the theoretical time.

11.3 Pump by Gradient Flow Measurement

Requirements

Acetone

HPLC grade water or equivalent

Restricted capillary (5-meter Length x 0.17 mm ID)

Mobile phase preparation

Mobile phase A (A=B if the pump is Quaternary): Take HPLC grade water 1000 ml filter with 0.45-micron nylon filter is cleaned and dried mobile phase bottle and sonicate it for about 10 min

Note

If the pump is Quaternary, mobile phase A is to be kept in reservoir A and B both. Mobile phase B (C=D if the pump is Quaternary): Take acetone 5 ml in clean and dried 1000 ml volumetric flask, dilute up to mark with HPLC grade water. Filter it with 0.45-micron nylon filter paper and degas by sonication for about 10 min.

Note

If the pump is Quaternary, mobile phase B is to be kept in reservoir C and D both.

System Requirements

Table 11.2

Testing conditions	Applied conditions
Column: Restricted capillary	ID No.:
Flow Rate: 2.0 ml/min.	
Wavelength: 265 nm	

Injection volume: 10 ml	
Runtime: 30 minutes for binary (20 min if the pump is quaternary)	
Column oven/Sample cooler temperature	

Sample/Standard Preparation

Take HPLC grade water or equivalent as a standard/sample

Procedure

Ensure the column should not be attached with the instrument If it is there then remove it and connect with the restricted capillary tube.

Purge the system with HPLC grade water, which is previously filtered through 0.45 m

Take the HPLC grade water previously filtered through 0.45 m nylon filter and use as mobile phase for Pump-A (Use for Pump A and Pump-B if the pump is quaternary)

Prepare 0.5 per cent v/v solution of Acetone in HPLC Water and filter in 0.45 m nylon filter and use as mobile phase for the Pump – B (Pump–C and Pump-D if the pump is quaternary).

Set the system gradient flow program as per Table 11.3 for binary and Table 11.4 for the quaternary pump.

Inject HPLC grade water or equivalent as a sample for five times and record peak height/area in the respective table.

Plot a linearity curve of peak height per cent (Theoretical) Vs observed peak height (μv) using the least square method. Calculate the squared correlation coefficient (r2) and RSD of peak height and record the observations in the observation table for the binary pump.

Table 11.3

Flow	*Time in Minutes*	*Per cent Channel–A*	*Per cent Channel–B*
2.0 ml	0	100	0
2.0 ml	5.0	100	0
2.0 ml	5.5	0	100
2.0 ml	6.0	100	0
2.0 ml	11.0	100	0
2.0 ml	11.5	30	70
2.0 ml	12.0	100	0

Flow	Time in Minutes	Per cent Channel–A	Per cent Channel–B
2.0 ml	17.0	100	0
2.0 ml	17.5	60	40
2.0 ml	18.0	100	0
2.0 ml	23.0	100	0
2.0 ml	23.5	90	10
2.0 ml	24.0	100	0
2.0 ml	29.0	100	0

Table 11.4: Gradient Flow Programming Table for the Quaternary Pump

Time	Flow	Per cent A	Per cent B	Per cent C	Per cent D	Curve
0	2.00	50	50	0	0	11
2.00	2.00	0	0	50	50	11
6.00	2.00	50	50	0	0	11
10.00	2.00	45	45	10	0	11
12.00	2.00	50	50	0	0	11
14.00	2.00	45	45	0	10	11
16.00	2.00	50	50	0	0	11
18.00	2.00	50	50	0	0	11

For binary pump, Inject three standard trial before starting the actual sequence for system equilibration and also observe the pressure graph. If it is found satisfactory then start the sequence.

Table 11.5: Observation Table for the Binary Pump

Sl.No.	Peak Height (100 per cent) Between 5.0 to 6.0 minutes	Peak Height (70 per cent) Between 11.0 to 12.0 minutes	Peak Height (40 per cent) Between 17.0 to 18.0 minutes	Peak Height (10 per cent) Between 23.0 to 24.0 minutes	Linearity the Squared Correlation Coefficient (r2) (r2 = NLT0.99)
1.					
2.					
3.					
4.					
5.					
Mean					
per cent RSD Þ (NMT2.0 per cent)					

11.7 HPLC Calibration

UV-Vis/PDA detector by reference energy check.

Calibration requirements: None

Mobile phase preparation: HPLC grade water or Methanol

System requirements: NA

Sample/standard preparation: NA

Calibration procedure:

Flush the system with mobile phase either HPLC grade water or HPLC grade methanol.

Check the reference energy for lamp in UV-Visible/PDA detector lamp at 225 nm.

Table 11.6

Sl.No.	Detector	Observed Value	Limit
1	UV-Visible		NLT 15 nA
2	PDA		NLT 10000 counts

General Note

Lamp shall be replace, when reference energy of UV-Visible/PDA detector lamp at 225 nm is less 15 nA/10000 counts respectively or fails to ignite or when the reference energy level of the lamp cause a decrease in sensitivity to the point where the baseline is too noisy and/or replacements of the lamp after one year interval, whichever is earlier.

*Reset the lamp hours counter to zero (0) at the time of replacement of lamp.

11.8 HPLC Calibration

UV-VIS/PDA detector by linearity measurement.

Calibration Requirements

Caffeine

Balance

Water HPLC grade or equivalent

Acetonitrile HPLC grade.

Mobile phase preparation: Water and Acetonitrile (85: 15)

System Requirements

Table 7

Testing conditions	Applied conditions
Column: C18 (25 cm x 4.6 mm), 5 m	Column No.:
Flow rate: 1 ml/min	
Wave length: 272 nm	
Injection volume: 20ml	
Run time: 10 min (Retention time: About 6.0 min)	
Column/Sample cooler temperature	

Sample/Standard Preparation

Prepare five concentration solution of caffeine, which are 25 ppm, 50 ppm, 100 ppm, 150 ppm, and 250 ppm.

Preparation of stock solution: Take 100.0 mg weight of caffeine in 100 ml volumetric flask, dissolve it and dilute up to mark with HPLC grade water.

25 ppm solution: Take 5 ml from stock solution with pipette in to 200 ml volumetric flask and dilute up to the mark with HPLC grade water.

50 ppm solution: Take 5 ml from stock solution with pipette in to 100 ml volumetric flask and dilute up to the mark with HPLC grade water.

100 ppm solution: Take 10 ml from stock solution with pipette in to 100 ml volumetric flask and dilute up to the mark with HPLC grade water.

150 ppm solution: Take 15 ml from stock solution with pipette in to 100 ml volumetric flask and dilute up to the mark with HPLC grade water.

250 ppm solution: Take 25 ml from stock solution with pipette in to 100 ml volumetric flask and dilute up to the mar with HPLC grade water.

Calibration Procedure

Set up the instrument as per the testing conditions.

Ensure that, the column is conditioned before injecting the sample.

Inject the all five linearity standard solution in duplicate and record the area of principal peak in observation table.

Plot a linearity curve of concentrations Vs corresponding mean area, using least square method. Calculate the squared correlation coefficient (r2), and record the observations in observation Table 11.7.

Table 11.8: Observation Table

Sl.No.	Caffeine Std Conc.	Actual Conc.	Area	Mean Area	Squared Correlation Coefficient (r2 = NLT 0.99)
1	25 ppm				
2	50 ppm				
3	100 ppm				Intercept: NMT 3.0 per cent of 100 ppm area
4	150 ppm				
5	250 ppm				

Acceptance Criteria

Linearity – squared correlation coefficient (r2) = NLT 0.99

Y intercept should be NMT 3.0 per cent of 100 ppm area.

Table 11.9

Sl.No.	Peak Height (100 per cent) Between 5.0 to 6.0 minutes	Peak Height (70 per cent) Between 11.0 to 12.0 minutes	Peak Height (40 per cent) Between 17.0 to 18.0 minutes	Peak Height (10 per cent) Between 23.0 to 24.0 minutes	Linearity the Squared Correlation Coefficient (r2) (r2 = NLT0.99)
1.					
2.					
3.					
4.					
5.					
Mean					
Per cent RSD Þ (NMT2.0 per cent)					

Observation table for the Quaternary pump.

For Quaternary Pump, Inject three standard Injection run as a trial injection before running the actual sequence for HPLC system equilibration and also observed pressure graph. if it is found satisfactory than start the sequence.

Table 11.10

Sl.No.	Peak Height (A + B) (100 per cent) at Approx. 6 minutes	Peak Height (C) (10 per cent) at Approx. 12 minutes	Peak Height (D) (10 per cent) at Approx. 16 minutes
1.			
2.			
3.			
4.			
5.			
Mean Þ			
Per cent RSD Þ (NMT 2.0 per cent)			

Calculation

(Mean of second peak height/Mean of first peak height) X 100 =______ per cent (Limit: Between 9.5 per cent and 10.5 per cent)

(Mean of third peak height/Mean of first peak height) X 100 =______ per cent (Limit: Between 9.5 per cent and 10.5 per cent)

Acceptance Criteria

Linearity – squared correlation coefficient (r2)= NLT 0.99 for binary.

per cent RSD of peak height obtained for five replicate injections of the sample should be NMT 2.0 per cent for Quaternary.

11.9 HPLC Calibration: Auto sampler by carry over check.

Calibration Requirements

Caffeine

Balance

HPLC grade water.

HPLC grade Acetonitrile (ACN).

Mobile Phase Preparation

Water: ACN (85:15) Mix 850 ml of HPLC grade water and 150 ml of acetonitrile and filter through 0.45 micron nylon filter, degas for 10 min by sonicator.

System Requirements

Note: Needle should not be rinsed in this study.

Table 11

Testing conditions	
Column: C18 (25 cm x 4.6 mm), 5m Make – Hichrom or equivalent	Applied conditions
Flow rate: 1.0 ml/min	Column No.:
Wavelength: 272 nm	
Injection volume: 20 ml	
Run time: 10 min (Retention time: About 6.0 min)	
Column/Sample cooler temperature	

Sample/Standard Preparation

Sample preparation(Conc. 1000 ppm):Weigh accurately 100 mg of caffeine and transfer it in 100 ml clean and dried volumetric flask, dissolve in and dilute up to the mark with water.

Calibration Procedure

Setup the HPLC system with the above conditions and keep the flow for 30 minutes to saturate the column with the mobile phase and inject 20 ml water as blank.

After getting the baseline from the blank inject the sample in duplicate and record the chromatogram.

Again inject 20 ml water as blank, if any peak observed at the principal peak retention time.

Calculate the per cent carryover using formula specified for carryover check and record the observation in observation table.

Compare the results obtained for compliance with the limits, as given in observation Table 12.

Table 11.12: Observation

No. of Injection	*RT and Area of Caffeine in Standard*	*Area at RT of Caffeine in Blank (B)*	*Per cent Carry over*	*Limits*
RT	Area			NMT0.2 per cent
1				
2				
Mean area				

Calculation: Calculate the per cent Carryover using following formula:

Per cent Carryover = {Area at RT of Caffeine in Blank (B)/Mean Area of sample preparation (A) } *100

Acceptance criteria: per cent Carryover should be NMT 0.2 per cent.

11.10 HPLC Calibration: Auto sampler by linearity check.

Calibration Requirements

Caffeine

Balance

HPLC grade water or equivalent.

HPLC grade acetonitrile (ACN).

Mobile phase preparation (85:15): Mix 850 ml of HPLC grade water and 150 ml of acetonitrile and filter through 0.45 micron nylon filter. Degas it for 10 min by sonicator.

System requirements:

HPLC Calibration: Auto sampler by linearity check.

Calibration requirements:

Caffeine

Balance

HPLC grade water or equivalent.

HPLC grade acetonitrile (ACN).

Mobile phase preparation (85:15): Mix 850 ml of HPLC grade water and 150 ml of acetonitrile and filter through 0.45 micron nylon filter. Degas it for 10 min by sonicator.

System requirements:

Table 13

Testing conditions	Applied conditions
Column: C18 (25 cm x 4.6 mm), 5m Make – Hichrom or Equivalent	Column No.:
Flow rate: 1.0 ml/min	
Injection volume: As per Table – 1	
Wavelength: 272 nm	
Run time: 10 min (Retention time: About 6.0 min)	
Column/Sample cooler temperature	NA

Sample/Standard Preparation

Sample preparation (Concentration 15 ppm): Weighed accurately 30 mg of caffeine and transferred it to 100 ml clean and dried volumetric flask, dissolved in and dilute up to the mark with water. Further diluted 5 ml of above solution to 100 ml with water and mixed well.

Calibration Procedure

Setup the HPLC system as per above Cinematographic conditions and saturate the column with the mobile phase for about 30 minutes.

Fill 5 vials with sample solutions respectively and place into vial positions 1, 29, 58, 87, 116 for Waters Alliance system

*Note: Exact vial position depends on the shape of sample tray. The ideal selections are given as following, which should be maintained. If total injection volume is more than the vial capacity then place the vial in next position or previous position.

50 ml capacity loop: Start by injecting 5 ml and proceed with10 ml, 20 ml, 40 ml, and 50 ml from the vials from different locations in five replicate injection.

100 ml capacity loop: Also for 100 ml loop capacity same injection pattern will follow additionally, 100 ml from the vials from different locations in five replicate injection.

200 ml capacity loop: Inject 20 ml, 80 ml, 120 ml, 160 ml, and 200 ml from the vials from different locations in five replicate injection.

500 ml/2000 μl capacity loop: Inject 20 ml, 80 ml, 200 ml, 400 ml, and 500 ml from the vials from different locations in five replicate injection.

Record the area due to principal peak. Take the mean area and calculate the per cent RSD for five replicate injection area and record the results in given the observation table.

Plot a linearity curve of concentrations Vs corresponding mean area, using least square method. Calculate the squared correlation coefficient r2, and record the observations.

Table 11.14: Observation Table

*Loop Capacity	Area of Caffeine	Area of Caffeine	Squared Correlation Coefficient (r2 = NLT 0.99)			
50 μl	5 μl	10 μl		20 μl	40 μl	50 μl
100 μl	5 μl	20 μl		50 μl	80 μl	100 μl
200 μl	20 μl	80 μl		120 μl	160 μl	200 μl
500 μl	20 μl	80 μl		200 μl	400 μl	500 μl
2000 μl	20 μl	80 μl		200 μl	400 μl	500 μl
Run – 1						

*Loop Capacity	*Area of Caffeine*	*Area of Caffeine*	*Squared Correlation Coefficient (r2 = NLT 0.99)*			
Run – 2						
Run – 3						
Run – 4						
Run – 5						
Mean						
per cent RSD (NMT 2.0 per cent)						

based on the capacity of loop size study shall be designed so as to cover minimum and maximum capacity of the loop).

Acceptance criteria: Linearity – squared correlation coefficient (r2) = NLT0.99

11.11 HPLC Calibration

UV-VIS/PDA detector by wavelength accuracy measurement.

Calibration requirements:

Caffeine Code no

Balance Code

Water HPLC grade or equivalent and Acetonitrile HPLC grade.

Mobile phase preparation:

Use filtered and degassed HPLC grade methanol as mobile phase.

System requirements:

Table 15

Testing conditions	Applied conditions
Column: Hypersil BDS C18 or equivalent (25 cm x 4.6 mm),5m,	Column No.:
Flow Rate:1.0 ml/min	
Wavelength: 201 nm to 209nm, 240 nm to 248nm 268 nm to 276 nm, for UV and PDA	
Injection volume: 20 ml	
Run time: 5 min. (10 min for PDA detector)	
Column/Sample cooler temperature	NA

Sample/Standard Preparation

Prepare the sample for checking of wavelength accuracy by taking cleaned dried 250 ml volumetric flask and take about 50 mg of caffeine, dissolve and dilute up to the mark with HPLC grade methanol, take 10 ml of this solution in 100 ml volumetric flask and dilute up to the mark with HPLC grade methanol. (Conc. about 20 ppm)

Note: Before calibration clean the system with isopropyl alcohol (IPA) or with the required solvent.

Calibration Procedure

Set up the instrument as per the given testing conditions (section 3).

Ensure that, the column is conditioned before injecting the sample.

Run the system for 201 nm to 209 nm, 240 nm to 248 nm and 268 nm to 276 nm by the increment of 1-nm wavelength and record the chromatograms.*i.e.* prepare the instrument methods with different wavelength if the detector is UV.

Note: In case of PDA detector select the range 201 nm to 209 nm, 240 nm to 248 nm and 268 nm to 276 nm and carryout only one sample single injection. Extract the data of 201 nm to 209 nm, 240 nm to 248 nm and 268 nm to 276 nm by the increment of 1-nm wavelength.

Record the observe area of each chromatogram in observation table.

Find the highest area for 201 nm to 209 nm, 268 nm to 276 nm and minimum area for 240 nm to 248 nm from the observation table and consider its respective wavelength as actual wavelength.

* Acceptance criteria: Maximum area should be at 205± 2 nm.

**Acceptance criteria: Minimum area should be at 245 ± 2 nm.

***Acceptance criteria: Maximum area should be at 272 ± 2 nm.

11.12 HPLC Calibration

Auto sampler for RI detector by linearity measurement.

Calibration requirements:

Column: Symmetry C18, 3.5 µm, (4.6 mm x 7.5 cm) column.

Caffeine

Balance

HPLC grade water or equivalent.

HPLC grade Methanol or equivalent.

Mobile phase preparation:

Water: Methanol (75:25): Mix 750 ml of HPLC grade water and 250 ml of methanol and filter through 0.45 micron nylon filter, degas for 10 min by sonicator.

Note: This Auto sampler calibration parameter shall be performed when HPLC system consist only with RI detector.

System requirements:

Table 11.16

Testing conditions	
Column: Symmetry C18, 3.5 µm, (4.6 mm x 7.5 cm) or equivalent.	Applied conditions
Flow rate: 1.0 ml/min	Column No.:
Column oven temperature: 35°C	
Flow cell temperature: 35°C	
Injection volume: As per table -1	
Run time: 5 min (Retention time: About 2.8 min)	
Sensitivity for Waters make HPLC	
Polarity + (positive)	
*Sensitivity:64	
*Response: 5	
Rinsing solvent: HPLC grade water: methanol (20:80)	

Diluent preparation:

Mix 80 ml of HPLC grade water and 20 ml of methanol and filter through 0.45 micron nylon filter, degas for 10 min by sonicator.

Sample/Standard Preparation

Preparation of stock solution (Conc. 2000 ppm):

Weigh accurately about 200 mg of caffeine in a clean and dried 10 ml volumetric flask, dissolve in and dilute up to the mark with Diluent. Take 1.0 ml of above solution in 10 ml volumetric flask and dilute up to the mark with diluent.

Calibration Procedure

Inject the sample preparations in duplicate and record the area of the principal peak in the given Table 11.17.

Plot a linearity curve of Injection volume Vs corresponding mean area, using least square method. Calculate the squared correlation coefficient (r2), and record the observations in given Table 11.17.

Table 11.17

Sl.No.	*Injection Volume*	*Area*	*Mean Area*
1	5 mL		
2	10 mL		
3	20 mL		
4	50 mL		
5	100 mL		

Note

Location shall decide the calibration range for loop capacity and injection volume based on there working range for the calibration of Auto sampler.

Acceptance Criteria

Linearity – squared correlation coefficient (r2) = NLT 0.99

11.13 HPLC Calibration

RI Detector by linearity measurement.

Calibration requirements:

Column: Symmetry C18, 3.5 µm, (4.6 mm x 7.5 cm) column.

Caffeine

Balance

HPLC grade water or equivalent.

HPLC grade Methanol or equivalent.

Mobile phase preparation:

Water: Methanol (75:25): Mix 750 ml of HPLC grade water and 250 ml of methanol and filter through 0.45 micron nylon filter, degas for 10 min by sonicator.

Note

This Auto sampler calibration parameter shall be performed when HPLC system consist only with RI detector.

System requirements:

Table 18

Testing conditions	Applied conditions
Column: Symmetry C18, 3.5 µm, (4.6 mm x 7.5 cm) or equivalent.	Column No.:
Flow rate: 1.0 ml/min	
Column oven temperature: 35°C	
Flow cell temperature: 35°C	

Table 19

Testing conditions	Applied conditions
Injection volume: 20 µl	
Run time: 5 min (Retention time: About 2.8 min)	
Sensitivity for Waters make HPLC	
Polarity + (positive)	
*Sensitivity:64	
*Response: 5	
Rinsing solvent: HPLC grade water: methanol (20:80)	

Diluent Preparation

Mix 80 ml of HPLC grade water and 20 ml of methanol and filter through 0.45m nylon filter, degas for 10 min.

Sample/standard preparation:

Preparation of stock solution (Conc. 1000 ppm): Weigh accurately about 100.0 mg of caffeine in a clean and dried 100 ml volumetric flask, dissolve in and dilute up to the mark with diluent.

100 ppm-Linearity Solution 1: Dilute 1 ml of stock solution to 10 ml with diluent and mix well.

200 ppm-Linearity Solution 2: Dilute 2 ml of stock solution to 10 ml with diluent and mix well.

400 ppm-Linearity Solution 3: Dilute 4 ml of stock solution to 10 ml with diluent and mix well.

500 ppm-Linearity Solution 4: Dilute 5 ml of stock solution to 10 ml with diluent and mix well.

1000 ppm-Linearity Solution 5: Stock solution of 1000 ppm.

Calibration Procedure

After stabilization of the system and column inject the diluent as blank.

Inject the all five linearity preparations in duplicate and record the areas of the principal peak in the observation table

Plot a linearity curve of concentrations Vs corresponding mean area, using least square method. Calculate the squared correlation coefficient (r2) and record the observations in given table.

Observation table: Refer SOP for HPLC Calibration

Acceptance criteria:

Linearity – squared correlation coefficient (r2) = NLT 0.99

11.14 HPLC Calibration

Fluorescence detector by linearity measurement.

Calibration Requirements

HPLC grade water or equivalent.

Anthracene

Column

HPLC grade acetonitrile and HPLC grade water or equivalent.

Balance

Mobile phase preparation:

Water: ACN (30:70): Mix 300 ml of HPLC grade water and 700 ml of acetonitrile and filter through 0.45 micron filters and degas about 10 min by Sonicator.

System requirements:

Table 20

Testing conditions	Applied conditions
Column: C18 (3.5 mm,4.6 mm x 75 mm), Symmetry column or equivalent.	Column No.:
Flow rate: 1.0 ml/min.	
Injection volume: 20 ml	
Run time:10 min	
Wavelength: Ex=249 nm,Em=402 nm	
Polarity:+	
Testing conditions	Applied conditions
Attenuation: 64	

Response: STD	
Gain:1, Unit: mv	
Bandwidth:18	
Column temperature: 30°C	

Sample/standard preparation:

Preparation of stock solution (Conc. 100 ppb):

Prepare the standard by weighing, accurately about 50 mg of anthracene and transfer into clean dried 50 ml volumetric flask, dissolve in and dilute up to the mark with acetonitrile (stock solution). Take 1.0 ml of the stock solution in 100 ml volumetric flask and dilute up to the mark with acetonitrile. Further Take 5.0 ml the solution in 500 ml volumetric flask and dilute up to the mark with mobile phase.

Linearity Solution Preparation

20 ppb-Linearity Solution 1: Dilute 2 ml of stock solution to 10 ml with mobile phase and mix it well.

40 ppb-Linearity Solution 2: Dilute 4 ml of stock solution to 10 ml with mobile phase and mix it well.

60 ppb-Linearity Solution 3: Dilute 6 ml of stock solution to 10 ml with mobile phase and mix it well.

80 ppb-Linearity Solution 4: Dilute 8 ml of stock solution to 10 ml with mobile phase and mix it well.

100 ppb-Linearity Solution 5: Stock solution of 100 ppb.

Calibration Procedure

Ensure that, the column is conditioned before injecting the solution.

Set the instrument as per system requirement.

Inject the all five sample preparation in duplicate.

Record the area due to anthracene peak. Calculate mean area and record the results in the observation table.

Plot a linearity curve of concentrations Vs corresponding mean area, using least square method. Calculate the squared correlation coefficient (r2), and record the observations in observation table.

Acceptance criteria

Linearity – squared correlation coefficient (r2) = NLT 0.99

11.15 HPLC Calibration

Fluorescence detector by wavelength accuracy measurement

Calibration Requirements

HPLC grade water or equivalent.

Column: NA

Mobile phase preparation: NA

System requirements:

Table 11.21

Parameters for emission wavelength calibration.	Parameter's Setting for emission wavelength calibration.
Type	Sample scan
Grating to scan	Excitation scan
Scan wavelength range	340 to 360
Emission wavelength	397
Gain	1
EUFS	100000
Pace	100
Mark each nm	Off

Table 11.22

Parameters for excitation wavelength calibration.	Parameter's Setting for excitation wavelength calibration.
Type	Sample scan
Grating to scan	Emission scan
Scan wavelength range	375 to 430
Excitation wavelength	350
Gain	1
EUFS	100000
Pace	100
Mark each nm	Off

Calibration Procedure

Connect the restricted capillary to Fluorescence detector in place of column.

Keep the HPLC grade water in reservoir.

Purge the measuring cell of Fluorescence detector with water and allow to run the water through cell about 10 min.

Fill the water in measuring cell of Fluorescence detector.

Scan the sample (water fill in flow cell) from 340 nm to 360 nm and measure the emission wavelength (for emission wavelength 397 nm). Record the observation in table.

For excitation wavelength calibration scan the sample (filled water in flow cell) from 375 nm to 430 nm and measure the highest peak listed first (for excitation wavelength 350 nm) peak. Record the observation in table.

Calibration Parameter	*Acceptance Criteria*	*Observed value*
Emission wavelength	397 ± 3 nm	
Excitation wavelength	350 ± 3 nm	

Note: If the excitation wavelength and emission wavelength test fails, then there may be something other than pure water in flow cell or flow cell may be dirty. Therefore rinse the flow cell with water again and repeat the test.

Acceptance Criteria

Emission wavelength 397 ± 3 nm.

Excitation wavelength 350 ± 3 nm.

Chapter 12

Calibration of GC

12.1 Flow Gas Accuracy

Sample/standard preparation: NA

Calibration Procedure

Switch on instrument as per the instrument SOP.

Open the carrier gas like Nitrogen or Helium from the cylinder or use gas generator for respected carrier gas.

Hydrogen and Zero air shall be switched off whenever checking carrier gas flow.

Setup the instrument to check the flow rate and attach the silicon tube inlet of flow meter.

Adjust the pressure between 5.5 and 6.5 kg/cm^3 by opening and closing of the valve located at the cylinder (if applicable).

Attach another end of silicon tube to the injector end.

Set the flow 1 mL/min and observe the flow reading in the flow meter. Calculate the reciprocal flow rate with the help of inbuilt stopwatch and record the observations in observation table.

Follow the same procedure to calibrate the flow for other gases as per table and record the results in observation table.

Verify the result against the acceptance criteria and in any case of non-compliance, follow the steps to rectify it.

Table 12.1

Sl.No.	Set Flow	Observed Flow	Acceptance Criteria (Difference between set flow and observed flow should meet the criteria given below
	For Capillary column		
1	1 ml/min		0.9 – 1.10 ml/min
2	2 ml/min		1.80 -2.20 ml/min
3	5 ml/min		4.5-5.50ml/min
	For packed column		
1	10 ml/min		9.0-11.0 ml/min
2	20 ml/min		18.0-22.0 ml/min
3	30 ml/min		27.0-33.0 ml/min
4	40 ml/min		36.0-44.0 ml/min
5	50 ml/min		45.0-55.0 ml/min

12.2 Detector Calibration (FID) by using Packed Column

System Requirements

Table 12.2

Parameter	Test Conditions	Applied Conditions
Column	S.S column	
Carrier gas	Nitrogen	
Flow rate	30 ml/min	
Detector	FID	
Column temperature	180 °C	
Injector Temperature	250 °C	
Detector temperature	250 °C	
Injection volume	1 ml	
Run time	20 min	

Sample/Standard Preparation

- ☆ Take 5.0 ml of ethyl alcohol in 50.ml of volumetric flask and dilute up to mark with
- ☆ DMF. (Stock solution 1)
- ☆ Take 5 ml of Isopropyl alcohol in 50 ml of volumetric flask dilute up to mark with
- ☆ DMF. (Stock solution 2)
- ☆ Preparation of final concentration

- ☆ 2000 ppm : Take 1 ml of stock solution 1 and 5.0 ml of stock solution 2 in 50.0 ml of volumetric flask dilute up to mark with DMF.
- ☆ 4000 ppm: Take 2 ml of stock solution 1 and 5.0 ml of stock solution 2 in 50.0 ml of volumetric flask dilute up to mark with DMF.
- ☆ 6000 ppm: Take 3 ml of stock solution 1 and 5.0 ml of stock solution 2 in 50.0 ml of volumetric flask dilute up to mark with DMF.
- ☆ 8000 ppm : Take 4 ml of stock solution 1 and 5.0 ml of stock solution 2 in 50.0 ml of volumetric flask dilute up to mark with DMF.
- ☆ 10000 ppm: Take 5 ml of stock solution 1 and 5.0 ml of stock solution 2 in 50.0 ml of volumetric flask dilute up to mark with DMF.

Calibration Procedure

- ☆ Set up the instrument as per the condition.
- ☆ Inject the samples three times from each solution and calculate the ratio of IPA and Ethyl alcohol. Record the observation in Table 12.3.
- ☆ Calculate the RSD per cent of the mean ratio. It should not be more than 5 per cent
- ☆ Plot a linearity curve of concentration vs corresponding mean area, using the least square method. Calculate the correlation coefficient (r2) and record the observation in Table 12.3.

Table 12.3

Sl.No.	*Ethyl Alcohol Conc*	*Area of IPA (X)*	*Area of Ethyl Alcohol(Y)*	*Area Ratio of Ethyl Alcohol to IPA*
1	2000			
Mean				
RSD per cent				
2	4000			
Mean				
RSD per cent				
3	6000			
Mean				
RSD per cent				
4	8000			
Mean				
RSD per cent				
5	10000			
Mean				
RSD per cent				

Table 12.4

Sl.No.	Ethyl Alcohol Std Conc	Mean Ratio Area	Acceptance Criteria
1	2000		Correlation coefficient should be NLT 0.990
2	4000 ppm		
3	6000 ppm		
4	8000 ppm		
5	10000 ppm		

12.3 Detector Calibration (FID) by Using Capillary Column-Liquid Injection Mode

System Requirement

Table 12.5

Parameter	Test Condition	Applied Condition
Column	30 m x 0.53 mm x 5.0 μ	
Carrier gas	Nitrogen or Helium	
Oven temperature	60 °C -3 min @ 30°C/min-210-2 min	
Detector temperature	220 oC -FID	
Head pressure	25 kPa	
Injection volume	1 μl	
Split ratio	40:1	
Injector temperature	180 °C	

Sample/Standard Preparation

- Take 5.0 ml of iso propyl alcohol (IPA) in 50.ml of volumetric flask and dilute up to mark with DMF. (Stock solution)
- **Internal Standard preparation:** Take 5 ml of Tetrahydrofuran (THF) to 50 ml of volumetric flask and dilute up to mark with DMF

Preparation of Final Concentration of IPA

- 5000 ppm : Take 2.5 ml of stock solution and 5.0 ml of THF (IS) in 50.0 ml of volumetric flask dilute up to mark with DMF.
- 8000 ppm: Take 4 ml of stock solution and 5.0 ml of THF (IS) in 50.0 ml of volumetric flask dilute up to mark with DMF.

- ☆ 10000 ppm: Take 5 ml of stock solution and 5.0 ml of THF (IS) in 50.0 ml of volumetric flask dilute up to mark with DMF.
- ☆ 12000 ppm : Take 6 ml of stock solution and 5.0 ml of THF (IS) in 50.0 ml of volumetric flask dilute up to mark with DMF.
- ☆ 15000 ppm: Take 7.5 ml of stock solution and 5.0 ml of THF (IS) in 50.0 ml of volumetric flask dilute up to mark with DMF.

Calibration Procedure

- ☆ Set up the instrument as per the condition.
- ☆ Inject the samples three times from each solution and calculate the ratio of IPA and THF. Record the observation in Table 12.6
- ☆ Calculate the RSD per cent of the mean ratio. It should not be more than 5 per cent
- ☆ Plot a linearity curve of concentration vs corresponding mean area, using the least square method. Calculate the correlation coefficient (r^2) and record the observation in Table 12.7

Table 12.6: Observation

Sl.No.	*IPA Conc*	*Area of IPA (X)*	*Area of THF (Y)*	*Ratio*
1	5000 ppm			
Mean				
RSD per cent				
2	8000 ppm			
Mean				
RSD per cent				
3	10000 ppm			
Mean				
RSD per cent				
4	12000 ppm			
Mean				
RSD per cent				
5	15000 ppm			
Mean				
RSD per cent				

Table 12.7

Sl.No.	IPA std	Mean Area Ratio	Acceptance Criteria
1	5000 ppm		Correlation coefficient should be NLT 0.990
2	8000 ppm		
3	10000 ppm		
4	12000 ppm		
5	15000 pmm		

12.4. GC Calibration–Flame Ionization Detector by Linearity Measurement (By Using Capillary Column-(Head Space mode-HSS)

Table 12.8

Parameter	Test Condition	Applied Condition
Column	30 m x 0.53 mm x 5.0 μ	
Carrier gas	Nitrogen or Helium	
Oven temperature	60 °C -3 min @ 30oC/min-210-2 min	
Detector temperature	220 oC -FID	
Head pressure	25 kPa	
Injection volume	1 ml	
Split ratio	40:1	
Detector range	1	
Attenuation	-3	

Table 12.9: HSS Condition

Vial temperature	85 °C
Syringe temperature	100 °C
Equilibration time	20 min
Injector temperature	200 °C
Transfer line temperature	90 °C
Withdrawn time	0.2 min
Pressurize time	30min
GC cycle time	20 min
Carrier gas pressure	14 psi

Sample Standard Preparation

Internal Standard preparation: Take 0.3 ml of Tetrahydrofuran (THF) to 100 ml of volumetric flask and dilute up to mark with water.

Preparation of Stock Solution

- Take 5 ml of IPA in a 100 ml of volumetric flask and dilute up to mark with water.

Preparation of Final Solutions

- 2500 ppm : Take 2.5 ml of stock solution and 5.0 ml of THF (IS) in 50.0 ml of volumetric flask dilute up to mark with DMF.
- 4000 ppm: Take 4 ml of stock solution and 5.0 ml of THF (IS) in 50.0 ml of volumetric flask dilute up to mark with DMF.
- 5000 ppm: Take 5 ml of stock solution and 5.0 ml of THF (IS) in 50.0 ml of volumetric flask dilute up to mark with DMF.
- 6000 ppm : Take 6 ml of stock solution and 5.0 ml of THF (IS) in 50.0 ml of volumetric flask dilute up to mark with DMF.
- 7500 ppm: Take 7.5 ml of stock solution and 5.0 ml of THF (IS) in 50.0 ml of volumetric flask dilute up to mark with DMF.
- Pipette5 ml each of linearity solution in three HSS vial respectively. Seal the vials with septa followed by the crimp.

Calibration Procedures

- Set up the instrument as per the condition.
- Inject the samples three times from each solution and calculate the ratio of IPA and THF. Record the observation in Table 12.10.
- Calculate the RSD per cent of the mean ratio. It should not be more than 5 per cent
- Plot a linearity curve of concentration vs corresponding mean area, using the least square method. Calculate the correlation coefficient (r^2) and record the observation in Table 12.11.

Table 12.10

Sl.No.	IPA Conc	Area of IPA (X)	Area of THF (Y)	Ratio
1	2500 ppm			
Mean				
RSD per cent				
2	4000 ppm			
Mean				
RSD per cent				

Sl.No.	IPA Conc	Area of IPA (X)	Area of THF (Y)	Ratio
3	5000 ppm			
Mean				
RSD per cent				
4	6000 ppm			
Mean				
RSD per cent				
5	7500 ppm			
Mean				
RSD per cent				

Table 11.11

Sl.No.	IPA std	Mean Area Ratio	Acceptance Criteria
1	2500 ppm		Correlation coefficient should be NLT 0.990
2	4000 ppm		
3	5000 ppm		
4	6000 ppm		
5	7500 ppm		

12.5 Calibration of Auto Injector by Linearity Measurement (by using capillary column)

System Requirement

Table 11.12

Parameter	Test Condition	Applied Condition
Column	Porapak Q-80 S.S. column	
Carrier gas	Nitrogen	
Flow rate	30 ml/min	
Injector temperature	250 °C	
detector	FID	
Injection volume	0.5 ml, 1.0 ml, 2.0 ml and 2.5 ml	
Run time	20 min	

Sample/Standard Preparation

- ☆ Take 5.0 ml of ethyl alcohol in 50.ml of volumetric flask and dilute up to mark with DMF. (Stock solution 1)
- ☆ Take 5 ml of Isopropyl alcohol in 50 ml of volumetric flask dilute up to mark with DMF. (Stock solution 2)
- ☆ Take 3.0 ml of stack solution 1 and 5.0 ml of stock solution 2 in 50 ml of volumetric flask dilute up to mark with DMF.

Calibration Procedures

- ☆ Set up the instrument as per the condition.
- ☆ Inject the samples 0.5 ml, 1.0 ml, 1.5 ml, 2.0 ml and 2.5 ml every three time and calculate the ratio of IPA and ethyl alcohol. Record the observation
- ☆ Calculate the RSD per cent of the mean ratio. It should not be more than 5 per cent
- ☆ Plot a linearity curve of concentration vs corresponding mean area, using the least square method. Calculate the correlation coefficient (r^2) and record the observation

12.6 Calibration of Column Oven Temperature

Table 12.13

Parameter	*Applied condition*	*Test condition*
Column oven temperature	Temperature	Temperature
	A) 35 °C	A) ....°C
	B) 100 °C	B) ....°C
	C) 150 °C	C) ....°C
	D) 200 °C	D) .... °C
	E) 250 °C	E) .... °C
	F) 300 °C	F) ....°C

Sample/Standard Preparation: NA

GC Calibration Procedure

- ☆ Insert the probe of the temperature indicator device in the column oven and close it.
- ☆ Set the column oven temperature 35 °C and keep idle to stabilize the temperature and take the reading.
- ☆ Repeat the same procedure for 35 °C, 100 °C, 150 °C, 250 °C, 300 °C and record the observation in table 11.4.

Table 12.14

Sl.No.	Set Temperature	Observed temperature	Acceptance Criteria
1	35 °C		Difference between set temperature and observed temperature should be ± 2 °C
2	100 °C		
3	150 °C		
4	250 °C		
5	300 °C		

12.7 Calibration of Vial Temperature in GC

System requirement: NA

- ✰ Sample and standard Preparation: NA
- ✰ Calibration procedure-:
- ✰ Set the vial temperature at 35 °C and keep idle to stabilize the temperature and take the readings.
- ✰ Repeat the procedure for 90 °C and 130 °C. Record the observation in Table 12.15.

Table 12.15

Sl.No.	Set Temperature	Observed Temperature	Acceptance Criteria
1	90 °C		Difference between set temperature and observed temperature should be ± 2 °C
2	130 °C		

Chapter 13

Method Validation

13.1 Introduction

Validating analytical methods is crucial to guarantee their fitness for purpose. This involves following a structured plan that encompasses scope, performance characteristics, and acceptance limits. Analytical methods must undergo validation or revalidation before their incorporation into routine analyses, particularly for batch release. The overarching philosophy of 21st-century current good manufacturing practices (cGMPs) and robust modern quality systems emphasizes that quality should be inherent in the product, and testing alone cannot guarantee product quality. From an analytical perspective, this means that analytical methods employed for product testing should possess inherent quality attributes. To incorporate quality attributes into the analytical method, bench-level scientists must apply fundamental quality principles. This paradigm shift necessitates that bench-level scientists possess scientific and technical expertise, product knowledge, process knowledge, and/or risk assessment skills to effectively execute the quality functions of analytical method validation.

Additionally, the following procedures are essential:

Comprehensive Training: Bench-level scientists should receive appropriate training to grasp the principles of method validation, validate analytical methods, and understand the principles involved with method validation.

Thorough Documentation and Data Interpretation: Proper documentation, understanding, and interpretation of data are paramount.

Cross-Functional Understanding: Bench-level scientists should develop a cross-functional understanding of how their activities impact the product and ultimately the customer (patient).

Management bears the responsibility of verifying that the skills acquired through training are applied in routine analyses performance.

In 1990, Europe, the United States, and Japan harmonized the submission requirements for new pharmaceuticals. This led to the creation of the International Conference on Harmonisation (ICH), which is a forum for constructive dialogues between regulatory authorities and industry. One of the first topics that the ICH addressed was the validation of analytical procedures. The ICH has developed guidelines for the validation of analytical procedures. These guidelines are intended to ensure that analytical procedures are accurate, precise, specific, and reproducible. They also provide guidance on the documentation of analytical procedures. The United States Food and Drug Administration (FDA) and the United States Pharmacopeia (USP) have also developed guidelines for the validation of analytical procedures. These guidelines are similar to the ICH guidelines, but they may also include additional requirements. The Brazilian regulatory agency, Agência Nacional de Vigilância Sanitária (ANVISA), has also developed guidelines for the validation of analytical procedures. These guidelines are based on the ICH guidelines, but they may also be tailored to the specific needs of the Brazilian market. There are four main types of analytical procedures:

Identification tests: These tests are used to identify a substance.

Quantitative tests: These tests are used to measure the amount of a substance in a sample.

Limit tests: These tests are used to determine whether the amount of an impurity in a sample is below a specified limit.

Performance tests: These tests are used to measure the performance of a product, such as its dissolution rate or its stability.

Validation Characteristics

The validation characteristics that are required for an analytical procedure will depend on the type of procedure and the intended use of the data. However, some of the most common validation characteristics include:

Accuracy: This is the closeness of the measured value to the true value.

Precision: This is the degree of reproducibility of the measured value.

Specificity: This is the ability of the procedure to measure the desired substance without interference from other substances.

Linearity: This is the ability of the procedure to produce a linear response over the range of interest.

Range: This is the range of concentrations over which the procedure can be used to produce accurate and precise results.

The USP 1225 chapter covers only the most common categories of tests which validation data should be required. Those categories can be observed on the following lines:

Category I: Analytical procedures for quantitation of major components in bulk drug substances or active ingredients (including preservatives) in finished pharmaceutical products/pesticide residue in a given matrices given by SANTE guidelines.

Category II: Analytical procedures for determination of impurities in bulk drug substances or degradation compounds in finished pharmaceutical products/pesticide residue. These procedures include quantitative assays and limit tests.

Category III: Analytical procedures for determination of performance characteristics (*e.g.*, dissolution, drug release/waiting period of pesticide residue for fixing MRL or dissipation pattern, *etc.*).

Category IV: Identification tests.

In Table 13.1, the required validation characteristics for various types of analytical procedures according to USP can be observed. The required validation characteristics for each type of analytical procedures according to ICH are illustrated in Table 13.2. The discussion of the validation of analytical procedures is directed to the four most common types of analytical procedures:

- ☆ Identification tests.
- ☆ Quantitative tests for impurities' content.
- ☆ Limit tests for the control of impurities.
- ☆ Quantitative tests of the active moiety in samples of drug substance or drug product or other selected component(s) in the drug product. Even though there are many other analytical procedures, such as dissolution testing for drug products or particle size determination for drug substance, these are no addressed in the initial text for validation of analytical procedures. The validation of these additional analytical procedures is equally important to those listed herein, and it may be addressed in subsequent documents.

Table 13.1: Data Elements Required for Validation According to USP

Analytical Performance Characteristics	*Category I*	*Category II*		*Category III*	*Category IV*
		Quantitative	*Limit tests*		
Accuracy	Yes	Yes	*	*	No
Precision	Yes	Yes	No	Yes	No
Specificity	Yes	Yes	Yes	*	Yes
Detection Limit	No	No	Yes	*	No
Quantitation Limit	No	Yes	No	*	No
Linearity	Yes	Yes	No	*	No
Range	Yes	Yes	*	*	No

*It may be required, depending on the nature of the specific test.

Table 13.2: Data Elements Required for Validation According to ICH

Type of Analytical Procedure Characteristics	*Identification*	*Testing for Impurities*		*Assay*
		Quantitative	*Limit*	
Accuracy	-	+	-	+
Precision				
Repeatability	-	+	-	+
Interm. precision	-	+(1)	-	+(1)
Specificity (2)	+	+	+	+
Detection Limit	-	-(3)	+	-
Quantitation Limit	-	+	-	-
Linearity	-	+	-	+
Range	-	+	-	+

(-) characteristic is not normally evaluated. (+) characteristic is normally evaluated. (1) in some cases where reproducibility has been performed, intermediate precision does not need to be done. (2) lack of specificity in one analytical procedure could be compensated by other supporting analytical procedure(s). (3) it may be needed in some cases.

13.2 Strategy for Analytical Method Validation

The validation of a specific method must be demonstrated through laboratory experiments by routinely analyzing samples. The preparation and execution have to follow a validation protocol, preferably written in a step-by-step instruction format. Possible steps for a complete method validation are listed in Table 13.4. This proposed procedure assumes that the instrument has been selected, and the method has also been developed. It meets criteria such as easy of use, ability to be automated and controlled by computer systems, costs per analysis, sample throughput, turnaround time, environmental, health, and safety requirements. Successful acceptance of the validation parameters and performance criteria by all involved parties requires a cooperative effort of several departments, including analytical development, quality control, regulatory affairs, and individuals requiring the analytical data. The operating procedure or Validation Master Plan (VMP) must clearly define the rules and responsibilities of each involved department in the validation of analytical methods. The validation experiments should be carried out by an expert analyst to avoid errors due to lack of experience. The analyst should also be very familiar with the technique and operation of the instrument. Before an instrument starting validating a method, its performance specifications must be verified by using generic chemical standards. Satisfactory results for a method can be only obtained with equipment that is performing well. A special attention must be paid to equipment characteristics which are critical for the method. For instance, if the detection limit is critical for a specific method, the specification of the instrument for baseline noise and some detectors must be verified.

Strategies and Steps in Analytical Method Validation

1. Develop a validation protocol, an operating procedure, or a validation master plan for the validation
2. For a specific validation project, define owners and responsibilities
3. Develop a validation project plan
4. Define the application, purpose, and scope of the method
5. Define the performance parameters and acceptance criteria
6. Define validation experiments
7. Verify relevant performance characteristics of equipment
8. Qualify materials, *e.g.* standards and reagents for purity, accurate amounts, and sufficient stability
9. Perform pre-validation experiments
10. Adjust method parameters and/or acceptance criteria if necessary
11. Perform full internal (and external) validation experiments
12. Develop standard operation procedures for executing the method in the routine
13. Define criteria for revalidation
14. Define type and frequency of system suitability tests and/or analytical quality control checks for the routine
15. Document validation experiments and results in the validation report

13.3 Documentation

13.3.1 Validation Protocol

The level of detail required in a method validation protocol depends on the company's culture. Some companies prefer simple protocols, while others require exhaustive documentation. In addition to the items listed below, each parameter to be validated may be described in detail, including:

- ☆ Solution preparation: How the solutions will be prepared.
- ☆ Experimental design: The specific steps involved in the experiment.
- ☆ Calculations: How the data will be analyzed and calculated.
- ☆ Software: Any software needed for the analysis (*e.g.*, Excel).

If a full-length protocol is required, it must be written and approved before starting validation work. Otherwise, deviations from the protocol may occur, which need to be documented in the final report. Many companies have templates for validation reports, making data entry easier. Once populated, the report is reviewed to ensure all parameters and criteria were met. Any deviations require justification and potentially an action plan (*e.g.*, reanalysis, procedure change, revalidation). If the analytical method is not yet approved, a final draft should be attached to the protocol. Before starting any experimental work, qualified personnel must write the protocol and have it approved by quality assurance.

Here are some of the items that must be specified in the validation protocol:

- ☆ Analytical method: The specific method for analysis.
- ☆ Test to be validated: The specific test being validated.
- ☆ Test parameters: Details about the test, including types and numbers of solutions and tests.
- ☆ Acceptance criteria: Criteria for each parameter based on internal SOPs (product or method-specific adaptations may be necessary and are acceptable, if justified).
- ☆ Batches: List of drug substance and/or drug product batches.
- ☆ Excipients: For drug products, the grade/quality of excipients used.
- ☆ Reference materials: List of reference materials used in the validation experiments.
- ☆ Instruments and apparatus: Information about the equipment used.
- ☆ Responsibilities: Who is responsible for different aspects of the validation process (*e.g.*, author, chemists, project leader, quality assurance).

13.3.2 Validation Report: Early-phase Validation Reports

- Use predefined tables to compare results with acceptance criteria for analytical methods (identified by code).
- Explain each experiment with detailed sample and standard preparation steps for late-phase validation.

Documentation

- Reference materials (standards with certificate of analysis).
- Calibrated and qualified instruments used in validation.
- Batches of drug substances/pesticides, impurities, solutions, or intermediates.
- Batches of drug substances, drug products, and excipient grades/qualities.

Test Results

- List test parameters and acceptance criteria with results and pass/fail indications.

Validation Outcome

- Indicate whether the method validation was successful.
- Propose any changes to the analytical method.
- Resubmit the final method for quality assurance approval.

Validation Report Content

- Objective and scope: Applicability and type of the method.
- Methodology summary: Briefly explain the analytical procedure.
- Chemicals and standards: List chemicals, reagents, standards, and QC samples with purity, grade, source, or preparation instructions.
- Quality checks: Procedures for checking the quality of standards and chemicals.
- Safety precautions: Important safety considerations for the method.
- Method parameters: All specific parameters and conditions of the analytical procedure.
- Critical parameters: Key parameters identified during robustness testing.
- Equipment: List of equipment with functional and performance requirements.
- Detailed procedures: Step-by-step instructions for conducting the experiments, including sample preparation.

- Statistical procedures: Explanation of statistical methods used and representative calculations.
- Routine analysis quality control: Procedures for ensuring data quality during routine analyses.
- Results presentation: Representative plots, chromatograms, spectra, and calibration curves.
- Method performance data: Data demonstrating the method's adherence to acceptance limits.
- Measurement uncertainty: Expected uncertainty of the results.
- Revalidation criteria: Conditions under which the method should be revalidated.
- Method developers and validators: List the responsible personnel.
- References: Relevant references used in the report.
- Summary and conclusions: Key findings and conclusions of the validation process.
- Approval: Signatures and dates of responsible reviewers and approvers.

Revalidation

Changes to method parameters may be necessary throughout its lifespan if performance falls outside acceptable limits. To avoid confusion, define operating ranges for each parameter based on experience or during development. Verify these ranges during validation and include them as method characteristics. This streamlines the decision-making process for revalidation. Revalidation is required under the following circumstances:

- Parameter change outside operating range: For example, changing column temperature from its 30°C-40°C range to 41°C necessitates revalidation.
- Scope change: This includes alterations to the sample matrix, operating conditions, or instrument characteristics not covered by the initial validation. An HPLC method validated on a pump with a 5mL delay volume requires revalidation if the new pump has a 0.5mL delay volume. System suitability or QC sample failures: If the cause lies outside instruments or other known factors, partial or full revalidation may be necessary.

Changes requiring a documented change control system:

- New samples with new compounds or matrices
- New analysts
- New instruments

- New locations
- New chemicals/reference standards
- Modified analytical parameters

Steps to determine revalidation need:

1. Evaluate if the change is within the method's scope. If so, no revalidation is necessary.
2. Define parameters for revalidation if the change falls outside the scope.
3. Perform validation experiments.
4. Re-evaluate and redefine system suitability test parameters if necessary.

By clearly defining operating ranges and establishing a documented change control system, you can ensure efficient decision-making regarding analytical method revalidation.

13.4 Parameters for Method Validation (Figures of merit)

The level of validation required for a test depends on its nature. For example, methods described in pharmacopeias might not need full validation but should be verified instead. Different tests require different validation parameters. As a project progresses and more information about the analytical method and product becomes available, the method gets gradually updated. While each company has its own approach and criteria for different assays, these approaches must comply with internal quality assurance guidelines and any relevant regulations. This section provides a detailed description of each parameter to be validated (figures of merit).

13.4.1 Specificity

Specificity should be investigated during validation of identification tests, impurity determination, and assays. The specific procedures used depend on the analytical procedure's intended purpose. It's not always possible to achieve complete discrimination for a specific analyte, so combining two or more analytical procedures (*e.g.*, adding a chiral HPLC method for optically active substances) may be necessary. For dosage forms, identity confirmation is crucial. This can be achieved by comparing the assay and related substances or content uniformity results to an external reference standard. This identity test ensures the presence of the correct active ingredient(s) in the correct ratio for formulations with multiple variants. This method can also be used for post-packaging analysis. Generally, the sample and standard chromatograms should match within ±10 per cent for retention time and normalized peak area. Performing a forced decomposition study alongside the specificity test is the simplest approach for any HPLC method. To gather more information (mass-to-charge ratio of parent ions, initial fragmentation pattern, and peak purity), using a mass spectrometry (MS) detector in series with a Photo Diode Array

(PDA) detector is recommended. Specificity is confirmed when an API peak is "pure" (according to PDA and/or MS) and there is no interference from a placebo solution (placebo dissolved in sample preparation solvent) at the API peak's retention time.

13.5 Linearity

The linearity of an analytical procedure is its ability (within a given range) to obtain test results which are directly proportional to the concentration (amount) of analyte in the sample. It may be directly demonstrated on the analyte, or on spiked samples using at least five concentrations over the whole working range. Moreover a visual evaluation of the analyte signal as a function of the concentration, appropriate statistical calculations are recommended, such as a linear regression. The parameters slope and intercept, residual sum of squares and the coefficient of correlation should be reported. A graphical presentation of the data and residuals is recommended. The terminology for this validation characteristic is somewhat misleading because linearity in the inner sense, *i.e.*, a linear relationship between analyte concentration and test results is certainly preferable, but not essential. A better term would have been analytical response. Some analytical procedures have intrinsic non-linear response functions, such as quantitative TLC, fluorescence detection, *etc.*, but they can of course be validated. The primary objective is to validate or verify the calibration model. Consequently, the requirements and the relevant parameters depend on the intended mode of calibration. The response function of a given analytical procedure is an intrinsic property of the respective analyte. That means, with respect to validation, that the answer is of a qualitative kind: Can the intended calibration be applied, yes or no? Therefore, solutions of the analyte itself are sufficient and there is no need to repeat linearity. Potential influences by the matrix, *i.e.*, the linearity of the analytical procedure would be better addressed in accuracy. Often, the fundamental response function is known for a given type of analytical procedure, such as a linear function for LC with UV detection, according to the Lambert–Beer law. In some cases, validation of linearity can be regarded more as a verification of the assumed response function, *i.e.*, the absence of (unacceptable) deviations. Primarily, this should be performed by means of graphical evaluation of the deviations of the experimental data from the assumed response model (residual analysis), known as residual plots. The evaluation of numerical parameters is only sensible after verification of the response function, *i.e.*, if only random errors exist. Frequently, the linearity is evaluated graphically, in addition to or as an alternative to mathematical evaluation. The evaluation is made by visually inspecting a plot of signal height or peak area as a function of analyte concentration. Because deviations from linearity are sometimes difficult to detect, two additional graphical procedures can be used. The first is to plot the deviations from the regression line versus the concentration or versus the logarithm of the concentration if the concentration range covers several decades. For linear ranges, the deviations should be equally distributed between positive and negative values. Another approach is to divide signal data by their respective concentrations, yielding the relative responses. A graph is plotted with the relative responses on the y-axis and the corresponding concentrations on the x-axis, on a log scale. The obtained line should be horizontal over the full linear range. At higher concentrations, there will typically be a negative deviation from linearity. Parallel horizontal lines are drawn on the graph corresponding, for example, 95 per cent and 105 per cent of the horizontal line. The method is linear up to the point where the plotted relative response line intersects the 95 per cent line. Figure 13.1 shows a comparison of the two graphical evaluations on a sample of caffeine using HPLC. A plot of the data should be included in the report. In addition, an

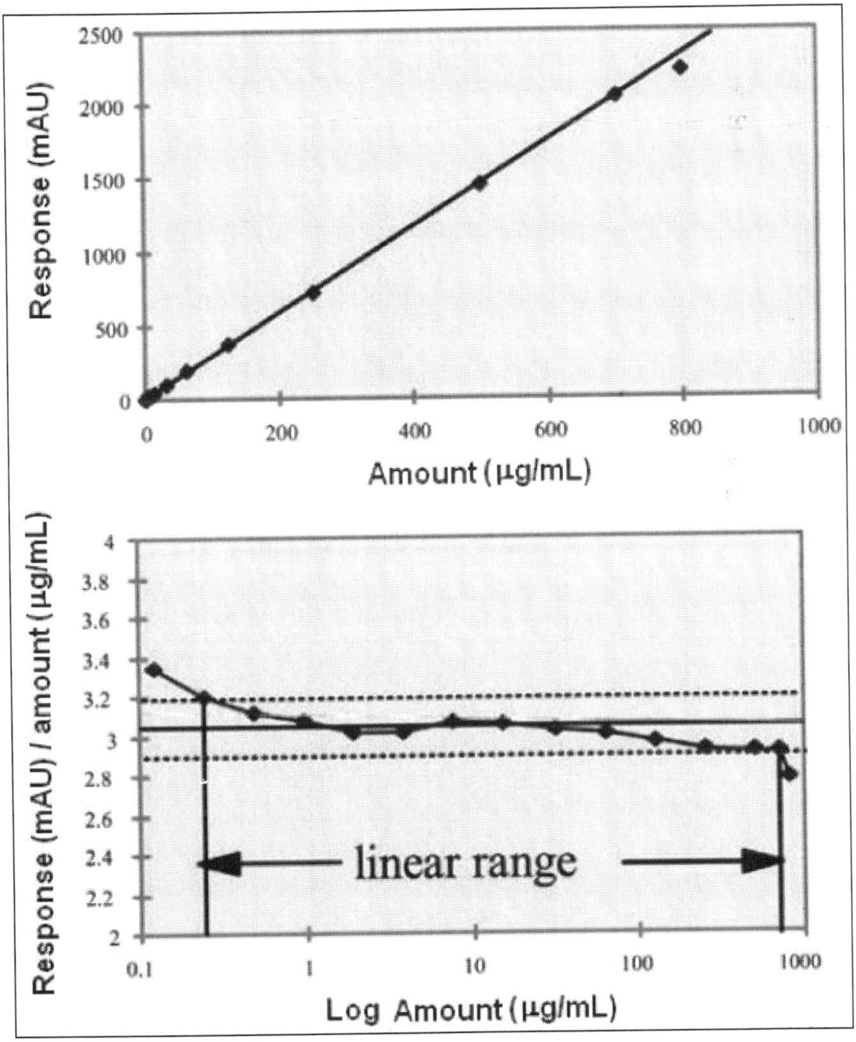

Figure 13.1: Linearity Plot of an Analyte Sample.

analysis of the deviation of the actual data points from the regression line may also be helpful for evaluating linearity. In order to establish linearity, a minimum of five concentrations is recommended. Other approaches should be justified. Plotting the sensitivity (response/amount) gives clear indication of the linear range. Plotting the amount on a logarithmic scale has a significant advantage for wide linear ranges.

13.6 Range

The range of an analytical method is the interval between the upper and lower levels (including these levels) that have been demonstrated to be determined with precision, accuracy, and linearity using the method as written. The range is normally expressed in the same units as the test results. For assay test, it requires a minimum of specified range to be 80 to 120 per cent of the test concentration, and for the determination of an impurity, the range to extend from the limit of quantitation, or from 50 per cent of the specification of each impurity, whichever is greater to 120 per cent of the specification.

13.7 Limit of Blank

EP17 defines LoB as the highest apparent analyte concentration expected to be found when replicates of a sample containing no analyte are tested. Note that although the samples tested to define LoB are devoid of analyte, a blank (zero) sample can produce an analytical signal that might otherwise be consistent with a low concentration of analyte. LoB is estimated by measuring replicates of a blank sample and calculating the mean result and the standard deviation (SD).

$$LoB = mean_{blank} + 1.645\,(SD_{blank}) \tag{12.1}$$

Assuming a Gaussian distribution of the raw analytical signals from blank samples, the LoB represents 95 per cent of the observed values. (Note: Typical modern analysers don't routinely display the actual analytical signal but automatically convert it to a concentration value. The raw analytical signal is preferable for establishing LoB as analysers may report all signal values below a certain fixed limit as "zero concentration"). The remaining 5 per cent of blank values represent a response that could actually be produced by a sample containing a very low concentration of analyte. Statistically, this false positivity is known as a Type I (or α) error (Figure 13.2). Conversely, while a sample that actually contains analyte is expected to exceed the LoB, it must also be recognised that a proportion of very low concentration samples will produce responses less than the LoB, representing Type II (or β) error (Figure 13.2). Thus, EP17 acknowledges that the overlap of the analytical responses of blank and low concentration is a statistical reality, and uses LoB as a reasonable starting point for estimating the LoD.

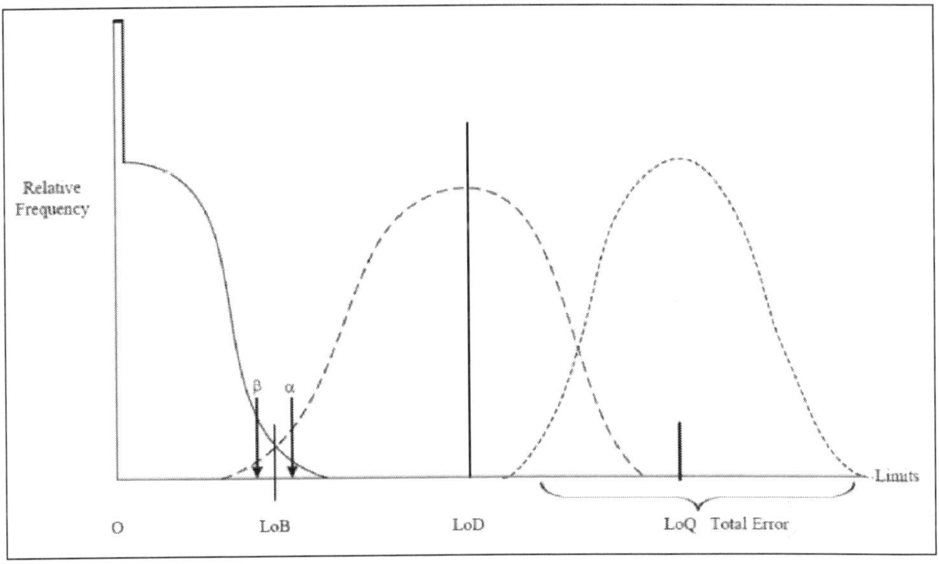

Figure 13.2: Relationship between LoB, LoD and LoQ.

The solid line defines the LoB and represents the distribution of results for a blank specimen. As modern analysers seldom report results of less than zero, the frequency of "zero" results is artificially increased as illustrated. The LoB shown excludes a small proportion of blank results ("á"). The dashed line defines the LoD and represents the scatter (imprecision) of results for a specimen of low concentration. The LoD is then set so that only a small proportion ("â") of these results will fall below LoB. The dotted line defines the LoQ and represents the distribution of results for a specimen of low concentration meeting the target for total error (imprecision *and* bias). It may be that this LoQ total target error is met by a specimen at the LoD concentration in which case LoQ = LoD. Otherwise, LoQ will have to be set after testing a specimen of higher concentration. Reproduced with permission, from NCCLS publication EP17-A, Protocols for Determination of Limits of Detection and Limits of Quantitation; Approved Guideline.

13.8 Limit of Detection

The limit of detection is the point at which a measured value is larger than the uncertainty associated with it. It is the lowest concentration of analyte in a sample that can be detected, but not necessarily quantified. The limit of detection is frequently confused with the sensitivity of the method. The sensitivity of an analytical method is the capability of the method to discriminate small differences in concentration or mass of the test analyte. In practical terms, sensitivity is the slope of the calibration curve that is obtained by plotting the response against the analyte concentration or mass. In chromatography, the detection limit is the injected amount that results in a peak with a height at least two or three times as high as the baseline noise level. Furthermore this signal/ noise method describes three more methods (ICH):

- ☆ Visual inspection: The detection limit is determined by the analysis of samples with known concentrations of analyte and by establishing the minimum level at which the analyte can be reliably detected.
- ☆ Standard deviation of the response based on the standard deviation of the blank: Measurement of the magnitude of analytical background response is performed by analyzing an appropriate number of blank samples and calculating the standard deviation of these responses.
- ☆ Standard deviation of the response based on the slope of the calibration curve: A specific calibration curve is studied using samples containing an analyte in the range of the limit of detection. The residual standard deviation of a regression line, or the standard deviation of y-intercepts of regression lines, may be used as the standard deviation.

LoD is the lowest analyte concentration likely to be reliably distinguished from the LoB and at which detection is feasible. It is therefore greater than LoB (Figure 13.2). A traditional and typical approach to estimate LoD consists of measuring replicates, usually n=20, of a zero calibrator or blank sample, determining the mean value and SD, and calculating LoD as the mean +2 SD. Variations of this approach use the mean plus 3, 4, or even 10 SDs to provide a more conservative LoD. The assumption is that if analyte is present, it will produce a signal greater than the analytical noise in the absence of analyte. This is a simple and quick method. The weakness is that there is no objective evidence to prove that a low concentration of analyte will indeed produce a signal distinguishable from a blank (zero concentration) sample. As Needleman and Romberg noted, "It defines only the ability to measure nothing".An alternative approach utilises analysis of samples containing small but known concentrations of the substance of interest (be it a drug, hormone or other analyte). The advantage of this empirical approach is that objective data is used to compare the analytical response of blank and low concentration samples to determine conclusively what concentration of analyte is necessary to distinguish its presence from its absence. Various analytical specifications (*e.g.* a minimum signal-to-noise ratio for a chromatographic method or a minimum absorbance requirement for a spectrophotometric procedure) can be applied to ensure that the LoD is meaningful and clearly distinguishable from a negative or blank sample. As defined in EP17, LoD is determined by utilising both the measured LoB and test replicates of a sample known to contain a low concentration of analyte.1 The mean and SD of the low concentration sample is then calculated according to:$LoD = LoB + 1.645(SD_{lowconcentrationsample})$

13.9 Limit of Quantitation

The limit of quantitation is the minimum injected amount that produces quantitative measurements in the target matrix with acceptable precision in chromatography, typically requiring peak heights 10 to 20 times higher than the baseline noise. If the required precision of the method at the limit of quantitation has been specified, the EURACHEM approach can be used. A number of samples with decreasing amounts of the analyte are injected six times. The calculated RSD per cent of the precision is plotted against the analyte amount. The amount that corresponds to the previously defined required precision is equal to the limit of quantitation. It is important to use not only pure standards for this test but also spiked matrices that closely represent the unknown samples. For the limit of quantitation, the ICH recommends, in addition to the procedures as described above, the visual inspection and the standard deviation of the response and the slope of the calibration curve. Any results of limits of detection and quantitation measurements must be verified by experimental tests with samples containing the analytes at levels across the two regions. It is equally important to assess other method validation parameters, such as precision, reproducibility and accuracy, close to the limits of detection and quantitation. Figure 13.2 illustrates both the limit of detection and the limit of quantitation.

13.10 Precision

Precision provides an indication of random errors and can be broken down into repeatability and intermediate precision. This procedure should only be performed when the entire analytical method procedure is finalized. Repeatability represents the simplest situation and involves analysis of replicates by the same analyst, generally one injection after the other. Repeatability tests are mandatory for all tests delivering numerical data. Repeatability is divided into two parts: injection repeatability and analysis repeatability (multiple preparations). Validation of the precision of an HPLC method occurs at three stages. The first stage is injection precision (injection repeatability) based on multiple injections of a single preparation of a sample on a particular sample on a given day. The set of criteria is given for area (per cent area normalization) methods (drug substance and drug product) based on per cent RSD of peak area. The second stage is analysis repeatability where multiple preparations and multiple injections of a sample are analyzed by the same chemist on the same day. The third stage is intermediate precision and is usually performed by different analysts, on a different system, on a different day on the same drug substance or drug product batch to determine the variability of the analytical test. The intermediate precision test may give indications to potential issues that may arise during method transfer. Relative standard deviation or coefficient of variation (Srel or per cent RSD) is used to assess whether the adequate precision has been obtained. If automation is utilized, then an intermediate precision test is required to compare results obtained through manual testing versus automated testing (if all solvent composition and analyte concentrations of all actives are identical in both methods).

13.11 Repeatability

In addition to the system precision, this short-term variability includes the contributions from the sample preparation, such as weighing, aliquoting, dilution, extraction, homogenisation, and *etc.* Therefore, it is essential to apply the whole analytical procedure (as described in the control test), rather merely to injecting the same sample solution six times. This is also the reason for using authentic samples because only then can the analytical procedure be performed exactly as in the routine application. There may be exceptions, but these should be demonstrated or cautiously justified. For example, analysing degradants near of the quantitation limit, where the variance contribution of the sample preparation can be neglected, injection precision and repeatability are identical. For some applications, where precision can be regarded as less critical, such as in early development or if the variability demands only a small part of specification range (less than approximately 10 per cent), or if the expected content of impurities is far away from the specification limit, artificially prepared (spiked) samples may be used, allowing several validation characteristics (linearity, precision and accuracy) to be addressed simultaneously. Repeatability can be calculated using Eq.(12.1) and Eq. (12.2) from a larger number of repeatedly prepared samples (at least 6).

$$s = \sum_{i=1}^{n} \frac{x - \bar{x}}{n - 1} \tag{12.2}$$

$$RSD\% = \frac{s}{\bar{x}} \times 100 \tag{12.3}$$

Where s is the standard deviation, xi is the individual measurement in a set of n measurements and x is the mean of all the measurements. per cent RSD is the standard deviation in percentage.

13.12 Intermediate Precision

Intermediate precision includes the influence of additional random effects according to the intended use of the procedure in the same laboratory and can be regarded as an (initial) estimate for the long-term variability. Relevant factors, such as operator, instrument, and days should be varied. Intermediate precision is obtained from several independent series of applications of the (whole) analytical procedure to (preferably) authentic, identical samples. In some cases of relative techniques, the preparation and analysis of the reference standard is an important variability contribution. Therefore, it is not appropriate to determine intermediate precision from the peak area of the sample alone (analysed on different days or even several concentrations only), as it is sometimes reported in validation literature. Apart from ignoring the contribution of the reference standard, any signal shift of the instrument will be falsely interpreted as random variability. In order to reflect the expected routine variability properly, the calibration must be performed exactly as described in the control test. Reproducibility, according to the ICH definition, is obtained by varying further factors between laboratories and is also particularly important in the assessment of "official" compendia methods or whether the method is applied at different sites. However, understood in the long-term perspective, both intermediate precision and reproducibility approach each other, at least in the same company. Reproducibility from collaborative trials can be expected to include additional contributions due to a probably larger difference of knowledge, experience, equipment, and *etc.* among the participating laboratories.

13.13 Reproducibility

Reproducibility represents the precision obtained between different laboratories. The objective is to verify that the method will provide the same results in different laboratories. The reproducibility of an analytical method is determined by analyzing aliquots from homogeneous lots in different laboratories with different analysts, and by using operational and environmental conditions that may differ from, but they are still within, the specified parameters of the method (inter-laboratory tests). Validation of reproducibility is important whether the method is to be used in different laboratories. The main factors affecting the reproducibility are:

- Differences in room temperature and humidity.
- Operators with different experience and thoroughness.
- Equipment with different characteristics, *e.g.*, delay volume of an HPLC system.
- Variations in material and instrument conditions, *e.g.*, HPLC, composition of mobile phase, pH, and flow rate of mobile phase.
- Variation in experimental details not specified by the method.
- Equipment and consumables of different ages.
- Columns from different suppliers or different batches.
- Solvents, reagents, and other materials with varying quality.

13.14 Accuracy

The accuracy of an analytical method is the extent to which test results generated by the method and the true value agree. Accuracy can also be described as the closeness of agreement between the value that is adopted, either as a conventional, true, or accepted reference value, and the value found. For the quantitative approaches, at least nine determinations across the specified range should be obtained, for example, three replicates at three concentration levels each. The percentage recovery or the difference between the mean and the accepted true value together with the confidence intervals are recommended. The expected recovery (Table 13.3) depends on the sample matrix, the sample processing procedure, and the analyte concentration. The AOAC Guidelines for Single Laboratory Validation of Chemical Methods for Dietary Supplements and Botanicals (AOAC, 2002) includes a table with estimated recovery data as a function analyte concentration.

Table 13.3: Analyte Recovery at different Concentrations

Active Ingredient (per cent)	*Analyte Ratio*	*Unit*	*Mean Recovery (per cent)*
100	1	100 per cent	98 – 102
≥ 10	10^{-1}	10 per cent	98 – 102
≥ 1	10^{-2}	1 per cent	97 – 103
≥ 0.1	10^{-3}	0.1 per cent	95 – 105
≥ 0.01	10^{-4}	100 ppm	90 – 107
≥ 0.001	10^{-5}	10 ppm	80 – 110
≥ 0.0001	10^{-6}	1 ppm	80 – 110
≥ 0.00001	10^{-7}	100 ppb	80 – 110
≥ 0.000001	10^{-8}	10 ppb	60 – 115

13.15 Robustness

Robustness tests examine the effect that operational parameters have on the analysis results. For the determination of robustness, a number of method parameters are varied within a realistic range, and the quantitative influence of the variables is determined. If the influence of the parameter is within a previously specified tolerance, the parameter is said to be within the robustness of the method range. Obtaining data on these effects helps to assess whether a method needs to be revalidated, for instance, when one or more parameters are changed to compensate the column performance over time. In the ICH document, it is recommended to consider the evaluation of robustness of the method during the development phase, and any results that are critical for the method should be documented. In summary, an analyst must have a critical look at three different types of parameters when robustness and ruggedness are investigated:

- Internal parameters (*e.g.*, temperature, pH, and *etc.*, in the case of HPLC).
- External parameters (*e.g.*, different analysts, instruments, laboratories, and *etc.*).
- Basic parameters (*e.g.*, stability of test solutions and *etc.*). From these lists, one point becomes very clear. A test for robustness is an

individual test and depends very much on the analytical technique and equipment applied. As a rule of thumb, it is recommended to examine at least those parameters which are part of the operational qualification of the respective equipment. Then, the set of parameters investigated in a robustness study can be arbitrarily extended to those specific to the method defined in the procedure. The usual way to performing robustness testing is first to define the parameters with reasonable maximum variation. Then, each parameter is successively varied whereas the others are held constant (at nominal setting). For instance, six parameters each one at two levels would require twelve experiments when one parameter is changed and the others are always set to nominal levels. If more parameters are included, more experiments must be conducted. This classical approach is called one-factor-a-time (OFAT) approach. Certainly, this kind of robustness testing has disadvantages, such as many experiments and time. In addition, only limited information is made available from such studies, since possible interactive effects which occur when more than one parameter (factor) is varied, cannot be identified. Nowadays, an experimental design approach (DOE: design of experiment) is often preferred for robustness testing. The aim of an experimental design is to obtain as much as possible relevant information in the shortest time from a limited number of experiments. Different designs can be used in robustness testing, *e.g.* including full– and fractional – factorial designs as well as Plackett–Burman designs. The latter have become very popular in method robustness testing during recent years. The choice of a design depends on the purpose of the test and the number of involved factors. Experimental designs in robustness testing can be employed for all analytical techniques.

13.16 Practical Explanation for LOD, LOQ, Accuracy, Repeatability using Excel Spread Sheets

Step 1

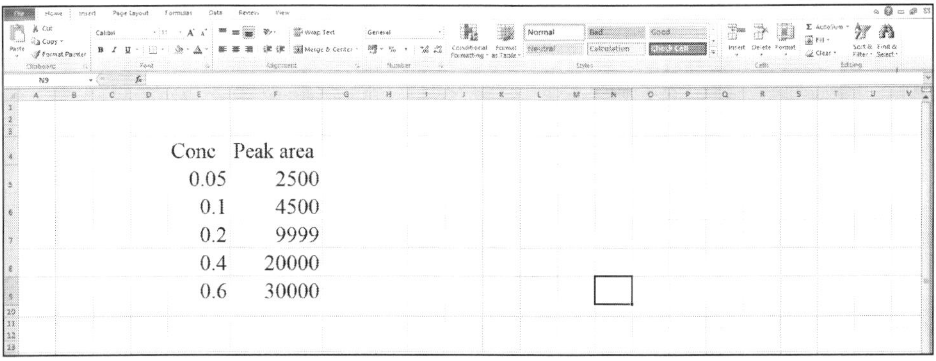

Conc	Peak area
0.05	2500
0.1	4500
0.2	9999
0.4	20000
0.6	30000

Step 2 Highlight the data and click on scatter diagram

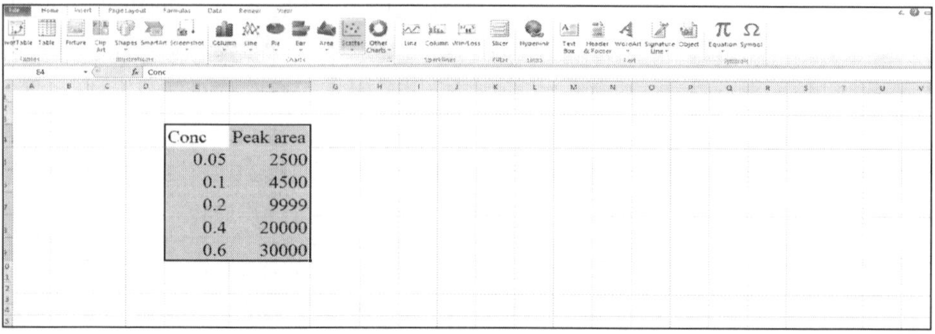

Step 3 After clicking the scatter diagram you will get graph as given below

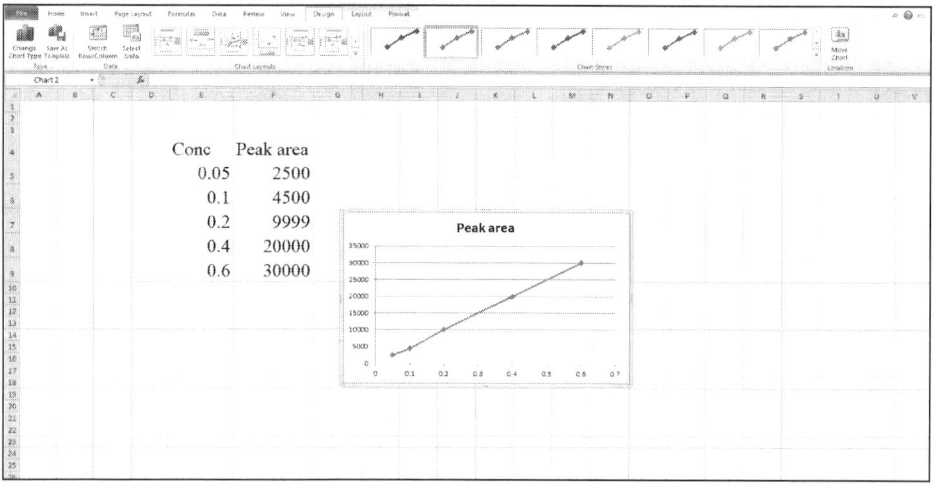

Step 4 Now put curser on the line and click right part of mouse you will get "add trend line" dialogue box and click "add trend line"

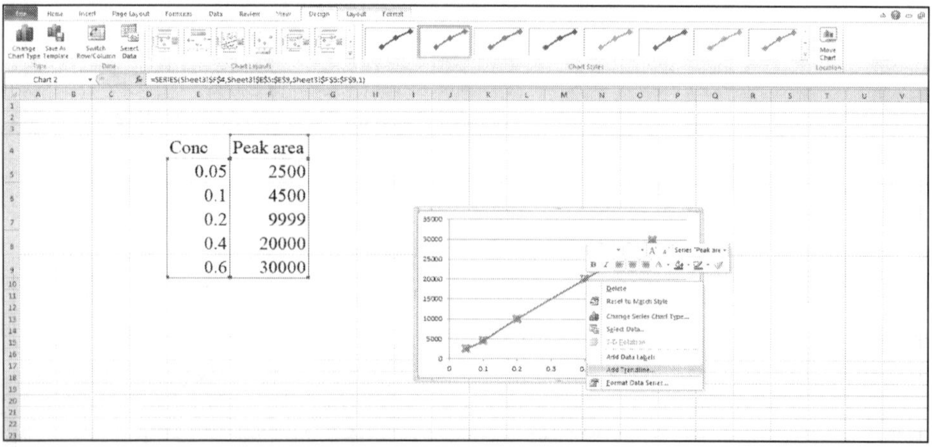

Step 5 Now dialogue box pops up and click "Linear" "Display equation on chart" and "Display R squared on Chart"

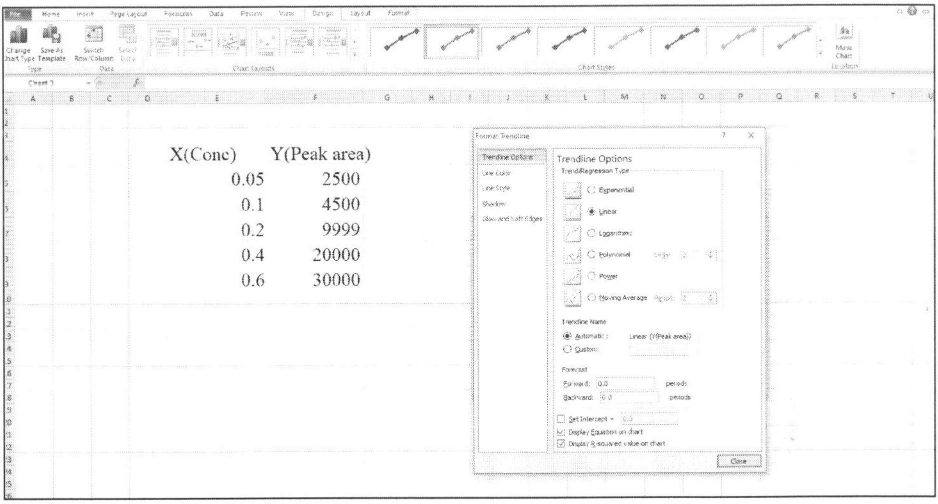

Step 6 Straight line equation with intercept and slope is displayed as given below

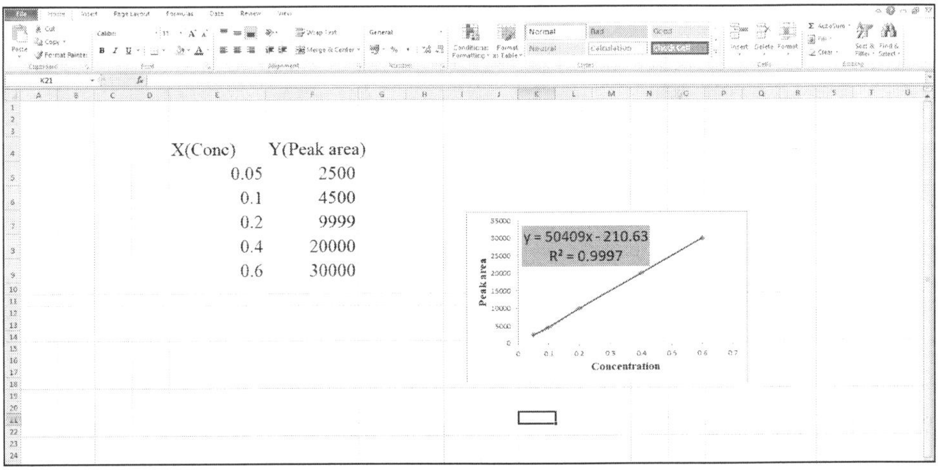

LOD = 3.3 × Residual SD/Slope LOQ = 10 × Residual SD/Slope

SD (Standard Deviation)

To get Residual standard deviations following steps are involved.

Step 7

To get data analysis tab on your excel sheet click on "file tab" and then on "option tab" as given below

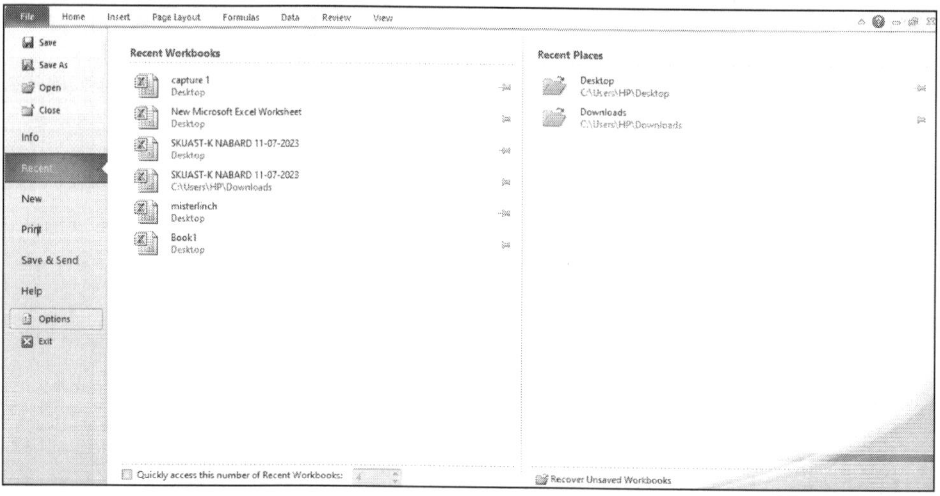

Step 8

Click on "Add-ins" and subsequently click "Analysis tool pak" and finally click "Go"

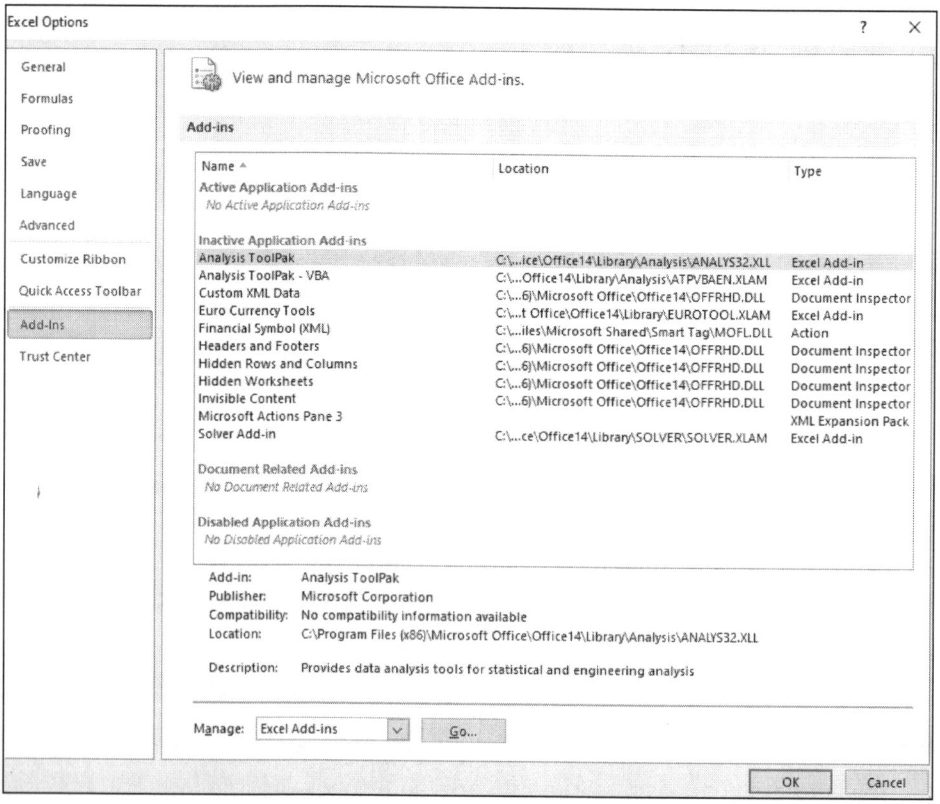

Step 9 Click "Analysis Toolpak" and then click "Ok"

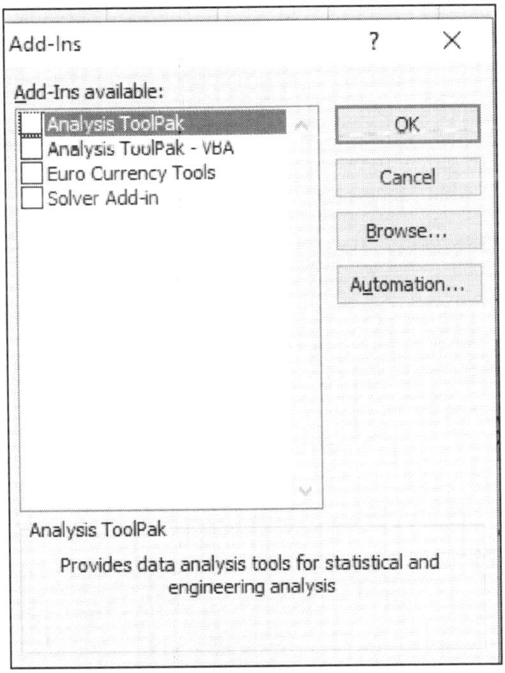

Step 8 Click "Data analysis" on extreme right, the following dialogue box will be seen. Here click "Regression" and "Ok"

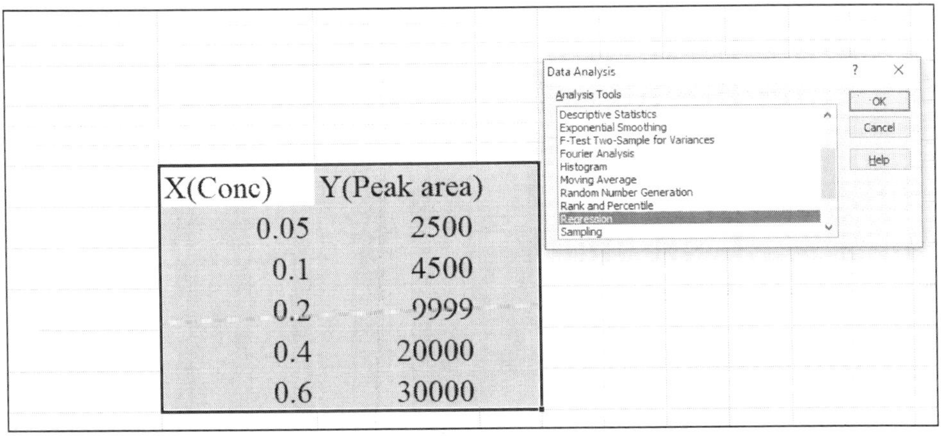

Step 9 When "Ok" tab is clicked following dialogue box will be observed

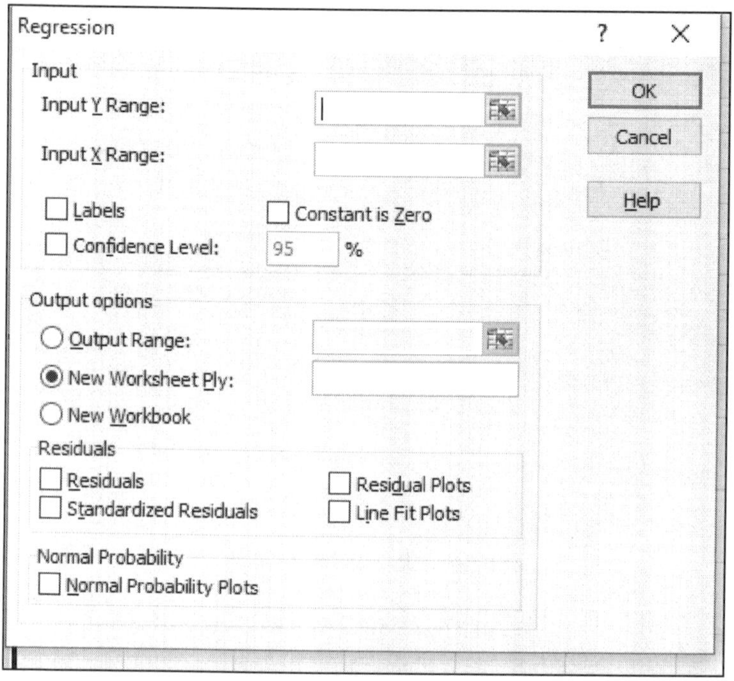

Ste 10 Click "input Y range" and highlight peak area and similarly click "input X range

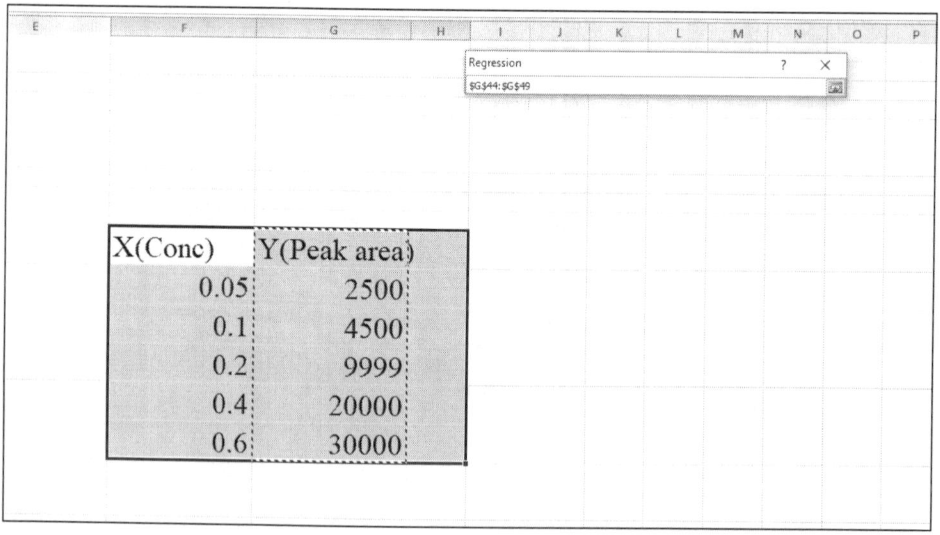

Step 11 Now click "output range" "Residuals" and "Residual plots"

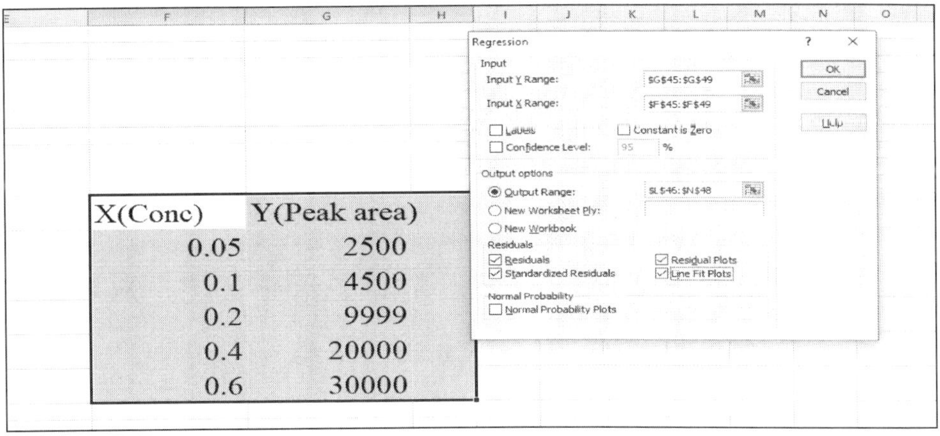

Step 12 Following statistics will pop out

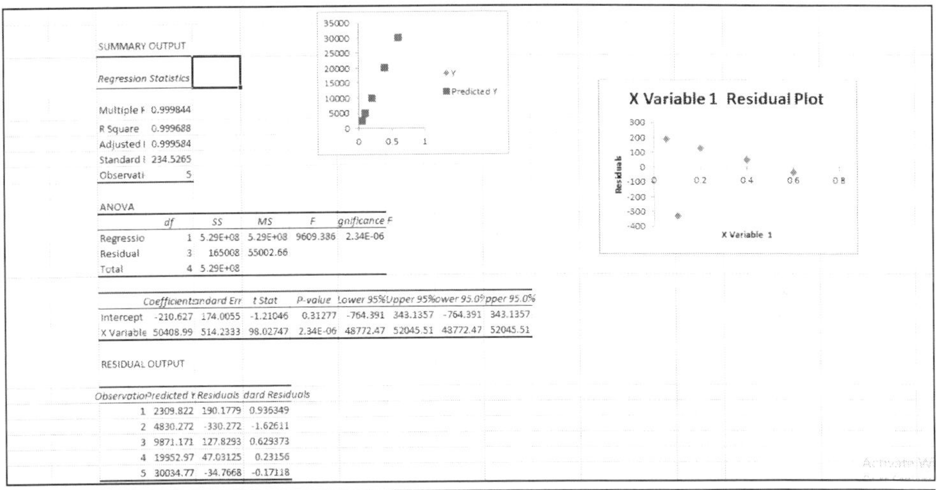

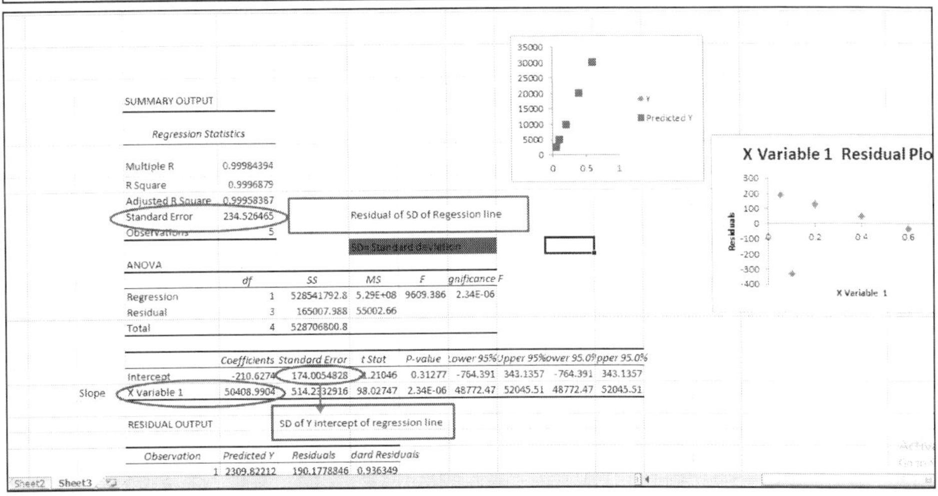

As already discussed in section 13.8 and 13.9 about LOD and LOQ, here ICH guidelines specifically discusses about LOD and LOQ.

Based on standard deviation of the response and the slope.

The detection limit (DL) may be expressed as

DL = 3.3 σ/S

where σ = the standard deviation of the response

S = the slope of the calibration curve

Measurement of the magnitude of analytical background response is performed by analysing an appropriate number of blank samples and calculating the standard deviation of these responses.

A specific calibration curve should be studied using samples containing an analyte in the range of DL. The residual standard deviation of a regression line or the standard deviation of the Y-intercepts of regression line may be sued as the standard deviation

Step 13 Here M63 is slope and M52 is standard error as shown in encirled ones.

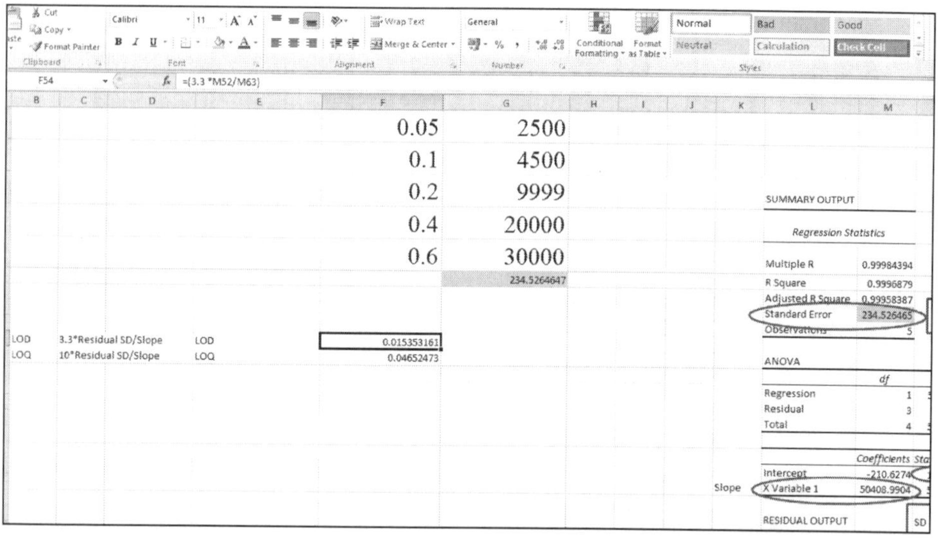

Repeatability: Here it the case of Perflourooctanic acid in ground water at five dates. The standard deviation as depicted in excel spread sheet $S = \sqrt{(x - x_{mean})}$,

Relative standard deviation (RSD) = Standard deviation (s)/mean,

$$S_{pool} = \frac{\sqrt{(n_1-1)\,[s1]^{2} + (n_2-1) + (n_3-1) + \cdots (n_k-1))}}{n_1+n_2+n_3+\ldots\ldots+N_k\text{-}k}$$

$$RSD_{pool} = \frac{\sqrt{(n_1-1)\,[RSD1]^{2} + (n_2-1)\,[RSD2]^{2} + (n_3-1)\,[RSD3]^{2} + \cdots (n_k-1)\,[RSDk]^{2}}}{n_1+n_2+n_3+\ldots\ldots+N_k\text{-}k}$$

	Repeatability				
PFOA (Perflouoroctanic acid in ground water)					
Measurements in2022	10 ng/L				
	15-Oct	18-Nov	30-Nov	15-Dec	30-Dec
	9.66	9.09	9.21	11.28	9.07
	10.85	9.11	9.24	11.12	10.04
	10.47	9.25	9.14	11.58	9.69
	10.12	9.31	9.11	10.97	10.94
	9.78		9.2	11.51	
Repeatability(1 day)					
Standard deviation					
Mean					
Relative standard deviation					
RSD%					

$$S = \frac{\sqrt{\sum_{i=1}^{n}(x - x_{mean})}}{n-1}$$

$RSD = s/_{mean}$

$$S_{pool} = \frac{\sqrt{(n_1-1)S1^2 + (n_2-1)S2^2 + (n_3-3)S3^2 + \cdots \ldots (n_k-1)Sk^2}}{n_1 + n_2 + n_3 + \ldots\ldots\ldots N_k\ \text{-}k}$$

$$RSD_{pool} = \frac{\sqrt{(n_1-1)RSD1^2 + (n_2-1)RSD2^2 + (n_3-3)RSD3^2 + \cdots \ldots (n_k-1)RSDk^2}}{n_1 + n_2 + n_3 + \ldots\ldots\ldots N_k\ \text{-}k}$$

Step 1 Calculation for standard deviation as depicted in excel spread sheet and calculate mean.

SUM =stdev(E7:E11)

	Repeatability				
PFOA (Perflouoroctanic acid in ground water)					
Measurements in2022	10 ng/L				
	15-Oct	18-Nov	30-Nov	15-Dec	30-Dec
	9.66	9.09	9.21	11.28	9.07
	10.85	9.11	9.24	11.12	10.04
	10.47	9.25	9.14	11.58	9.69
	10.12	9.31	9.11	10.97	10.94
	9.78		9.2	11.51	
Repeatability(1 day)					
Standard deviation		=stdev(E7:E11)			
Mean					
Relative standard deviation					
RSD%					

Step 2 Finally calculate RSD (Standard deviation/mean).

SUM =(F14/F15)

D	E	F	G	H	I
	15-Oct	18-Nov	30-Nov	15-Dec	30-Dec
	9.66	9.09	9.21	11.28	9.07
	10.85	9.11	9.24	11.12	10.04
	10.47	9.25	9.14	11.58	9.69
	10.12	9.31	9.11	10.97	10.94
	9.78		9.2	11.51	
Repeatability(1 day)					
Standard deviation		0.49166			
Mean		10.176			
Relative standard deviatior		=(F14/F15)			
RSD%					

Step 3 Now calculate repeatability for all five dates. Estimate pool repeatability as mentioned in the formulae and its calculation in excel sheet.

SUM =E19*E14^2

C / D	E	F	G	H	I
	Repeatability				
PFOA (Perflouoroctanic acid in ground water)					
Measurements in2022	10 ng/L				
	15-Oct	18-Nov	30-Nov	15-Dec	30-Dec
	9.66	9.09	9.21	11.28	9.07
	10.85	9.11	9.24	11.12	10.04
	10.47	9.25	9.14	11.58	9.69
	10.12	9.31	9.11	10.97	10.94
	9.78		9.2	11.51	
Repeatability(1 day)					
Standard deviation	0.49	0.11	0.05	0.26	0.78
Mean	10.176	9.19	9.18	11.292	9.935
Relative standard deviation(s)	0.048	0.012	0.006	0.023	0.079
RSD%	4.83	1.17	0.58	2.27	7.86
Repeatability pool	15-Oct	18-Nov	30-Nov	15-Dec	30-Dec
Degree of freedom (n-1)	4	3	4	4	3
(n-1)*s^2	=E19*E14^2				

SUM =sqrt(sum(E20:I20)/sum(E19:I19))

D	E	F	G	H	I
	9.78		9.2	11.51	
Repeatability(1 day)					
Standard deviation	0.49	0.11	0.05	0.26	0.78
Mean	10.176	9.19	9.18	11.292	9.935
Relative standard deviation(s)	0.048	0.012	0.006	0.023	0.079
RSD%	4.83	1.17	0.58	2.27	7.86
Repeatability pool	15-Oct	18-Nov	30-Nov	15-Dec	30-Dec
Degree of freedom (n-1)	4	3	4	4	3
(n-1)*s^2	0.97	0.03	0.01	0.26	1.83
Repeatability (multiple days)	=sqrt(sum(E20:I20)/sum(E19:I19))				
Relative stanadard deviation pool					
Mean					
RSD					
(n-1)*RSD^2					
RSD pool					
RSD pool%					

D	E	F	G	H	I
Repeatability(1 day)					
Standard deviation	0.49	0.11	0.05	0.26	0.78
Mean	10.176	9.19	9.18	11.292	9.935
Relative standard deviation(s)	0.048	0.012	0.006	0.023	0.079
RSD%	4.83	1.17	0.58	2.27	7.86
Repeatability pool	15-Oct	18-Nov	30-Nov	15-Dec	30-Dec
Degree of freedom (n-1)	4	3	4	4	3
(n-1)*s^2	0.97	0.03	0.01	0.26	1.83
Repeatability (multiple days)	0.42	0.39	0.44	0.55	0.78
Relative stanadard deviation pool					
Mean	10.176	9.19	9.18	11.292	9.935
RSD	0.048	0.012	0.006	0.023	0.079
(n-1)*RSD^2	=E19*E24^2				
RSD pool					
RSD pool%					

13.17 Standard Addition

Place 5 ml of unknown in each blank

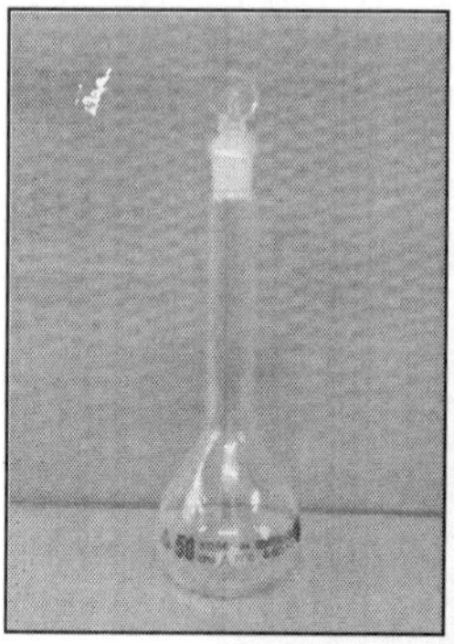
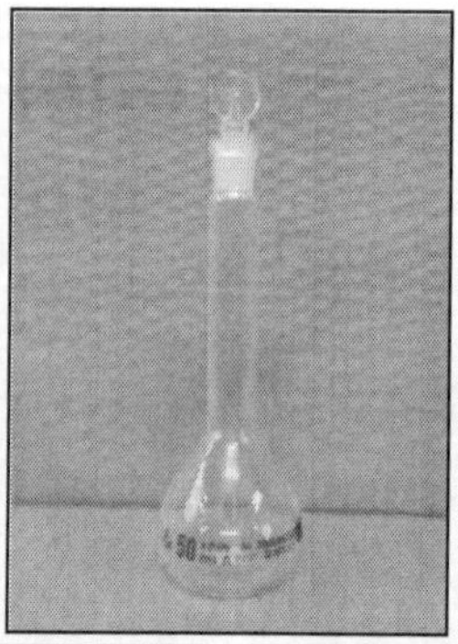
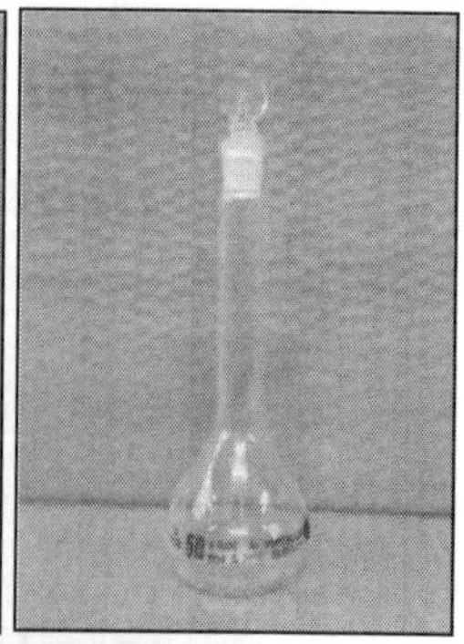
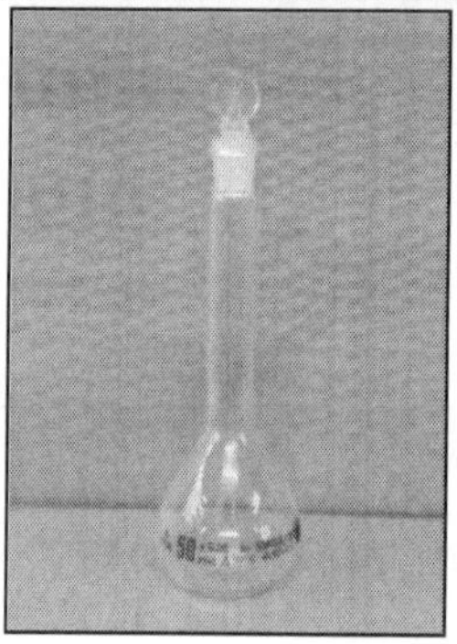
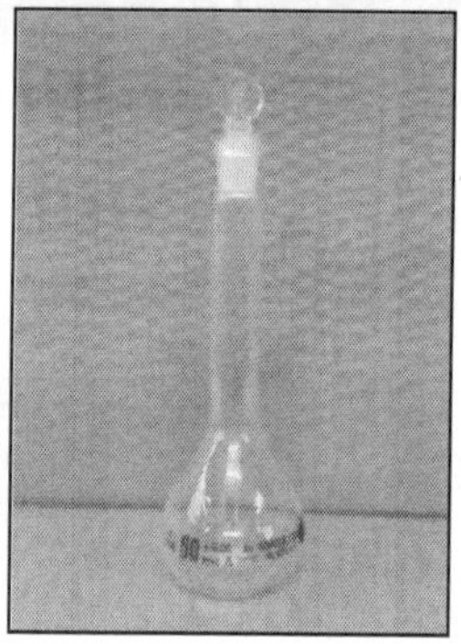

Pb^{2+} stock (ppb) 150.00

Sample vol (ml) 10.00 ml

Final vol (ml) 50.00

Now add 0 ml, 5 ml, 10 ml, 15 ml and 20 ml of standard solution in each (10 ml) of volumetric flask.

Fill each volumetric flask upto 50 ml.

Step 1 calculate the standard concentration in 50 ml of volumetric flask as shown in excel sheet.

=I5*$G$5/$G$7

C	D	E	F	G	H	I	J	K
						Volume of stock added	Standard concentration(ppb)	signal
		Pb^{2+} stock (ppb)		150		0	=I5*$G$5/$G$7	3.06
		Sample vol (ml)		10		5	15	4.56
		Final vol (ml)		50		10	30	6.06
						15	45	7.56
						20	60	9.06

Step 2 Now click on "insert tab" and click on "scatter diagram" as shown in excel sheet

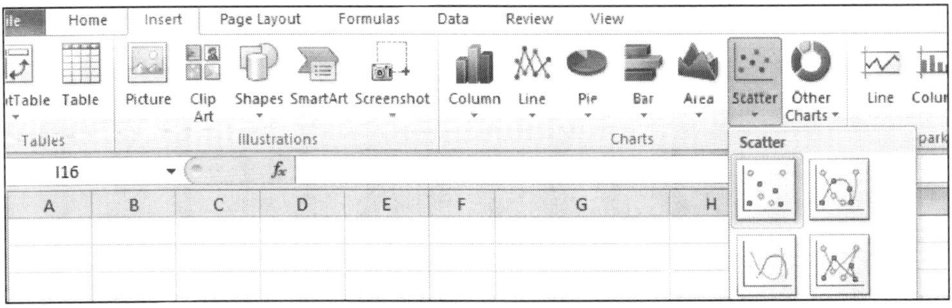

Step 3: Now click on "select data"

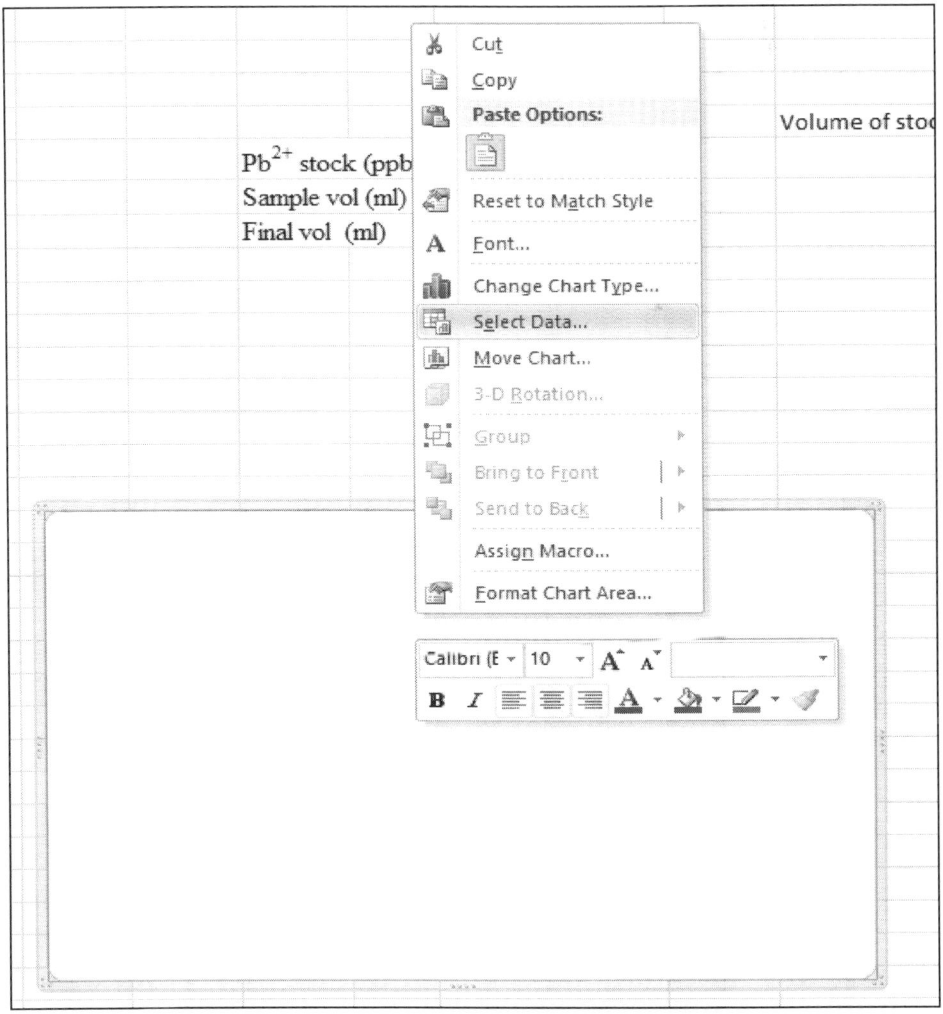

Step 4: when "select data" is clicked following dialogue box will pop out. Enter Standard concentration in "series X value" and signal in "series Y value"

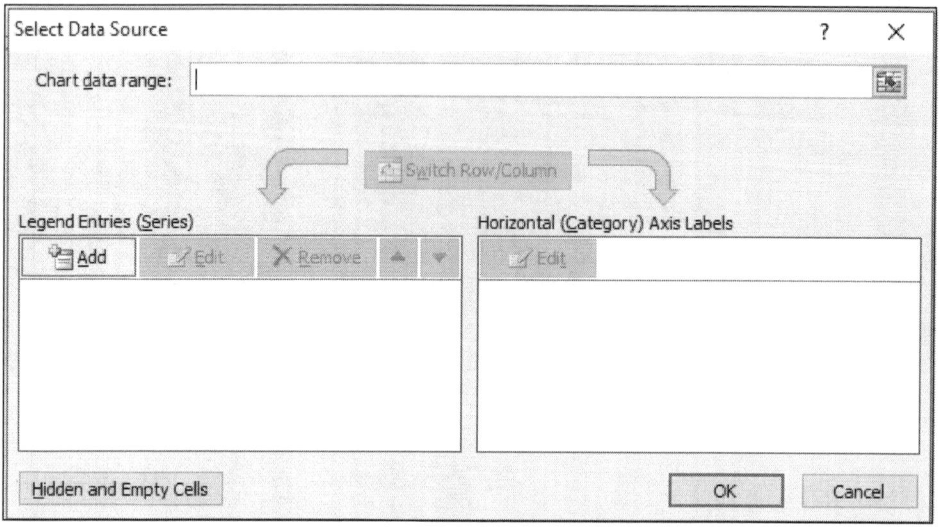

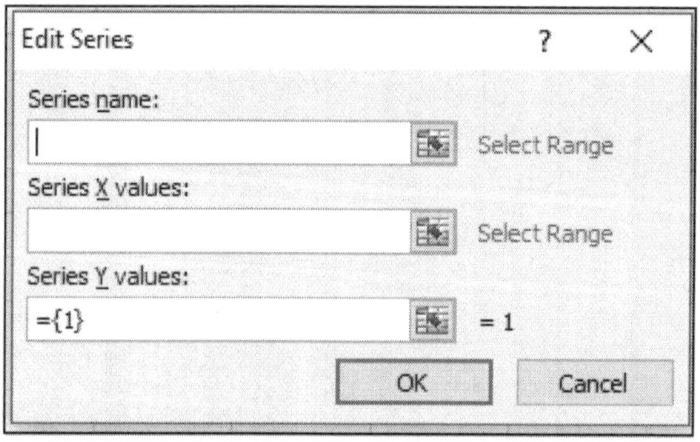

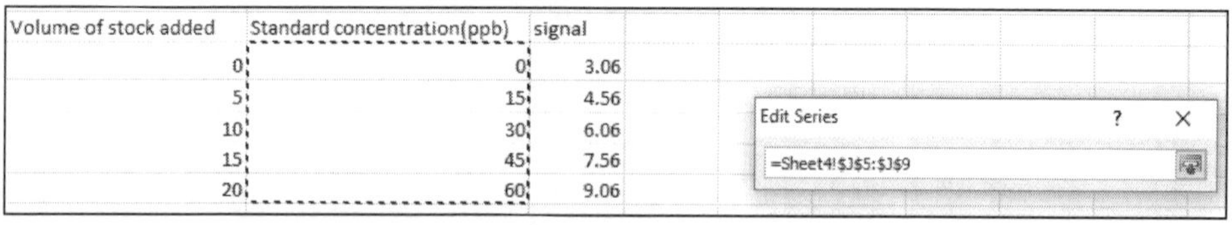

Volume of stock added	Standard concentration(ppb)	signal
0	0	3.06
5	15	4.56
10	30	6.06
15	45	7.56
20	60	9.06

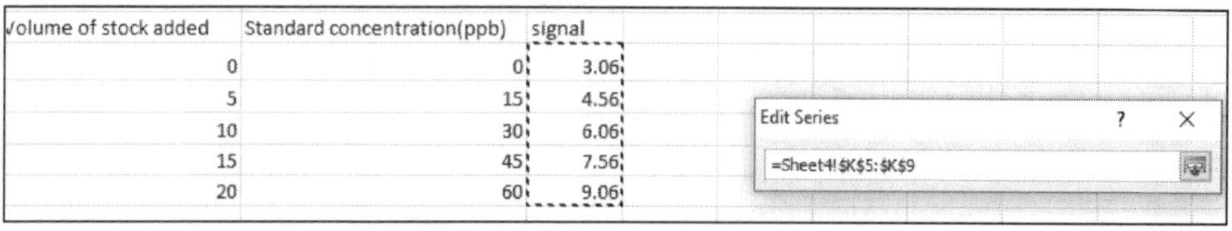

Volume of stock added	Standard concentration(ppb)	signal
0	0	3.06
5	15	4.56
10	30	6.06
15	45	7.56
20	60	9.06

Step 5 Now put curser on data point and click right part of mouse and click "add trend line" as given in following excel sheet

Step 6: Get the slope and intercept from the equation.

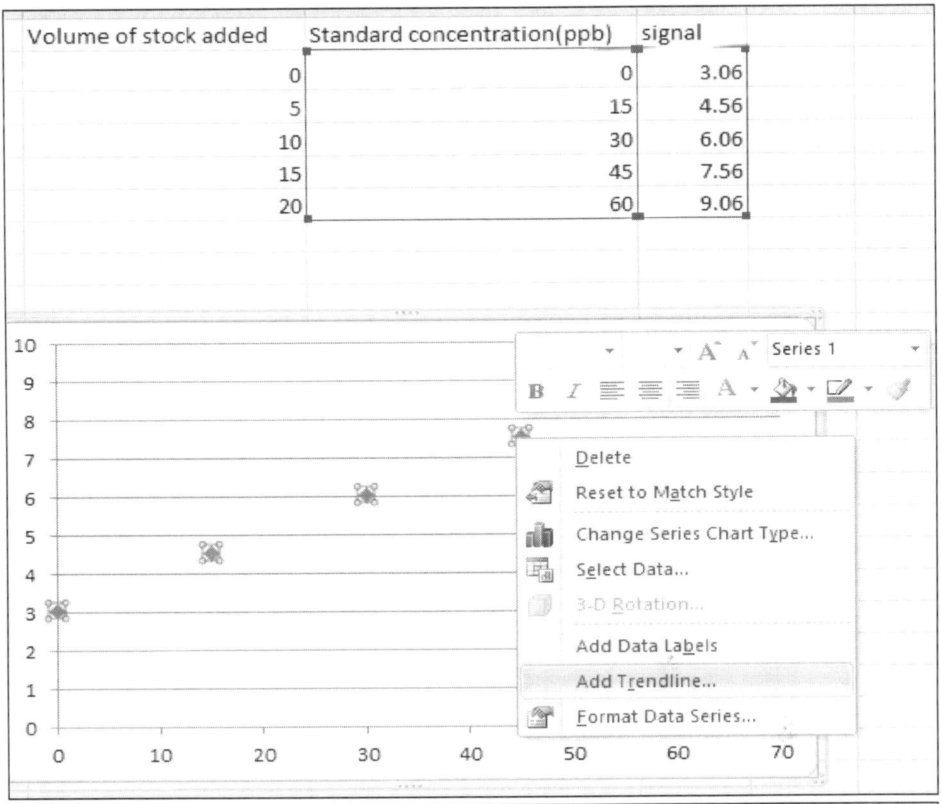

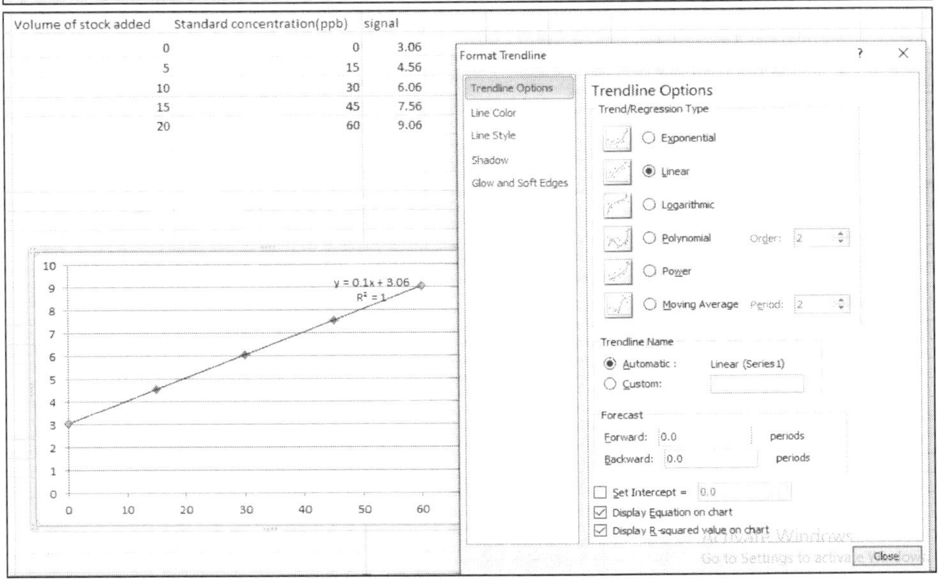

	Standard concentration(ppb)	signal
0	0	3.06
5	15	4.56
10	30	6.06
15	45	7.56
20	60	9.06
	=slope(K5:K9,J5:J9)	

olume of stock added	Standard concentration(ppb)	signal
0	0	3.06
5	15	4.56
10	30	6.06
15	45	7.56
20	60	9.06
n	0.1	
nt	=intercept(K5:K9,J5:J9)	

Step 7: When the line of the equation (st.line, y= ax+m) is extrapolated to negative y-axis the equation becomes as

0= ax+m, where y becomes zero

Therefore x = b/m (x is concentration of unknown sample) = 30.6 ppb

m	0.1
int	3.06
b/m	=J16/J14

Step 8 Finally multiply with dilution factor as given in excel sheet. [Pb in original sample] = 306 ppb

Pb^{2+} stock (ppb)	150		0		0
Sample vol (ml)	5		5		15
Final vol (ml)	50		10		30
			15		45
			20		60
		m			0.1
		int			3.06
		b/m			30.6
		Pb conc		=J17*G7/G6	

13.18 System Suitability Tests (SST)

SST ensures quality pharmaceutical analyses by verifying the performance of chromatography systems before analyzing samples. It checks crucial aspects like separation efficiency, precision, and consistency to guarantee reliable results. Following United States Pharmacopeia (USP) and International Conference on Harmonization (ICH) guidelines, SST is mandatory for all HPLC analyses in the pharmaceutical industry. It guarantees that equipment, analysis, and samples function together effectively before analyzing actual test samples. SST takes a comprehensive approach, viewing the entire chromatography system (equipment, electronics, analysis, samples) as one unit. By testing key aspects like column efficiency, precision, and repeatability, it ensures the system's overall suitability for the intended analysis. USP and ICH guidelines have specially emphasize on the use of SST to satisfy the latest regulatory requirements. One of the most important facts about the SST is that it should be performed not only before the experiment but also at different time interval throughout the all HPLC assays. The reason behind the SST analysis at different time interval is based on the concept that it should not be assumed that the system will behave properly throughout experiment. In addition to this, SST studies corresponding to compound of our interest is not enough to check system suitability because the system's separation capacity is not investigated thoroughly. But, System Suitability Samples (SSSs) or resolution test mixtures of components of interest and expected impurities is required. For this reasons SSTs are analysed before and during testing. The most important SST parameters which are investigated for different HPLC analysis is Resolution (R), repeatability (RSD– relative standard deviations–of peak response and retention time), column efficiency (N), and Tailing factor (T). These parameters are possessing para amount interest as they indicate system specificity, precision, and column stability.

Other parameters include capacity factor (k) and Signal-to-noise ratio (S/N) for impurity peaks. All most all HPLC data systems are based on the software data system which can calculate the measurement and report of these SST parameters. most "official" analytical methods has fixed the acceptance criteria or SST limits which may vary with different tests and are typically less stringent for biologics and trace impurities.

SST limits should emphasize the minimum criterion rather than optimal values. Analytical methods simply adopted the general limits from the CDER (Centre for Drug Evaluation and Research) guidance document. Different guidelines related to SST limits are shown (Table 13.3).

Table 13.3: SST Limits for HPLC Chromatogram According to different Guidelines

SST Limits	*CDER Guidelines*
Repeatability of peak response	≤1.0 per cent for 5
Resolution	>2.0 general
Tailing factor	≤2.0
Column efficiency	>2000 (plate count)
Capacity factor	>2

The system suitability parameters which are generally accepted by regulatory authorities and independent auditor are depicted below:

Peak retention time, Peak area, Amount, Peak height, Peak width at half height, Peak symmetry, Peak tailing, Capacity factor (k´), Plate numbers, Resolution between peaks, Selectivity relative to preceding peak. All these parameters have been already discussed.

13.19 Measure of Uncertainty

Before discussing measure of uncertainty let us delve some important concepts.

13.19.1 Measurement

A measurement tells us about a property of something. It might tell us how heavy an object is, or how hot, or how long it is. A measurement gives a number to that property. The result of a measurement is normally in two parts: a number and a unit of measurement, *e.g.* 'How long is it?. 2 metres.'

13.19.2 What is Uncertainty of Measurement?

Every measurement has an inherent uncertainty (doubt), which means we can't know the exact value with perfect certainty. This is true even for the most advanced instruments. Uncertainty of measurement represents the range of possible values within which the true value is believed to lie with a certain level of confidence. It's like saying a stick is two meters long with a "margin of error" of one centimeter.

13.19.3 Expressing Uncertainty of Measurement

Since there is always a margin of doubt about any measurement, we need to ask 'How big is the margin?' and 'How bad is the doubt?' Thus, two numbers are really needed in order to quantify an uncertainty. One is the width of the margin, or interval. The other is a confidence level, and states how sure we are that the 'true value' is within that margin. For example: We might say that the length of a certain stick measures 20 centimetres plus or minus 1 centimetre, at the 95 per cent confidence level. This result could be written: 20 cm ±1 cm, at a level of confidence of 95 per cent. The statement says that we are 95 per cent sure that the stick is between 19 centimetres and 21 centimetres long. There are other ways to state confidence levels.

13.19.4 Error versus Uncertainty

It is important not to confuse the terms 'error' and 'uncertainty'. Error is the difference between the measured value and the 'true value' of the thing being measured. Uncertainty is a quantification of the doubt about the measurement result. Whenever possible we try to correct for any known errors: for example, by applying corrections from calibration certificates. But any error whose value we do not know is a source of uncertainty.

13.19.5 Basic Statistics

The two most important statistical calculations are to find the average or ***arithmetic mean,*** and the ***standard deviation*** for a set of numbers.

If repeated measurements give different answers, you may not be doing anything wrong. Different readings in repeated measurements can be normal. It could be because of the world being naturally variable (like wind speed) or your tool not being perfectly stable (like a stretching tape measure). If there is variation in readings when they are repeated, it is best to take many readings and take an average. An average gives you an estimate of the 'true' value. An average or arithmetic mean is usually shown by a symbol with a bar above it, *e.g.* x/ ('xbar') is the mean value of x. Figure 12.3 shows an illustration of a set of values and their mean value. Example 1 shows how to calculate an arithmetic mean.

Broadly speaking, the more measurements you use, the better the estimate you will have of the 'true' value. The ideal would be to find the mean from an infinite set of values. The more results you use, the closer you get to that ideal estimate of the mean. But performing more readings takes extra effort, and yields 'diminishing returns'. What is a good number? Ten is a popular choice because it makes the arithmetic easy. Using 20 would only give a slightly better estimate than 10. Using 50 would be only slightly better than 20. As a rule of thumb usually between 4 and 10 readings is sufficient.

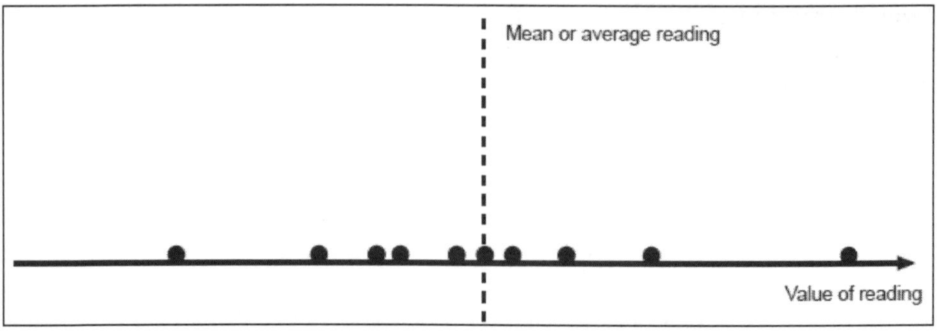

Figure 13.3: 'Blob Plot' Illustrating an Example Set of Values and Showing the Mean.

Suppose you have a set of 10 readings. To find the average, add them together and divide by the number of values (10 in this case). The readings are: 16, 19, 18, 16, 17, 19, 20, 15, 17 and 13.

The sum of these = 170

The average of the 10 readings is: 170/10 = 17

13.19.6 Spread Standard Deviation

When repeated measurements give different results, we want to know how widely spread the readings are. The spread of values tells us something about the uncertainty of a measurement. By knowing how large this spread is, we can begin to judge the quality of the measurement or the set of measurements. Sometimes it is enough to know the range between the highest and lowest values. But for a small set of values this may not give you useful information about the spread of the readings in between the highest and the lowest. For example, a large spread could arise because a single reading is very different from the others. The usual way to quantify spread is standard deviation. The standard deviation of a set of numbers tells us about how different the individual readings typically are from the average of the set. As a 'rule of thumb', roughly two thirds of all readings will fall between plus and minus (±) one standard deviation of the average. Roughly 95 per cent of all readings will fall within two standard deviations. This 'rule' applies widely although it is by no means universal. The 'true' value for the standard deviation can only is found from a very large (infinite) set of readings. From a moderate number of values, only an estimate of the standard deviation can be found. The symbol s is usually used for the estimated standard deviation.

Calculating an Estimated Standard Deviation

It is rarely convenient to calculate standard deviations by hand, with pen and paper alone. But it can be done as follows: Suppose you have a set of n readings (let's use the same set of 10 as above). Start by finding the average: For the set of readings we used before, 16, 19, 18, 16, 17, 19, 20, 15, 17 and 13, the average is 17. Next, find the difference between each reading and the average,

i.e. -1 +2 +1 -1 0 +2 +3 -2 0 -4,

and square each of these,

i.e. 1 4 1 1 0 4 9 4 0 16.

Next, find the total and divide by n-1 (in this case n is 10, so n-1 is 9), *i.e.*

$$\frac{1 + 4 + 1 + 1 + 0 + 4 + 9 + 4 + 0 + 16 = 40/9}{9} = 4044$$

The estimated standard deviation, s, is found by taking the square root of the total, *i.e.*

$s = \sqrt{4.44} = 2.1$

The complete process of calculating the estimated standard deviation for a series of n measurements can be expressed mathematically as:

$$s = \frac{\sqrt{\sum_{i=1}^{n}(x - [\bar{x})]^2}}{(n - 1)} \quad (12.4)$$

where xi is the result of the ith measurement and x_{mean} is the arithmetic mean of the n results considered.

13.19.7 Errors

There are two types of errors Random error and systematic error

Random Error

Random error affects your measurements in unpredictable ways: your measurements are equally likely to be higher or lower than the true values. In the graph below, the black line represents a perfect match between the true scores and observed scores of a scale. In an ideal world, all of your data would fall on exactly that line. The green dots represent the actual observed scores for each measurement with random error added.

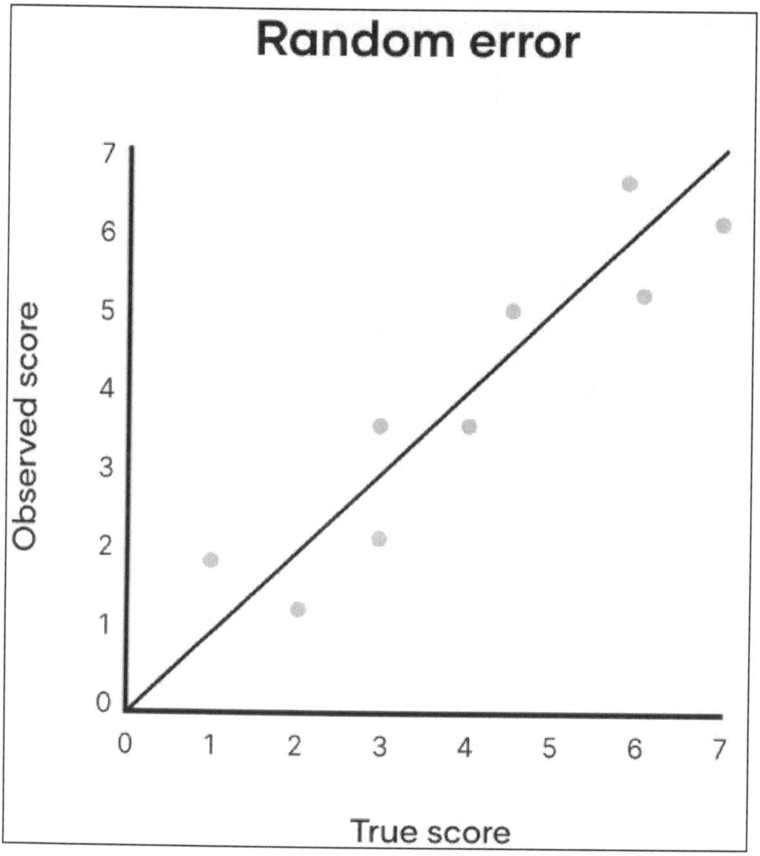

Sources of Random Errors

Some common sources of random error include:

1. Natural variations in real world or experimental contexts.
2. Imprecise or unreliable measurement instruments.
3. Individual differences between participants or units.
4. Poorly controlled experimental procedures.

Systematic Error

Systematic error means that your measurements of the same thing will vary in predictable ways: every measurement will differ from the true measurement in the same direction, and even by the same amount in some cases.

Systematic error is also referred to as bias because your data is skewed in standardized ways that hide the true values. This may lead to inaccurate conclusions.

Types of Systematic Errors

Offset errors and scale factor errors are two quantifiable types of systematic error.

An offset error occurs when a scale isn't calibrated to a correct zero point. It's also called an additive error or a zero-setting error. A scale factor error is when measurements consistently differ from the true value proportionally (*e.g.*, by 10 per cent). It's also referred to as a correlational systematic error or a multiplier error. In the graphs below, the black line shows when your observed value is the exact true value, and there is no random error.

The blue line is an offset error: it shifts all of your observed values upwards or downwards by a fixed amount (here, it's one additional unit).

The purple line is a scale factor error: all of your observed values are multiplied by a factor–all values are shifted in the same direction by the same proportion, but by different absolute amounts.

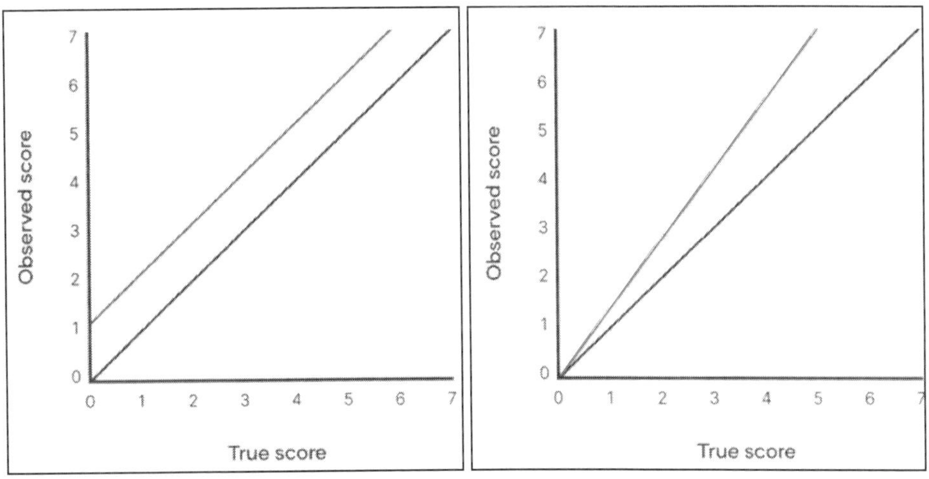

Sources of Systematic Errors

The sources of systematic error can range from your research materials to your data collection procedures and to your analysis techniques. This isn't an exhaustive list of systematic error sources, because they can come from all aspects of research. Response bias occurs when your research materials (*e.g.*, questionnaire) prompt participants to answer or act in inauthentic ways through leading questions

13.8.8 Distribution - The 'Shape' of the Errors

The spread of a set of values can take different forms, or probability distributions. In a set of readings, sometimes the values are more likely to fall near the average than further away. This is typical of a normal or Gaussian distribution. You might see this type of distribution if you examined the heights of individuals in a large group of men. Most men are close to average height; few are extremely tall or short.

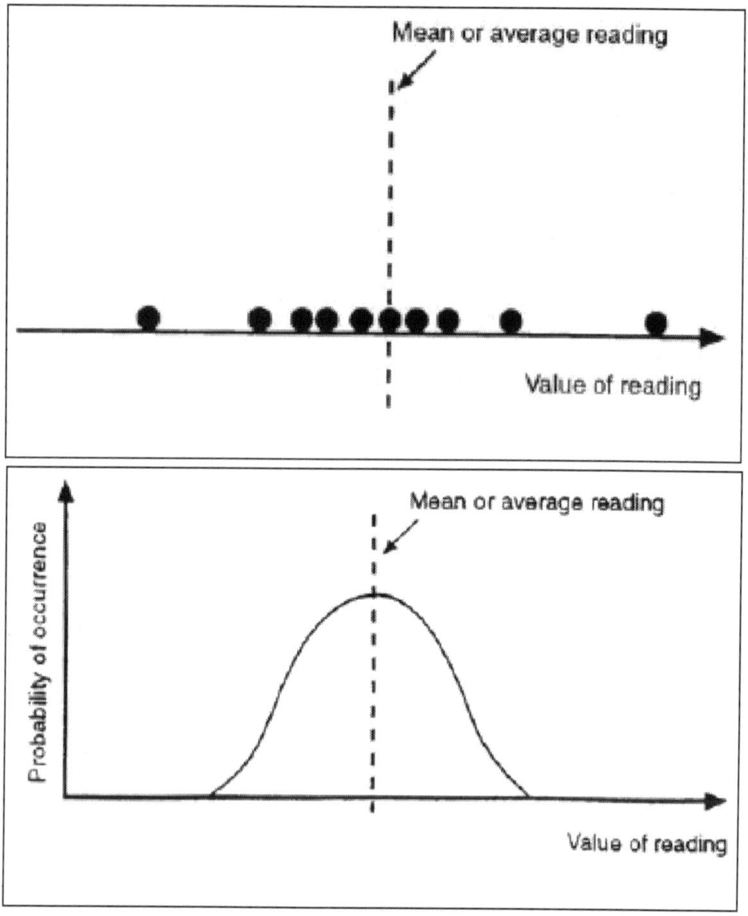

Figure 13.2: 'Blob Plot' of a Set of Values Lying in a Normal Distribution.

13.19.9 Uniform or Rectangular Distribution

When the measurements are quite evenly spread between the highest and the lowest values, a rectangular or uniform distribution is produced. This would be seen if you examined how rain drops fall on a thin, straight telephone wire, for example. They would be as likely to fall on any one part as on another. Figure 4 shows a set of 10 'random' values in an approximately rectangular distribution. A sketch of a rectangular distribution is shown in Figure 13.3.

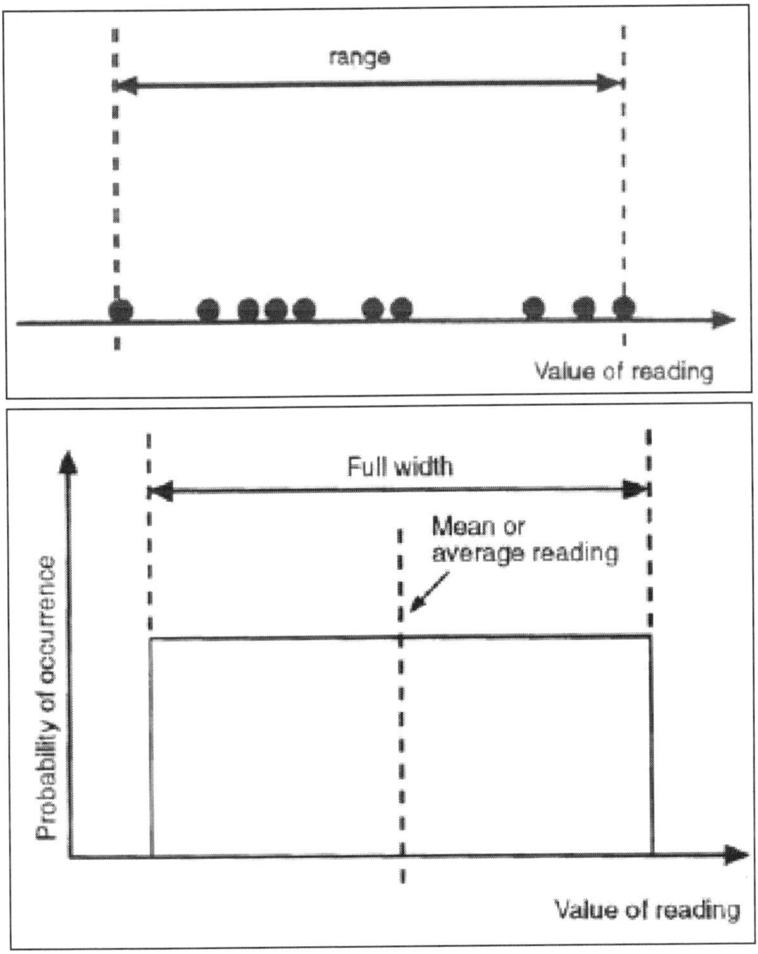

Figure 13.3: 'Blob Plot' of a Set of Values Lying in a Rectangular Distribution.

13.19.10 How to Calculate Uncertainty of Measurement

To calculate the uncertainty of a measurement, firstly you must identify the sources of uncertainty in the measurement. Then you must estimate the size of the uncertainty from each source. Finally the individual uncertainties are combined to give an overall figure. There are clear rules for assessing the contribution from each uncertainty, and for combining these together. There two ways to estimate uncertainties 'Type A' and 'Type B' evaluations. In most measurement situations, uncertainty evaluations of both types are needed. Type A evaluations - uncertainty estimates using statistics (usually from repeated readings). Type B evaluations - uncertainty estimates from any other information. This could be information from past experience of the measurements, from calibration certificates, manufacturer's specifications, from calculations, from published information, and from common sense. There is a temptation to think of 'Type A' as 'random' and 'Type B' as 'systematic', but this is not necessarily true.

Eight main steps to evaluating uncertainty

13.19.11 Standard Uncertainty

All contributing uncertainties should be expressed at the same confidence level, by converting them into standard uncertainties. A standard uncertainty is a margin whose size can be thought of as 'plus or minus one standard deviation'. The standard uncertainty tells us about the uncertainty of an average (not just about the spread of values). A standard uncertainty is usually shown by the symbol u (small u), or u(y) (the standard uncertainty in y).

13.19.11.1 Calculating Standard Uncertainty for a Type A Evaluation

When a set of several repeated readings has been taken (for a Type A estimate of uncertainty), the mean, x/, and estimated standard deviation, s, can be calculated for the set. From these, the estimated standard uncertainty, u, of the mean is calculated from:

$$u = {}^{s}/_{\sqrt{n}} \qquad 12.5$$

where n was the number of measurements in the set. (The standard uncertainty of the mean has historically also been called the standard deviation of the mean, or the standard error of the mean.)

13.19.11.2 Calculating Standard Uncertainty for a Type B Evaluation

Where the information is more scarce (in some Type B estimates), you might only be able to estimate the upper and lower limits of uncertainty. You may then have to assume the value is equally likely to fall anywhere in between, *i.e.* a rectangular or uniform distribution. The standard uncertainty for a rectangular distribution is found from:

$${}^{a}/_{\sqrt{3}} \qquad 12.6$$

where a is the semi-range (or half-width) between the upper and lower limits. Rectangular or uniform distributions occur quite commonly, but if you have good reason to expect some other distribution, then you should base your calculation on that. For example, you can usually assume that uncertainties 'imported' from the calibration certificate for a measuring instrument are normally distributed.

13.19.12 Converting Uncertainties from One Unit of Measurement to Another

Uncertainty contributions must be in the same units before they are combined. As the saying goes, you cannot 'compare apples with pears'. For example, in making a measurement of length, the measurement uncertainty will also eventually be stated in terms of length. One source of uncertainty might be the variation in room temperature. Although the source of this uncertainty is temperature, the effect is in terms of length, and it must be accounted for in units of length. You might know that the material being measured expands in length by 0.1 per cent for every degree rise in temperature. In that case, a temperature uncertainty of ±2 °C would give a length uncertainty of ±0.2 cm in a piece of the material 100 cm long. Once the standard uncertainties are expressed in consistent units, the combined uncertainty can be found using one of the following techniques.

13.19.13 Combining Standard Uncertainties

Individual standard uncertainties calculated by Type A or Type B evaluations can be combined validly by 'summation in quadrature' (also known as 'root sum of the squares'). The result of this is called the combined standard uncertainty, shown by uc or uc(y). Summation in quadrature is simplest where the result of a measurement is reached by addition or subtraction. The more complicated cases are also covered below for the multiplication and division of measurements, as well as for other functions.

13.19.14 Summation in Quadrature for Addition and Subtraction

The simplest case is where the result is the sum of a series of measured values (either added together or subtracted). For example, you might need to find the total length of a fence made up of different width fence panels. If the standard uncertainty (in metres) in the length of each fence panel was given by a, b, *etc.*, then the combined standard uncertainty (in metres) for the whole fence would be found by squaring the uncertainties, adding them all together, and then taking the square root of the total,

$$\textit{Combined uncertainty} = \sqrt{a^2 + b^2 + \cdots} \quad 12.7$$

13.19.15 Summation in Quadrature for Multiplication or Division

For more complicated cases, it can be useful to work in terms of relative or fractional uncertainties to simplify the calculations. For example, you might need to find the area, A, of a rectangular carpet, by multiplying the length, L, by the width, W (*i.e.* A=L × W). The relative or fractional uncertainty in the area of the carpet can be found from the fractional uncertainties in the length and width. For length L with uncertainty u(L), the relative uncertainty is u(L)/L. For width W, the relative uncertainty is u(W)/W. Then the relative uncertainty u(A)/A in the area is given by

$$u\frac{(A)}{A} = \sqrt{u\left(\frac{L}{L}\right)2 + u\left(\frac{W}{W}\right)2} \quad 12.8$$

For a case where a result is found by multiplying three factors together, equation (5) would have three such terms, and so on. This equation can also be used (in exactly the same form) for a case where the result is a quotient of two values (*i.e.* one number divided by another, for example, length divided by width). In other words, this form of the equation covers all cases where the operations are multiplication or division.

13.20 Correlation

The equations given above in Section 12.9 to calculate the combined standard uncertainty are only correct if the input standard uncertainties are not inter-related or correlated. This means we usually need to question whether all the uncertainty contributions are independent. Could a large error in one input cause a large error in another? Could some outside influence, such as temperature, have a similar effect on several aspects of uncertainty at once - visibly or invisibly? Often individual errors are independent. But if they are not, extra calculations are needed.

13.20.1 Coverage Factor k

Having scaled the components of uncertainty consistently, to find the combined standard uncertainty, we may then want to re-scale the result. The combined standard uncertainty may be thought of as equivalent to 'one standard deviation', but we may wish to have an overall uncertainty stated at another level of confidence, *e.g.* 95 percent. This re-scaling can be done using a coverage factor, k. Multiplying the combined standard uncertainty, uc, by a coverage factor gives a result which is called the expanded uncertainty, usually shown by the symbol U, *i.e.*

$$U = kuc \quad 12.9$$

A particular value of coverage factor gives a particular confidence level for the expanded uncertainty. Most commonly, we scale the overall uncertainty by using the coverage factor k = 2, to give a level of confidence of approximately 95 percent. (k = 2 is correct if the combined standard uncertainty is normally distributed. A particular value of coverage factor gives a particular confidence level for the expanded uncertainty. Most commonly, we scale the overall uncertainty by using the coverage factor k = 2, to give a level of confidence of approximately 95 percent. (k = 2 is correct if the combined standard uncertainty is normally distributed.

13.21 Let us do some Practical Problem to Measure Uncertainty

Standardization of NaOH

NaOH is standardized against Potassium hydrogen phthalate (KHP) using titration.

KHP is dried and weighed

NaOH is then titrated against KHP solution

Conc of NaOH (mol/L) =

$$\frac{1000 * molecular\ mass\ of\ KHP * purity\ of\ KHP}{molar\ mass\ of\ KHP * voulme\ of\ titration}$$

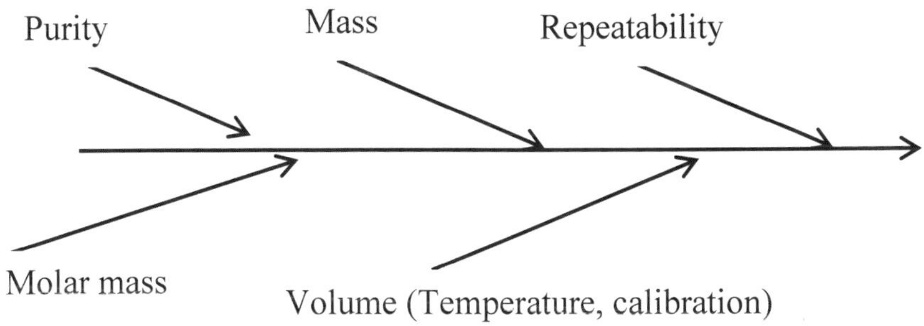

Fish Bone Diagram (Sources of uncertainty).

a) Uncertainty Due to Repeatability (type A)

Titration of NaOH with KHP was performed five times and the results were as follows

Sl.No.	Conc. NaOH mol/L
1	0.1024
2	0.1009
3	0.0997
4	0.0994
5	0.1016
Average	0.1008
Standard Deviation	0.001263
Std. uncertainty	0.000565

Here standard uncertainty = $\frac{\text{standard deviation}}{\sqrt{n}}$

$0.001263/\sqrt{5} = \underline{0.000565}$

b) Uncertainty Due to Repeatability (type B)

i) *Mass of KHP (balance)*

0.3888KHP g is weighed on the balance. The uncertainty in weighing arise due to uncertainty in linearity. The calibration certificates quotes linearity of balance as 0.15 mg *i.e.* 0.00015 g

Uncertainty due to balance linearity is = $\frac{0.00015}{\sqrt{3}}$ (rectangular distribution already discussed in section 12.8)

= 0.000086

The contribution of balance is counted twice, one for tare and one for final weight.

Total uncertainty due to balance (due to weighing) =

$\sqrt{0.000086 * 0.000086 * 2}$ = **0.0001225** g

ii) Purity of KHP

Purity of KHP as given by manufacturer is between 99.95 per cent to 100.05 per cent. The COA (certificate of Analysis) says that purity of KHP is 1.000 ± 0.0005. Since the supplier gives no further information on the uncertainty, therefore we assume rectangular distribution.

Uncertainty in purity of KHP = $\frac{0.0005}{\sqrt{3}}$ = **0.00029**

iii) Molar mass of KHP (atomic weight KHP = 204.22 g/L)

Element	No. of Atoms	Atomic Weight	Quoted Uncertainty	Uncertainty/ Atom	Uncertainty *No. of Atoms
C	8	12.0107	±0.0008	0.000462	0.003695
H	5	1.00794	±0.00007	0.000040	0.000202
O	4	15.994	±0.0003	0.000173	0.000693
K	1	39.0983	±0.0001	0.000058	0.000058

Here uncertainty per atom is calculated as per rectangular distribution =

$= \frac{12.0107}{\sqrt{0.0008}}$ = 0.000462 for Carbon.

Since there are 8 carbons in KHP therefor uncertainty will be = 0.000462 * 8 = 0.003695

Total uncertainty in molar mass of KHP is calculated as follows:

$$\sqrt{(0.003695 * 0.003695) + (0.000202 * 0.000202) + (0.000693 * 0.000693) + (0.000058 * 0.000058)}$$

= **0.0037653** g/mol

iv) Uncertainty Due to Volumetric Glass Ware

A burette of 20 mL is used for titration. As per COA, volume of burette is 20 ± 0.03 mL, without mentioning any information. There we assume rectangular distribution

Uncertainty due to calibration of burette = $\frac{0.003}{\sqrt{3}}$ = 0.01732 mL

Although coefficient of expansion is negligible, thermal expansion of water needs to be considered.

Thermal expansion of water is $2.1 \times 10^{-4}\ ^{o}C^{-1}$

If the temperature variation in the lab is ± 3°C, volume expansion due to temperature variation is $2.1 \times 10^{-4}\ ^{o}C^{-1} \times 20 \times 3 = 0.0126$ mL [Here burette capacity is 20 ml and 3 is temperature variation].

Uncertainty due to volume expansion = $\frac{0.0126}{\sqrt{3}}$ = 0.00727

Combined uncertainty due to volume =

$$\sqrt{(0.01732 * 0.01732) + (0.00727 * 0.00727)}$$

Uncertainty due to volume = **<u>0.018786</u>** mL

Uncertainty Budget for NaOH Standardization

Description	*Value (x)*	*Std Uncertainty (u)*	*Relative Std Uncertainty (u/x)*
Mass of KHP	0.388 g	0.0001225	0.000315
Purity of KHP	1.0	0.00029	0.00029
Molar mass	204.22 g/mol	0.0038 g/mol	0.0000184
Volume of NaOH	19.0 ml	0.018786 mL	0.000987
Repeatability	0.1008 mol/l	0.000565 mol/l	0.005603

Combined uncertainty Uc =

$$\sqrt{(\blacksquare(average * (0.000315 * 0.000315) + (0.00029 * 0.00029) + (0.0000184 * 0.0000184)@ + (0.000987 * 0.000987 + (0.005603 * 0.005603)))}$$

=**<u>0.00058</u>** mol/l

Expanded uncertainty = combined uncertainty × coverage factor

Confidence Level	*68.27 per cent*	*90 per cent*	*95 per cent*	*95.45 per cent*	*99 per cent*	*99.73 per cent*
Coverage factor (k)	1.000	1.645	1.960	2.000	2.576	3.000

Selection of coverage factor: Generally the value of coverage factor can be considered as 2. However, when the combined uncertainty is based on observation with relatively less degree of freedom (less than about six), it should be derived from effective degree of freedom. Welch-Satterthawite equation is used to calculate the effective degree of freedom. Coverage factor corresponding to the degree of freedom (using students t-distribution) is to be used for calculating expanded uncertainty.

Welch-Satterthawite formula for effective degree of freedom

$$V_{eff} = \frac{\dfrac{Uc^4\ (y)}{\sum_{i=1}^{N} ui^4(y)}}{Vi}$$

Uc = Combined standard uncertainty

Vi = Degree of freedom for standard uncertainty

Ui = standard uncertainty parameter

Degree of freedom is number of values in final calculation of statistics that are free to vary.

Degree of freedom = No. of independent observation –No. of parameters estimated.

For repeatability experiments if 10 measurements (n) are taken, degree of freedom is n-1 = 9.

For type B uncertainties it is generally considered to be infinity.

Now we apply the formula to our example of NaOH

$$\text{Effective degree of freedom for NaOH standardization} = \frac{(0.00058)^4 \times 4}{(0.000565)^4 \times 4}$$

$$= 4.3$$

Effective degree of freedom is 4.3 (approx. 4.0), corresponding the coverage factor from student's t-distribution =2.78 for 95 per cent confidence level

Expanded uncertainty for NaOH conc = 2.78 x 0.00058 =0.0016 mol/l

The result can be expressed as follows

Conc of NaOH = 0.1008 ± 0.0016 mol/l at 95 per cent CL and a coverage factor of 2.

Chapter 14

Instrument Qualification

14.1 Introduction

Equipment qualification is a formal process that provides documented evidence that an instrument is fit for its intended use and kept in a state of maintenance and calibration consistent with its use. The entire qualification process consists of four parts

1. Design qualification (DQ)
2. Installation qualification (IQ)
3. Operational qualification (OQ)
4. Performance qualification (PQ)

14.1.1 Design Qualification (DQ)

Describes the user requirements and defines the functional and operational specifications of the instrument. It should ensure that instruments to be purchased have the necessary functions and performance that will enable them be suitable for the intended applications. Later on the DQ document will be used as a basis for tests in the OQ phase.

Table 14.1 lists elements with examples that should be included in the design qualification for the HPLC systems in the selected QA/QC laboratory. Please note that all instruments are required to have gradient pumps, thermostatted column compartments and diode-array detectors to ensure that they all can be used for all applications.

Table 14.1: Design Qualification Elements

Design Elements	*Examples*
Intended use	Analysis of drug compounds and impurities as an example)
User requirement specification for the HPLC analysis	Up to 100 samples/day ☆ Automated over-night analysis ☆ Limit of quantitation: 0.1 per cent ☆ Automated confirmation of peak identity and purity with diode-array detection ☆ Automated compound quantitation and printing of report
Functional specifications	☆ Pump Binary or higher gradient ☆ Detector UV/V is Diode-array, 190-900 nm ☆ Autosampler 100 samples, 0.5 ul to 5 ml sample volume Column compartment 15 to 60 Deg C, peltier controlled ☆ Computer System control, data acquisition for signals and spectra, peak integration and quantitation, spectral evaluation for peak purity and compound confirmation. Electronically save all chromatograms generated by the system.
Operational specifications	Detector: Baseline noise:<5 x 10^{-5} AU
	☆ Sampler: Precision inj. volume: <0.5 per cent ☆ RSD, sample carry over: <0.5 per cent ☆ Pump: precision of retent.time: <0.5 per cent ☆ RSD
User instructions	☆ Operational manual on paper ☆ Computer based tutorial
Validation/ qualification	Vendor must provide procedures and servicesfor IQ and OQ
Maintenance	☆ Vendor must deliver maintenance procedure and recommend schedule ☆ Instrument must include early maintenance feedback for timely exchange of most important maintenance parts ☆ Maintenance procedures must be supplied on Multimedia CD ROM
Training	Vendor must provide familiarization and training

14.1.2 Installation Qualification

Installation qualification establishes that the instrument is received as designed and specified, that it is properly installed in the selected environment, and that this environment is suitable for the operation and use of the instrument. Table 14.2 lists steps as recommended before and during installation. IQ should include analysis of a test sample. A successful run of such a sample verifies correct installation of all modules and electrical and fluid connections.

Before Installation

- Obtain manufacturer's recommendations for installation site requirements.
- Check the site for the fulfillment of the manufacturer's recommendations (utilities such as electricity, and environmental conditions such as humidity and temperature).
- Allow sufficient shelf space for the equipment, SOPs, operating manuals and software. During installation
- Compare equipment, as received, with purchase order (including software, accessories, spare parts)
- Check documentation for completeness (operating manuals, maintenance instructions, standard operating procedures for testing, safety and validation certificates)
- Check equipment for any damage
- Install hardware (computer, equipment, fittings and tubings for fluid connections, columns in HPLC and GC, power cables, data flow and instrument control cables)
- Switch on the instruments and ensure that all modules power up and perform an electronic self-test
- Identify and make a list with a description of all hardware, include drawings where appropriate.
- Run test sample and compare chromatogram print-out with reference chromatogram
- List equipment manuals and SOPs
- Prepare an installation report

14.1.3 Operational Qualification (OQ)

Operational qualification (OQ) is the process of demonstrating that an instrument will function according to its operational specification in the selected environment" It verifies that the HPLC system complies with key functional and operational requirements as specified in the design qualification. In modular HPLC systems it is recommended to perform system tests (holistic testing), rather than performing tests module by module (modular testing). Individual module tests should be performed as part of the diagnosis if the system fails. Table 14.2 includes test parameters, procedures and recommended user limits. At the beginning of the test, a leak test based on the instrument's flow rate accuracy is performed. If this test fails, most other tests are likely to fail.

Table 14.2: Test Parameters and Acceptance Criteria for Case Study

Parameter	*Procedure (*)*	*User Limit*
Leak testing	Flow test by volume or weight/time	± 5 per cent
Baseline drift	ASTM Method E19.09, 20 min	<2 x 10-3 AU
Baseline noise	ASTM Method E19.09, 20 x 1 min	<5 x 10-5 AU
Precision of injection volume	6 x injection of caffeine standard, RSD of peak areas	0.3 per cent RSD
Precision of flow rate	6 x injection of caffeine standard, RSD of retention times	0.5 per cent RSD
Detector linearity	inject 5 standards	>1.5 AU, 5 per cent
Wavelength accuracy	holmium oxide filter	± 1 nm
Temperature accuracy	comparison with external measuring device	±1 °C
Temperature precision ±	monitoring temperature over 20 min	0.25 °C
Autosampler carry over	Injection of blank solvent after large concentration	< 0.5 per cent
Mobile phase composition accuracy	Step gradients from 4 to 7 per cent B, step heights relative to 100 per cent, with acetone tracer	± 1 per cent

(*) see detailed in section 14.3

14.1.4 Performance Qualification

Performance Qualification (PQ) is the process of demonstrating that an instrument consistently performs according to a specification appropriate for its routine use. Important here is the word 'consistently'. The test frequency is much higher than for OQ. Another difference is that PQ should always be performed under conditions that are similar to routine sample analysis. For a chromatograph this means using the same column, the same analysis conditions and the same or similar test compounds. PQ should be performed on a daily basis or whenever the instrument is used. The test frequency not only depends on the stability of the equipment but on everything in the system that may contribute to the analysis results. For a liquid chromatograph, this may be the chromatographic column or a detector's lamp. The test criteria and frequency should be determined during the development and validation of the analytical method. In practice, PQ can mean system suitability testing, where critical key system performance characteristics are measured and compared with documented, preset limits. For example, a well-characterized standard can be injected 5 or 6 times and the standard deviation of amounts are then compared with a predefined value. PQ tests are applications specific. If the limits of detection and/or quantitation are critical, the lamp's intensity profile or the baseline noise should be tested. They should use the same column and chemicals as used for the real samples. Test should include

- precision of the amounts
- precision of retention times
- resolution between two peaks
- peak width at half height or
- peak tailing
- baseline noise
- wavelength accuracy of the UV/Visible wavelength detector, preferably using built-in holmium-oxide filters.

14.2 Documentation

On completion of equipment qualification, documentation should be available that consists of:

- Design qualification document
- IQ document (includes description of hardware and software)
- Procedures for OQ testing
- OQ test reports
- PQ test procedures and representative results OQ test reports should include test parameters, acceptance criteria and actual results.

14.2.1 Flow Rate Accuracy

Materials: Beaker (capacity of 5-10 mL) or weighing vessel Calibrated thermometer Calibrated chronometer Calibrated analytical balance

Settings: Mobile phase: degassed water for chromatography

Capillary *e.g.* 2000 x 0.12 mm capillary for back pressure Flow rate: x HPLC system: 0.5 mL/min and 5.0 mL/min or maximum

Flow rate used x HPLC system: 0.2 mL/min and 2.0 mL/min or maximum flow rate used

If high-pressure mixing systems are installed, this test has to be done on each solvent channel. Procedure: Set the flow rate at a desired level (at least test one flow rate, the lower operational level in use) and leave water to flow into the beaker or weighing vessel, previously weighed (empty). After a defined time (*e.g.* 5.0 min) weigh the beaker again containing the delivered water. Measure the temperature of the water used. Repeat the procedure at least 3 times. Calculate the volume delivered to the beaker and the flow rate:

$$V = \frac{m}{Q} \qquad 13.1$$

M = weight of water delivered in the measured time [g]

Q = density of water at actual temperature from table [g/mL]

V = volume delivered [mL]

$$f = \frac{V}{t} \quad 13.2$$

f = measured flow rate [mL/min]

t = elapsed time [min]

V = volume delivered [mL]

$$D = 100 * \frac{f - F}{F} \quad 13.3$$

D = deviation [per cent] F = nominal flow rate [mL/min] f = measured flow rate [mL/min]

14.2.2 Gradient Composition Accuracy and Ripple

Settings: Capillary *e.g.* 2000 x 0.12 mm for back pressure

Detection: 265 nm

Mobile phase A: water for chromatography R Mobile phase B: water for chromatography R containing 0.5 per cent acetone

Flow rate: 1.0 mL/min

Procedure: The test is carried out in the following way by using a gradient program depending on the number of solvent channels and the configuration of the system:

A-B

A-B and A-C

A-C, A-B and B-D

Table 14.3

Time [min]	*Per cent Mobile Phase A (water)*	*Per cent Mobile Phase B (Water/acetone mixture)*
0.0	100	0
0.1	90	10
10	90	10
10.1	50	50
20	50	50
20.1	10	90
30	10	90
30.1	0	100
40	0	100
40.1	100	0

Start the test by pumping water for at least 10 min to equilibrate the system. The zero per cent value at the start of the test is the baseline. All steps are measured at the beginning of the horizontal part of the line either by software (as absorption units) or manually on the paper print using a ruler. The height of the 100 per cent water/acetone mixture is used as H in the following calculation.

$$\%H = 100 * \frac{h}{H} \qquad 13.4$$

Per cent H = calculated composition

h = height of the measured line

H = height of the 100 per cent water/acetone mixture line (mobile phase B)

$$d = \%H - G \qquad 13.5$$

d = deviation

G = nominal gradient composition [per cent water/acetone mixture = mobile phase B]

The ripple of the gradient composition is the percentage of noise of the 50 per cent line from the gradient program.

$$\%R = 100 * \frac{N}{h/50} \qquad 13.6$$

$\%R$ = ripple

$h/50$ = height of the 50 per cent line

N = height of the noise, *i.e.* the difference between the minimum and the maximum of the

50 per cent line measured during 1 minute in the linear region

14.3.2 Injector

Volume precision and carry-over are the tests proposed for the periodic and motivated check of the LC injector.

Volume Precision and Carry-Over

Solutions: Solvent A: water for chromatography R: methanol R (40:60 V/V)

Reference solution (a): dissolve 15.0 mg methylparaben and 15.0 mg of propylparaben in solvent A and dilute to 100.0 mL with the same solvent.

Reference solution (b): Dilute 1.0 mL of reference solution (a) to 10.0 mL with solvent A. Settings:

Column: Lichrospher 100 RP8, 5 µm, 125 x 4 mm, without pre-column or equivalent

Mobile phase: water for chromatography R: methanol R (40:60 V/V)

Flow rate: 1.0 mL/min

Detection: 254 nm

Injection volume:

- ☆ HPLC ("short pathway" flow-cell, for example 5 or 10 mm): 20 µL
- ☆ UHPLC ("short pathway" flow-cell, for example 5 or 10 mm): 10 µL
- ☆ HPLC/UPLC ("long pathway" flow-cell, for example 60 or 85 mm): 2 µL

Procedure: Injection scheme: – 1 x solvent A (blank injection 1)

– 6 x reference solution (b)

– 1 x reference solution (a)

– 1 x solvent A (blank injection 2)

– 1 x reference solution (b)

Repeatability of peak areas: The relative standard deviation of the peak areas of methylparaben and propylparaben in the chromatograms of the six consecutive injections of reference solution (b) should be ≤ 1.0 per cent

Carry-over: The percentage of the peak area corresponding to propylparaben in the blank injection 2 does not exceed 2 per cent of the peak area of the propylparaben peak in the chromatogram obtained with reference solution (b) injected after the blank injection 2, this corresponds to 0.2 per cent.

14.3.3 Linearity and Accuracy

Injector linearity is determined by injecting five different injection volumes (Inj vol 1 to Inj vol 5) of a caffeine standard solution at an appropriate concentration (for example 0.5 mg/ml water for chromatography R) from the same vial. Appropriate volumes are chosen with maximum injectable volume as upper point. Injector accuracy can be inferred by dividing the y-intercept (Area) by the slope (Area/volume) of the linearity plot.

Solutions: Blank: water for chromatography R

Reference solution: caffeine standard solution in water for chromatography R

Settings: Column: suitable column or capillary (*e.g.* RP 18)

Mobile phase: acetonitrile R: water for chromatography R (15:85 V/V)

Oven temperature: 40 °C

Flow rate:

- ☆ HPLC 1.0 mL/min
- ☆ UHPLC 0.5 mL/min (or appropriate flow rate based on instrument specifications/operational use)

Detection: 273 nm

Procedure: Injection scheme: 1 x Inj vol 1 µL blank

1 x Inj vol 1 µL reference solution

1 x Inj vol 2 µL reference solution

1 x Inj vol 3 µL reference solution

1 x Inj vol 4 µL reference solution

1 x Inj vol 5 µL reference solution

14.3.4 Thermostating

Accuracy Materials: Calibrated (electronic) thermometer with a suitable probe.

Procedure: Select a temperature along the operational or required temperature range of the equipment (*e.g.* the lowest point can be checked, for example + 5° C, or typical temperature used). Wait until the system is equilibrated. By means of the calibrated thermometer, measure the actual temperature in the autosampler and compare it to the selected temperature. As an alternative procedure, fill a vial with water, wait until equilibration and measure the temperature of the water by using an appropriate probe. If applicable, repeat the measure from several points of the autosampler (mapping) following the instructions provided by the manufacturer.

Oven/Cooling Device

Thermostating accuracy and stability are the parameters tested in this example of periodic and motivated check of the oven/cooling device.

Thermostating Accuracy and Stability

Materials: Calibrated thermometer.

Procedure: 1) Set the column oven temperature to 40 °C, wait about 30 minutes to equilibrate the system, put a calibrated thermometer into the oven and take 6 temperature readings at 4 minutes' intervals. 2) If applicable, repeat this procedure setting the oven/cooler at other operational temperatures based on the working range of temperature of the equipment.

The accuracy is calculated as average of 6 temperature readings. The stability is verified using the results obtained in the accuracy test. The stability is calculated for each temperature as the difference between the highest and the lowest temperature of the six readings.

14.3.5 Multi-Wave Length Detector

The periodic and motivated check of the LC UV/visible and DAD detectors can be performed by testing the linearity, wavelength accuracy and drift. Linearity Solutions:

Std. 1: 0.5 µg/mL caffeine

Std. 2: 1.0 µg/mL caffeine

Std. 3: 5.0 µg/mL caffeine

Std. 4: 25.0 µg/mL caffeine

Std. 5: 50.0 µg/mL caffeine

Std. 6: water for chromatography R (blank)

For example, prepare a stock solution (Std. 0) by weighing about 10.0 mg of caffeine and fill up to 20.0 mL with water for chromatography R (dissolve using sonication or mix). Dilute this solution, for example following the subsequent scheme:

Std. 5: dilute 10.0 mL of stock solution to 100.0 mL

Std. 4: dilute 5.0 mL of stock solution to 100.0 mL

Std. 3: dilute 1.0 mL of stock solution to 100.0 mL

Std. 2: dilute 10.0 mL of Std. 3 to 50.0 mL

Std. 1: dilute 10.0 mL of Std. 3 to 100.0 mL Use water for chromatography R to dilute

Settings: Column: suitable column or capillary (*e.g.* RP 18)

Mobile phase: acetonitrile R: water for chromatography R (15:85 V/V)

Oven temperature: 40 °C

Flow rate:

- ✰ HPLC 1.0 mL/min
- ✰ UHPLC 0.5 mL/min (or appropriate flow rate based on instrument specifications/operational use)

Detection: 273 nm Injection volume:

- ✰ HPLC ("short pathway" flow-cell, for example 5 or 10 mm): 20 µL
- ✰ UHPLC ("short pathway" flow-cell, for example 5 or 10 mm): 10 µL
- ✰ HPLC/UPLC ("long pathway" flow-cell, for example 60 or 85 mm): 2 µL (or appropriate volume based on instrument specifications/operational use).

Procedure:

Injection scheme:

2 x blank

1 x Std. 1

1 x Std. 2

1 x Std. 3

1 x Std. 4

1 x Std. 5

Remark: As this test employs different test solutions to be injected, it covers also the check of correct positioning of vials in the autosampler.

14.3.6 Wavelength Accuracy

Wavelength accuracy (and related adjustment) can be carried out by built-in test procedures, following instructions of the instrument manual/manufacturer. In all other cases use the procedure described below.

Solutions: Caffeine Std. 5 from the linearity testing

Settings: Use the operational conditions described under linearity testing except for detection. Procedure: Inject the caffeine solution and record the spectrum by scanning from 190 to 290 nm (1 nm incremental steps, whenever possible). The maxima are at 205 nm and 273 nm, the minimum at 245 nm.

Drift Typically drift verification is carried out by built-in test procedures. Follow the instructions of the instrument manual/manufacturer.

14.3.7 Fluorescence Detector

The following three parameters are proposed for the performance of the periodic and motivated check of the LC fluorescence detector:

Wavelength Accuracy for Excitation

Method: Rinse and fill the measuring cell with de-ionized water Adjust the excitation wavelength to 350 nm and emission wavelength to 397 nm. Measure the excitation spectrum. Wavelength Accuracy for Emission

Method: Rinse and fill the measuring cell with de-ionized water Adjust the excitation wavelength to 350 nm and emission wavelength to 397 nm. Measure the emission spectrum.

14.3.8 Signal to Noise Ratio

Method

Rinse and fill the measuring cell with de-ionized water. Set-up the excitation wavelength at 350 nm and the emission wavelength at 397 nm. The signal is acquired over a total of 23 minutes. At 20.30 minutes the emission wavelength is switched to 450 nm. The signal is measured at the maximum wavelength of Raman band (*i.e.* 397 nm for 350 nm excitation) and the noise in a region (450 nm) where no Raman signal is present. Signal-to-noise is calculated dividing the height of the Raman band divided by the noise.

14.3.9 Electrochemical Detector

There are several types of electrochemical detectors available from different manufacturers, based on amperometry, coulometry or conductivity method of detection.

Amperometric detector

Drift and Noise

Settings: Cell potential of a dummy cell: 800 mV

Rise time filter: 0.1 s

Range: 0.1 nA

Temperature: 30 °C

Procedure: Drift: Measure the electric current and subtract 2.67 nA (theoretical value) Noise: Measure the noise over a period of 5 minutes.

Integrated Amperometric Detector

Drift and Noise Noise: Read the signal every minute.

Settings: Pressure: ≥1000 psi

Flow: 1.0 mL/min

Restriction capillary: PEEK, 150 cm x 0.13 mm

Column: CarboPac PA1 Guard

Eluent: 50 mM NaOH, filtered through 0.2 µm Nylon, degassed

Detector settings: Waveform "Amino Acids" (pH, Ag, AgCl Reference)

Run time: 20 min

Linearity Solutions:

1, 2, 4, 8, and 10 µM of Threonine in water for chromatography R

Settings:

Pressure: ≥1500 psi

Flow: 0.25 mL/min

Restriction capillary: PEEK, 150 cm x 0.13 mm

Column: CarboPac PA1 Guard 2 mm

Eluent: 50 mM NaOH, filtered through 0.2 µm Nylon, degassed

Detector settings: Waveform "Amino Acids" (pH, Ag, AgCl Reference)

Duration/run time: 3 min

Injection volume 25µL

Conductivity detector

Drift and Noise Pure water is pumped through the detector cell at 1.0 mL/min. Measure the baseline noise over an appropriate period.

Linearity Solutions: 5, 10, 25, 50 and 100 ppm solutions of Nitrate in water for chromatography R.

Settings: Column: Backpressure loop

Eluent: Deionized water

Flow rate: 1.0 mL/min

14.3.10 Refractive Index Detector

Signal to Noise ratio, drift over time and linearity are the parameters proposed for the periodic and motivated check of the RI (refractive index) detector.

Signal to Noise Ratio

Method 1

Solutions: D-fructose solution at 4.0 mg/mL (*e.g.* dilute 200.0 mg fructose in 20 mL with water for chromatography R, add 25 mL acetonitrile R and dilute to 50.0 mL with water for chromatography R)

Settings: Column: Spherisorb NH2, 5 µm, 250 x 4.6 mm or equivalent

Oven temperature: 38 °C

Cell temperature: 35 °C

Flow rate: 1.0 mL/min

Injection volume: 20 µL

Mobile phase: 0.253g sodium dihydrogen phosphate R in 220 mL of water for chromatography R and 780 mL of acetonitrile R

Procedure: After equilibration, inject three times a blank solution of mobile phase over a run time where the system is stable. Measure the baseline noise over an appropriate period. The baseline noise is accepted if the mean height of the three replicates is $<$ 1000 µV.

To calculate the signal to noise ratio, inject three times a solution of fructose at 0.4 mg/mL and calculate the mean of the three replicates.

Drift Over Time

Procedure: Calculate the slope of the signal within the specified range within a defined time range (*e.g.* 60 minutes). Alternatively, the limits can be expressed in ΔRI/min or in per cent of full scale of the selected range, following manufacturer methods/specifications.

Linearity

Method

Solutions: Appropriately dilute the stock solution of D-fructose at concentration of 4.0 mg/mL to obtain minimum 5 diluted solutions in a range from 0.4 to 2.0 mg/mL, for example: Std. 1: 2.0 mg/mL D-fructose

Std. 2: 1.6 mg/mL D-fructose

Std. 3: 1.2 mg/mL D-fructose

Std. 4: 0.8 mg/mL D-fructose

Std. 5: 0.4 mg/mL D-fructose

Use water for chromatography R to dilute.

Settings: Apply the same conditions as described for signal-to-noise ratio testing

Procedure: After equilibration, inject each diluted solution once (starting from the more diluted) and the blank. Integrate the peak corresponding to D-fructose and calculate the linearity of response by linear regression mode.

14.3.11 Circular Dichroism Detector

The following tests are proposed to perform the periodic and motivated check of CD (circular dichroism) detector.

Linearity and Signal to Noise Ratio

Solutions: Reference solution (a): dissolve 25.0 mg D (-) pantolactone in 50.0 mL of water

Reference solution (b): dilute 2.0 mL of reference solution (a) to 10.0 mL

Reference solution (c): dilute 4.0 mL of reference solution (a) to 10.0 mL

Reference solution (d): dilute 6.0 mL of reference solution (a) to 10.0 mL

Reference solution (e): dilute 8.0 mL of reference solution (a) to 10.0 mL

Reference solution (f): dilute 0.5 mL of reference solution (b) to 25.0 mL

Use water for chromatography R to dilute.

Settings: Column: C18, 5 µm, 150 x 4.0 mm

Mobile phase: acetonitrile R: water for chromatography R (10:90 V/V) Flow: 1.0 mL/min Detection: 225 nm I

Injection volume: 20 µL

Procedure: Check the linearity of CD- and UV-signal of D(-) pantolactone reference solution (a), (b), (c), (d), (e).

Measure the noise of the CD-signal of reference solution (f) between 0 – 10 min.

- Calculate the absolute amount (µg) in the cell
- Calculate the signal-to-noise ratio (S/N) for 0.01 µg in the cell

- Calculate the sensitivity with the calculated S/N and the specified S/N= 2 (0.01 x 2/S/N calculated).

Drift Over Time

Settings: Column: C18, 5 µm, 150 x 4.0 mm

Mobile phase: acetonitrile R: water for chromatography R (10:90 V/V)

Flow: 1.0 mL/min

Detection: 290 nm

Injection volume: 20 µL

Procedure: Inject water and stop the flow after 5 minutes. Measure the CD-signal for 1 hour. Measure with the cursor the drift of the baseline between 5 and 65 min.

Spectra Comparison Solutions:

Reference solution (a): dissolve 5.0 mg dexamethasone in 10.0 mL of mobile phase

Settings: Column: C18, 5 µm, 150 x 4.0 mm

Mobile phase: acetonitrile R: water for chromatography R (40:60 V/V)

Flow: 1.0 mL/min

Detection: 230 nm

Injection volume: 20 µL

Procedure: Inject water and stop the flow after 5 minutes. Measure the CD-signal for 1 hour. Measure with the cursor the drift of the baseline between 5 and 65 min.

Spectra Comparison Solutions: Reference solution (a): dissolve 5.0 mg dexamethasone in 10.0 mL of mobile phase Settings: Column: C18, 5 µm, 150 x 4.0 mm Mobile phase: acetonitrile R: water for chromatography R (40:60 V/V) Flow: 1.0 mL/min Detection: 230 nm Injection volume: 20 µL Procedure: Compare the maxima/minima obtained at the Installation of the detector (see table 14.4)

CD max	CD min	UV max
222 nm	224 nm	236 nm
230 nm	252 nm	
284 nm		

14.4 GC Performance Qualification Tests

US Pharmacopoeia (USP) general chapter <1058> on Analytical Instrument qualification (AIQ) was first implemented in 2008 and remained unchanged for nine years. During 2017, the USP implemented two updates to <1058>.

These updates have a significant impact on AIQ, and as the only major pharmacopeia with a chapter dedicated to AIQ, changes to USP <1058> are of global significance.

	Parameter to be checked	*Typical tolerance limit*
1. Inlet System	1.1 Injector leak test	Pressure drops <1.5 Kpa within 5 minutes.
	1.2 Pressure/flow accuracy and stability	Covered by overall test 1
	1.3 Repeatability of GC injections (overall tests 1)	
	Peal area: - In split mode - In splitless mode	RSD≤3.0 per cent RSD≤3.0 per cent
	1.4 Retention times:- Injector temperature accuracy and stability 1.5 Carryover (Covered by overall test 3)	Covered by overall test 2
	Head space injector 1.2 Repeatability of head space injector - Peak area -Retention time -Vial heater temperature	RSD≤5.0 per cent RSD≤2.0 per cent± 4°C
2. Oven	2.1 Repeatability of oven temperature characteristics	Covered by overall test 2
3. FID detector	3.1 Linearity (Covered by overall test 3) 3.2 Costant detector response 3.3 Noise 3.4 Drift	Covered by overall test 1 and 2 Plz see section– Plz see section –

Several tests are proposed to check various parameters at the same level (overall tests). In order to run the tests in a more commercial way, other suitable solutions can be used as for example "grob test" mixture available from different suppliers. This commercial should be appropriate to the column material used.

14.5 Inlet System

The following tests are proposed for the periodic and motivated check of the GC Inlet System.

14.5.1. Injector Leak Test

Method: If not otherwise specified by the instrument manufacturer, the leak test is carried out according to the procedure laid down in the instrument manual or by the built in automatic leak check procedure of the instrument.

- ☆ Otherwise use the test described below: Disconnect the column from the injector and close the injector outlet with a sealed cap.
- ☆ Close the septum purge and the bypass.
- ☆ Adjust the flow and pressure controller to the maximal possible value of the pressure gauge. Adjust the flow controller to zero.
- ☆ Read the pressure after 1 minute and record the value.
- ☆ Record the pressure after 5 minutes

Limits: Pressure drop ≤ 15 kPa within 5 minutes.

14.5.1. Inlet Pressure/Flow Accuracy and Stability

A direct measurement of these parameters was not deemed practical or necessary, but the optimal conditions of flow/pressure can be verified by the overall test 1.

Limits: refer to overall test 1.

14.5.1. Repeatability of GC Injection

The verification of this parameter is covered by the overall test 1. This test is to be performed in both split and split less mode.

Limits: refer to overall test 1.

14.5.3. Injector Temperature Accuracy and Stability

Due to the fact that the temperature cannot be reliably measured without opening and modifying the system and due to the difficulties of introducing a probe inside this module, the verification of this parameter is considered to be covered by the overall test 2. Limits: refer to overall test 2

14.5.1. Injector Carry Over

After having injected the solutions for the linearity test of the FID detector, in increasing order, inject the blank and measure the peaks that correspond to the major peaks (= analytes) in the linearity solutions. The verification of this parameter is covered by the overall test 3. Limits: refer to overall test 3

14.5.5. Head Space Injectors

Repeatability of Headspace Injections

Method: The GC-HS operating conditions below are provided as an example, adjustments may be needed depending on the equipment used.

Test solution: 0.5 per cent ethanol in water R (V/V) GC-Settings:

Column: 95 per cent Dimethyl/5 per cent diphenylpoly siloxane 30 m x 0.25 µm; 0.25 mm (HP-5 was found suitable)

Carrier gas: Helium Column

Flow: 1.2 ml/min

Injector temperature: 200 °C

Split ratio: 1:50

Oven temperature: 40 °C isotherm

Detector temperature: 250 °C

Run time: 1.5 folds the retention time of the main peak

Retention time: about 2.2 min

Headspace-Settings:

Carrier pressure: 9.9 psi

Vial pressure: 14.2 psi

Shake: low

Oven Temperature: 80 °C

Loop Temperature: 90 °C T

ransfer line Temperature: 100 °C

Vial size: 20 mL

Volume of sample solution/vial: 5 mL

Vial Equilibration Time: 15.0 min

Inject Time: 1.0 min

Loop Equilibration Time: 0.1 min

Loop Fill Time: 0.5 min

Vial Pressurization Time: 0.08 min

Carry out 6 consecutive injections of the test solution and calculate the RSD of the different peak areas and retention times.

Limits: Peak areas: the RSD should be ≤ 5.0 per cent

Retention time: the RSD should be ≤ 2.0 per cent

14.5.6 Vial Heater Temperature

The heater temperatures are to be set up at values which depend on the operating conditions of the methods applied.

Suitable calibrated temperature devices are to be used.

Put the calibrated device in the oven of the head space compartment. Set the temperature at the required values. When equilibration is achieved, record the value displayed on the calibrated device.

Limits: ± 4 °C from set point

14.5.7 Oven

Repeatability of the Oven Temperature Characteristics

Due to the fact that the temperature cannot be reliably measured without opening and modifying the system conditions and that even when introducing a probe inside the oven, its location would not reflect the real temperature conditions at all points, the verification of this parameter is covered by the overall tests 2A and 2B. Limits: refer to overall test 2

14.5.8 FID Detector

The following tests are proposed for the periodic and motivated check of the GC FID detector.

FID Detector Linearity

Increasing amounts of analyte are injected and a linear response should be obtained. The verification of this parameter is covered by the overall test 3.

Limits: refer to overall test 3. 3.2.

Constant FID Detector Response

The proper and reproducible functioning of the FID can be demonstrated by checking the peak areas obtained from a pre-defined standard solution. The verification of this parameter is covered by the overall test 1 or 2. Limits: refer to overall test 1 or 2.

14.5.9 FID Detector Noise and Drift

If the instrument has a built-in automatic system for the verification of the noise and drift, follow the manufacturer's instructions and apply the defined acceptance criteria. Otherwise, use the test described below:

Settings: Column installed

Suitable flow, depending on column length/diameter

No injection

Oven temperature: 40 °C

Detector on and heated at working temperature (270-300 °C)

Method: After stabilisation of the system, record the signal for 15 minutes.

Noise: evaluate 10 periods of 1 minute and calculate the mean value.

Drift: evaluate the slope of the baseline over the 15 minutes.

Limits: The acceptance criteria for these parameters have to be chosen in accordance with the instrument vendor's instructions and the intended use of the instrument. If no instructions are given, the user has to pre-define these acceptance criteria by taking into account the previous experience and the intended use of the instrument. No fixed values can be pre-defined in this guideline due to the high variety of integration systems used and consequently the acceptance criteria may be expressed in different units (voltage, current, arbitrary units per time).

Overall Test 1

The overall test 1 covers the following parameters:

- ✰ Pressure/flow accuracy and stability in the inlet system: Retention time repeatability
- ✰ Repeatability of injection: peak area precision
- ✰ In split mode
- ✰ In split less mode

The test may be combined with overall test 3.

Split mode:

Test solution: 1-octanol in n-hexane 1 per cent (V/V).

Settings: Column: 100 per cent Dimethylpolysiloxane 30m x 0.32mm ID x 0.25µm film (SPB-1 was found suitable)

Carrier gas: Helium

Velocity: 25cm/sec

Split: 1:100

Injection: 1µl

Injector temperature: 220 °C

Oven temperature: 100 °C isotherm

Detector temperature: 300 °C

Run time: 1.5 folds the retention time of the main peak

Retention time of 1-octanol: about 5 min

Split less mode:

Stock solution: 1-octanol in n-hexane 1 per cent (V/V)

Test solution: Dilute 10 ml of the stock solution with n-hexane to 100 ml (corresponds to 1µl/ml of 1-octanol in n-hexane)

Settings:

Column: 100 per cent Dimethylpolysiloxane 30m, 0.32mm ID, 0.25µm film (SPB-1 was found suitable)

Carrier: Helium

Velocity: 30cm/sec

Split less injection: purge valve closed during 2 min

Injection: 0.2µl of the test solution

Injector Temperature: 220 °C

Oven Temperature: Initial 60 °C for 4 min, 15 °C/min. up to 135 °C, final timc 1min

Detector temperature: 300 °C

Runtime: 1.5 folds the retention time of the main peak

Retention time of 1-octanol: about 8 min

Method: Carry out 6 consecutive injections of the test solution and calculate the RSD of the different peak areas and retention times.

Limits::

Retention time repeatability: the RSD of the retention times should be ≤ 2.0 per cent

Peak area precision (split and split-less mode): the RSD of the peak areas should be ≤ 3.0 per cent

Overall Test 2

The overall test 2 covers the following parameters:

- Injector, oven and detector temperature accuracy and stability: retention time repeatability

Two alternative tests are proposed:

Overall Test 2A

Test solution: 0.035 ml 1-octanol

0.035 ml 2-octanone

0.035 ml 2,6-dimethylanilin

0.035 ml n-tridecane

0.035 ml n-tetradecane

35 mg n-eicosane dissolved in 50 ml Dichloromethane

Settings: Column: 100 per cent Dimethylpolysiloxane 30m x 0.32mm ID x 0.25µm film (SPB-1 was found suitable)

Carrier gas: Helium

Velocity: 25 cm/s

Split: 1:100 Injection

volume: 1 µl

Injector temperature: 220 °C

Detector: FID Detector temperature: 300 °C

Gradient programme: 60 °C (4 min), 5 °C/min, 270 °C (3 min)

Method: Inject the solution twice and calculate the relative retention (RR) times in relation to n-eicosane (RR = 1)

The following table shows the approximately expected relative retention times.

Analyte	1-octanol	2-octanone	2,6-dimethylaniline	n-tridecane	ntetradecane
RRT	0.30	0.22	0.37	0.52	0.60

Limits: The RSD of each RR from two consecutive injections should be ≤ 1.0 per cent

Overall Test 2B

Test Solution: 1.0 per cent (m/m) n-Nonane and Hexadecane in Tetradecane.

Settings: Column: 100 per cent Dimethylpolysiloxane 25m x 0.32mm ID x 0.52µm film (Ultra-1 was found suitable)

Injection volume: 1 µl

Solvent: Tetradecane

Oven temperature: 110 °C

Gradient programme: 110 °C, 20 °C/min, 180 °C (final time: 3.5 min)

Detector temperature: 250 °C Injector temperature: 200 °C Detector:

FID Flow rates: as defined by the instrument manufacturer

Split ratio: 1:15

Split vent: 30 ± 3.0 ml/min

Septum purge: 3-5 ml/min

Method: Allow the system to equilibrate. Injection sequence:

1. Blank (Tetradecane)
2. 6 replicates of the test solution. Calculate the mean of the retention times and peak areas and the relative standard deviation of n-Nonane and n-Hexadecane.

Limits: Retention time repeatability: RSD of the peak retention times of the 6 replicates ≤ 2.0 per cent Retention time (Rt) accuracy: for this example, the retention time ranges shown in the table below are proposed. Nevertheless, individual ranges should be predefined by the laboratory depending on the column used (*e.g.* Rt ± 0.2 min).

Compound	Rt (min)
n-Nonane (C9)	1.3 – 1.7
Tetradecane (C14)	4.0 – 4.7
Tetradecane (C14)	4.0 – 4.7

Overall Test 3

This test is a modified version of the overall test 1 to be used for the verification of:

1. Detector linearity: linearity of the areas recorded
2. Injector carry-over: area recorded in the blank run

It is described for both split and split less mode and may be combined with overall test 1.

Split mode:

Test solution: 1-octanol in n-hexane 1 per cent (V/V)

Prepare further reference solutions by diluting the test solution as described below.

Settings: see overall test 1

Injection sequence: 5.0 ml of the test solution diluted to 25.0 ml with n-hexane (2 µl/ml): 2 injections

10.0 ml of the test solution diluted to 25.0 ml with n-hexane (4 µl/ml): 2 injections

15.0 ml of the test solution diluted to 25.0 ml with n-hexane (6 µl/ml): 2 injections

20.0 ml of the test solution diluted to 25.0 ml with n-hexane (8 µl/ml): 2 injections

if combined with overall test 1 for repeatability: test solution (10 µl/ml): 6 injections n-hexane as blank (carry over)

Split-less mode:

Stock solution: 1-octanol in n-hexane 1 per cent (V/V)

Test solution: Dilute 10 ml of the stock solution with n-hexane to 100 ml (corresponds to 1µl/ml of 1-octanol in n-hexane).

Prepare further reference solutions by diluting the test solution with n-hexane.

Settings: see overall test 1

Injection sequence: 5.0 ml of the test solution diluted to 25.0 ml with n-hexane (0.2 µl/ml): 2 injections

10.0 ml of the test solution diluted to 25.0 ml with n-hexane (0.4 µl/ml): 2 injections

15.0 ml of the test solution diluted to 25.0 ml with n-hexane (0.6 µl/ml): 2 injections

20.0 ml of the test solution diluted to 25.0 ml with n-hexane (0.8 µl/ml): 2 injections

if combined with overall test 1 for repeatability: test solution (1 µl/ml): 6 injections n-hexane as blank (carry over)

Limits:

Linearity: coefficient of correlation of the calibration line obtained with the reference solutions and the test solution: $r2 \geq 0.999$.

Carry-over: the percentage of the peak area corresponding to the analyte in the blank solution should be ≤ 0.2 per cent of the peak area of this analyte in the chromatogram obtained with the solution with the highest concentration within the sequence.

Chapter 15

Troubleshooting

15.1 Introduction

Troubleshooting is s systematic approach to problem-solving that is often used to find and correct issues with complex machines, electronics computer and software systems. The first step in most troubleshooting method is gathering information on the issue, such as an undesired behaviour or a lack of unexpected functionality.

Once the issue and how to reproduce it are understood, the next step is to eliminate unnecessary components to determine if the issue persists. This can help identify component incompatibility issue and the problem caused by third party products. Trouble-shooters look for common known issues for example retention variability.

15.2 Isolating HPLC Problems

In an HPLC system, problems can arise from many sources. First define the problem, then isolate the source. Use Table 15.1 to determine which component(s) may be causing the trouble. A process of elimination will usually enable you to pinpoint the specific cause and correct the problem.

Table 15.1

Problem	*Probable Cause*	*Remedy/Comments*
Problem No. 1: No Peaks/Very Small Peaks		
No problem 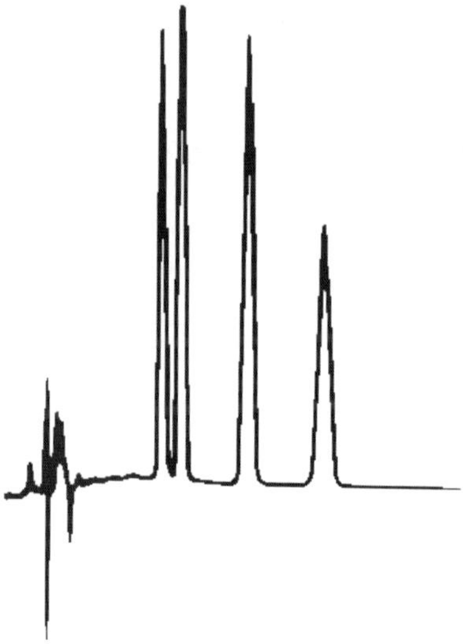	1. Detector lamp off. 2. Loose/broken wire between detector and integrator or recorder. 3. No mobile phase flow. 4. No sample/ deteriorated sample/wrong sample. 5. Settings too high on detector or recorder.	1. Turn lamp on. 2. Check electrical connections and cables. 3. See "No Flow" (Problem No. 2). 4. Be sure automatic sampler vials have sufficient liquid and no air bubbles in the sample. Evaluate system performance with fresh standard to confirm sample as source of problem. 5. Check attenuation or gain settings. Check lamp status. Auto-zero if necessary.
Problem 		
Problem 		

Problem	*Probable Cause*	*Remedy/Comments*
Problem No. 2: No Flow		
Normal 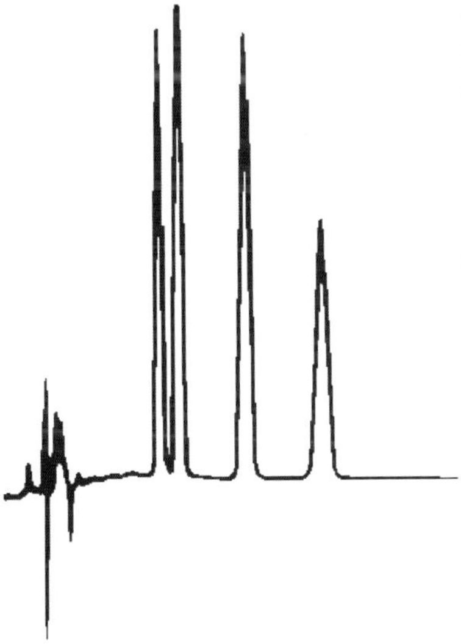	1. Pump off. 2. Flow interrupted/ obstructed. 3. Leak. 4. Air trapped in pump head. (Revealed by pressure fluctuations.)	1. Start pump. 1. Check mobile phase level in reservoir(s). Check flow throughout system. Examine sample loop for obstruction or air lock. Make sure mobile phase components are miscible and mobile phase is properly degassed. 3. Check system for loose fittings. Check pump for leaks, salt build-up, unusual noises. Change pump seals if necessary. 4. Disconnect tubing at guard column (if present) or analytical column inlet. Check for flow. Purge pump at high flow rate (*e.g.*, 5-10 mL/ min.), prime system if necessary. (Prime each pump head separately.) If system has check valve, loosen valve to allow air to escape. If problem persists, flush system with 100 per cent methanol or isopropanol. If problem still persists, contact system manufacturer.
Problem 		

Problem	Probable Cause	Remedy/Comments
Problem No. 3: No Pressure/Pressure Lower Than Usual		
Normal 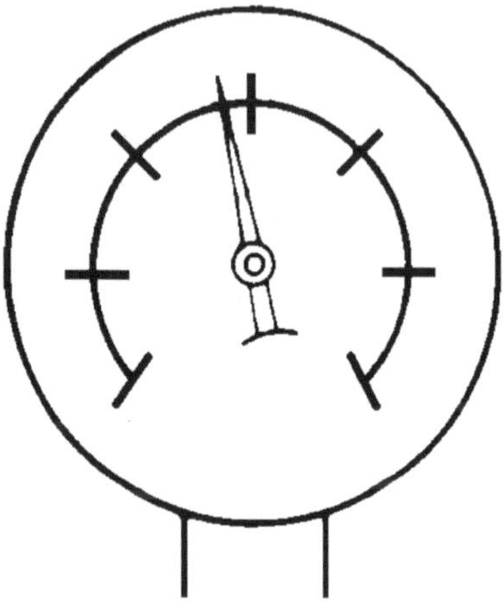	1. Leak. 2. Mobile phase flow interrupted/ obstructed. 3. Air trapped in pump head. (Revealed by pressure fluctuations). 4. Leak at column inlet end fitting. 5. Air trapped elsewhere in system. 6. Worn pump seal causing leaks around pump head. 7. Faulty check valve. 8. Faulty pump seals.	1. Check system for loose fittings. Check pump for leaks, salt build-up, unusual noises. Change pump seals if necessary2. Check mobile phase level in reservoir(s). Check flow throughout system. Examine sample loop for obstruction or air lock. Make sure mobile phase components are miscible and mobile phase is properly degassed. 3. Disconnect tubing at guard column (if present) or analytical column inlet. Check for flow. Purge pump at high flow rate (*e.g.*, 10 mL/ min.), prime system if necessary. (Prime each pump head separately.) If system has check valve, loosen valve to allow air to escape. 4t. Reconnect column and pump solvent at double the flow rate. If pressure is still low, check for leaks at inlet fitting or column end fitting. 5. Disconnect guard and analytical column and purge system. Reconnect column(s). If problem persists, flush system with 100 per cent methanol or isopropanol. 6. Replace seal. If problem persists, replace piston and seal. 7. Rebuild or replace valve. 8. Replace seals.

Problem	Probable Cause	Remedy/Comments
Problem 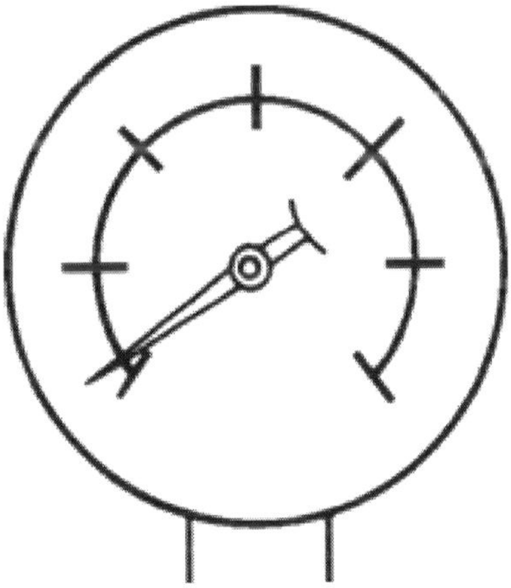		

Problem	Probable Cause	Remedy/Comments
Problem No. 4: Pressure Higher Than Usual		
Normal 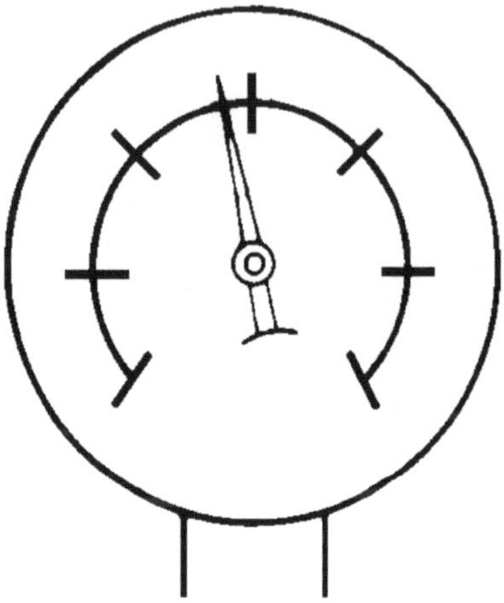	1. Problem in pump, injector, in line filter or tubing	1. Remove guard column and analytical column from system. Replace with unions and 0.010" I.D. or larger tubing to reconnect injector to detector. Run pump at 2-5 mL/min. If pressure is minimal, see Cause. If not, isolate cause by systematically eliminating system components, starting with detector, then in-line filter, and working back to pump. 2. Remove guard column (if present) and check pressure. Replace guard column if necessary. If analytical column is obstructed, reverse and flush the column, while disconnected from the detector (section 14.3). If problem persists, column may be clogged with strongly retained contaminants. Use appropriate restoration procedure. If problem still persists, change inlet frit (page 14.3) or replace column.
Problem 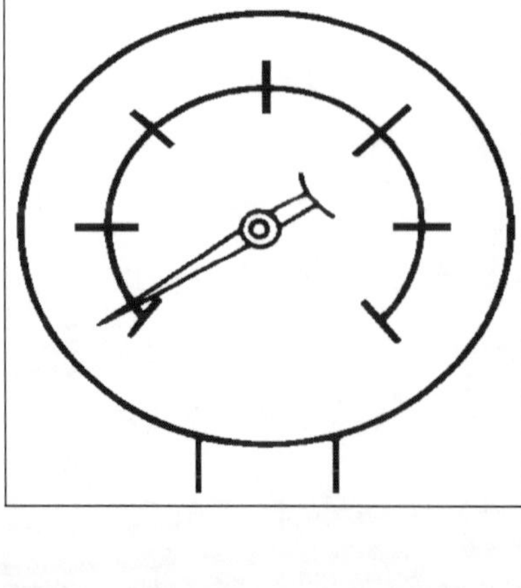		

Problem	Probable Cause	Remedy/Comments
Problem No. 5: Variable Retention Times		
No problem 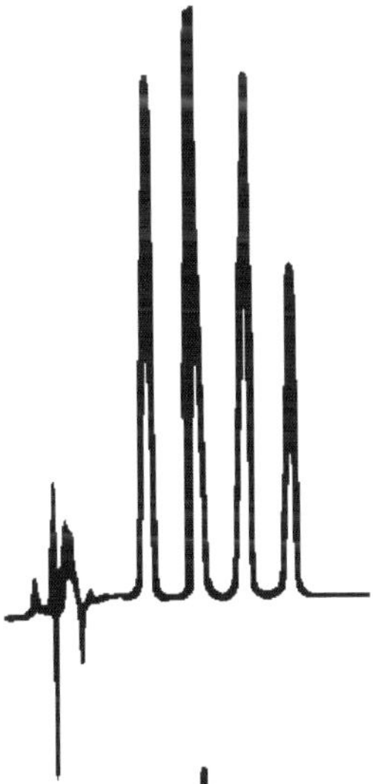 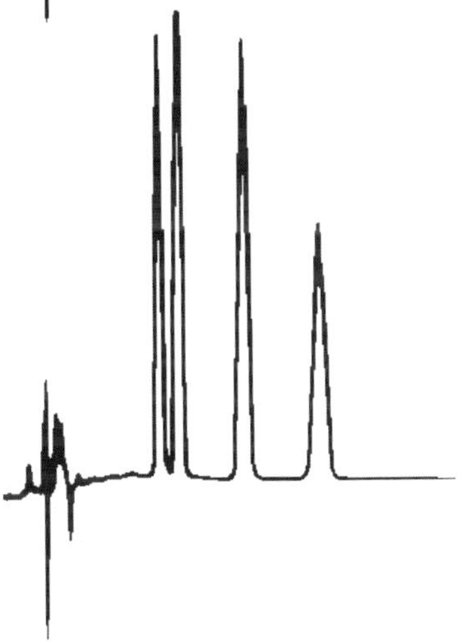	1. Leak. 2. Change in mobile phase composition. (Small changes can lead to large changes in retention times.) 3. Air trapped in pump. (Retention times increase and decrease at random times.) 4. Column temperature fluctuations (especially evident in ion exchange systems). 5. Column overloading. (Retention times usually decrease as mass of solute injected on column exceeds column capacity.)6. Sample solvent incompatible with mobile phase. 7. Column problem. (Not a common cause of erratic retention. As a column ages, retention times gradually decreases.	1. Check system for loose fittings. Check pump for leaks, salt build-up, and unusual noises. Change pump seals if necessary. 2. Check make-up of mobile phase. If mobile phase is machine mixed using proportioning values, hand mix and supply from one reservoir. 3. Purge air from pump head or check valves. Change pump seals if necessary. Be sure mobile phase is degassed. 4. Use reliable column oven. (Note: higher column temperatures increase column efficiency. For optimum results, heat eluant before introducing it onto column.)5. Inject smaller volume (*e.g.*, 10 µL vs. 100 µL) or inject the same volume after 1:10 or 1:100 dilutions of sample. 6. Adjust solvent. Whenever possible, inject samples in mobile phase. 7. Substitute new column of same type to confirm column as cause. Discard old column if restoration procedures fail

Problem	*Probable Cause*	*Remedy/Comments*
Problem 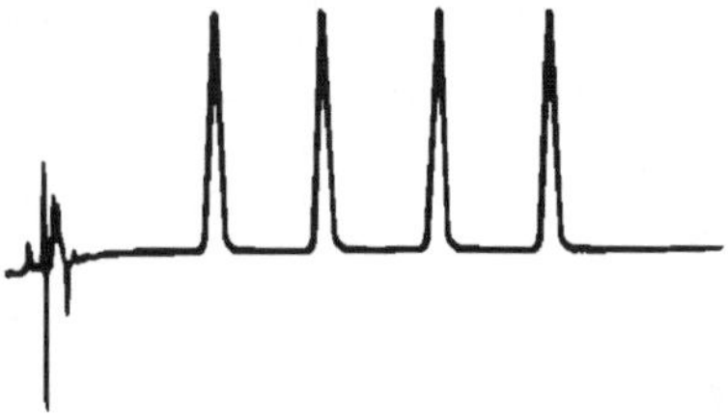		
Problem No. 6: Loss of Resolution		
No Problem 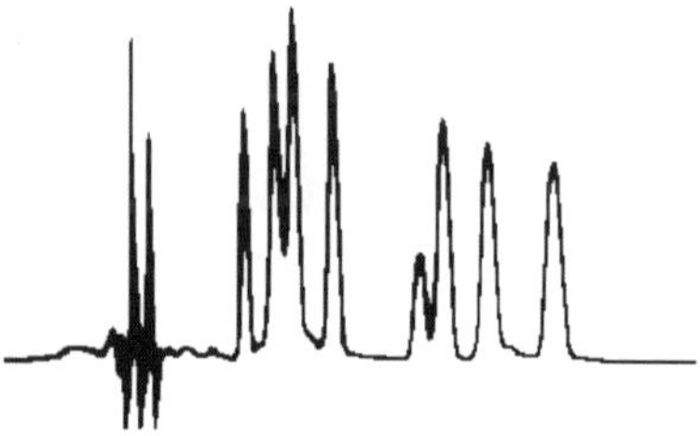	1. Mobile phase contaminated/ deteriorated (causing retention times and/or selectivity to change). 2. Obstructed guard or analytical column.	1. Prepare fresh mobile phase (page 2). 2. Remove guard column (if present) and attempt analysis. Replace guard column if necessary. If analytical column is obstructed, reverse and flush (section 14.3). If problem persists, column may be clogged with strongly retained contaminants. Use appropriate restoration procedure (Table 2, If problem still persists, change inlet frit or replace column.
Problem 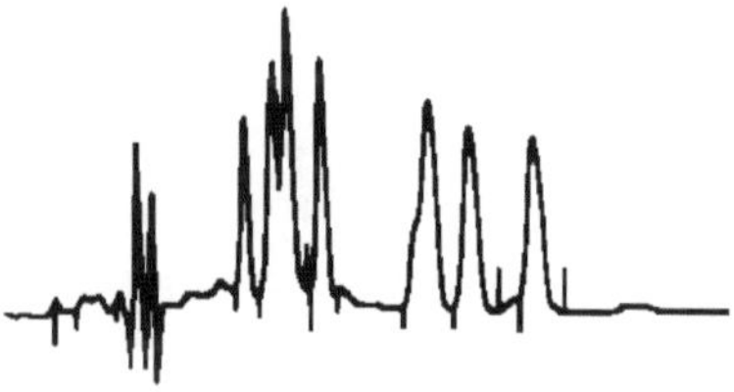		

Problem	Probable Cause	Remedy/Comments
Problem No. 7: Split Peaks		
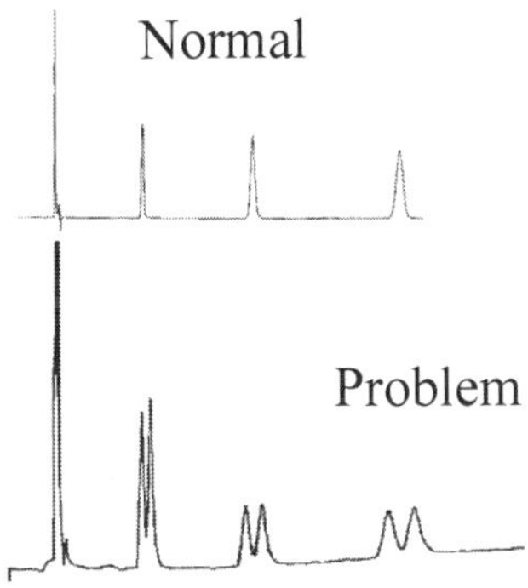	1. Contamination on guard or analytical column inlet 2. Partially blocked frit. 3. Small (uneven) void at column inlet. 4. Sample solvent incompatible with mobile phase	1. Remove guard column (if present) and attempt analysis. Replace guard column if necessary. If analytical column is obstructed, reverse and flush (page 14). If problem persists, column may be clogged with strongly retained contaminants. Use appropriate restoration procedure (Table 14.2). If problem still persists, inlet frit is probably (partially) plugged. Change frit (section 14.2) or replace column. 2. Replace frit (see above) 3. Repack top of column with pellicular particles of same bonded phase functionality. Continue using the column in reverse flow direction. 4. Adjust solvent. Whenever possible, inject samples in mobile phase.

<table>
<tr><th>Problem</th><th>Probable Cause</th><th>Remedy/Comments</th></tr>
<tr><td colspan="3">Problem No. 8: Peaks Tail on Initial and Later Injections</td></tr>
<tr><td>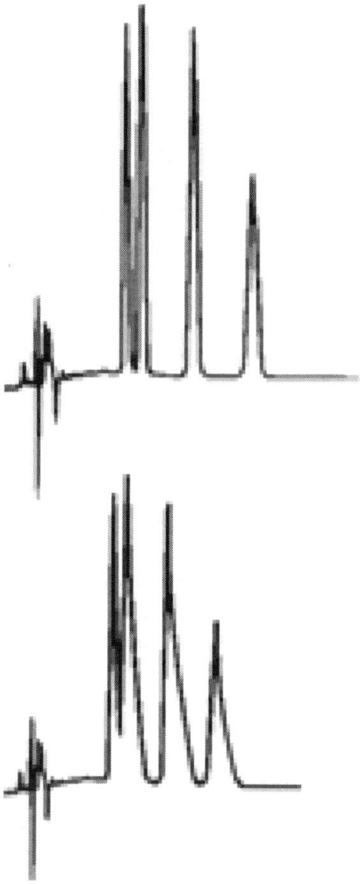</td><td>1. Sample reacting with active site.
2. Wrong mobile phase pH.
3. Wrong column type4. Small (uneven) void at column inlet.
5. Wrong injection solvent.</td><td>1. First check column performance with standard column test mixture. If results for test mix are good, add ion pair reagent or competing base or acid modifier.
2. Adjust pH for basic compounds, lower pH usually provides more symmetric peaks.
3. Try another column type (e.g., deactivated column for basic compounds).
5. Peaks can tail when sample is injected in stronger solvent than mobile phase. Dissolve sample in mobile phase.</td></tr>
</table>

Problem	Probable Cause	Remedy/Comments
Problem No. 9: Tailing Peaks		
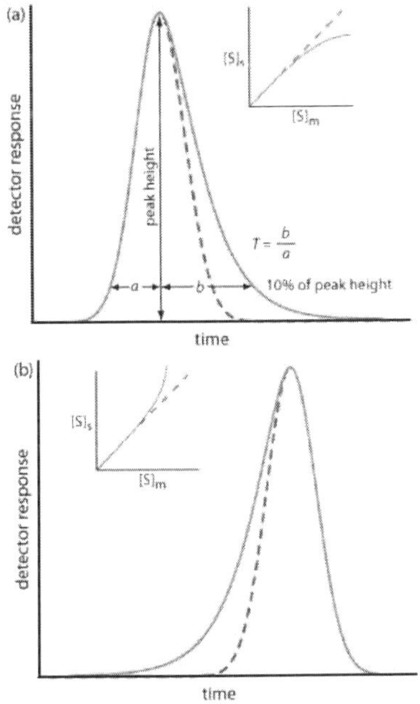	1. Guard or analytical column contaminated/ worn out. 2. Mobile phase contaminated/ deteriorated. 3. Interfering components in sample	Remove guard column (if present) and attempt analysis. Replace guard column if necessary. If analytical column is source of problem, use appropriate restoration procedure (Table 2. If problem persists, replace column Check make-up of mobile phase. Check column performance with standards.
Problem No. 10: Fronting Peaks		
No Problem 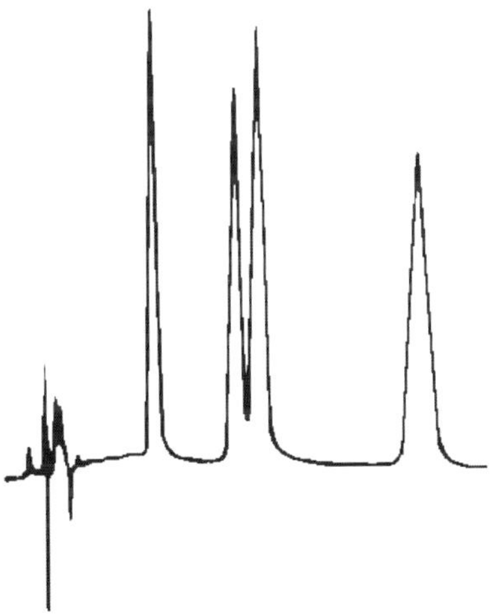	1. Column overloaded2. Sample solvent incompatible with mobile phase. 3. Shoulder or gradual baseline rise before a main peak may be another sample component.	1. Inject smaller volume (*e.g.*, 10 µL vs. 100 µL). Dilute the sample 1:10 or 1:100 fold in case of mass overload2. Adjust solvent. Whenever possible, inject samples in mobile phase. Flush polar bonded phase column with 50 column volumes HPLC grade ethyl acetate at 2-3 times the standard flow rate, then with intermediate polarity solvent prior to analysis3. Increase efficiency or change selectivity of system to improve resolution. Try another column type if necessary (*e.g.*, switch from nonpolar C18 to polar cyano phase).

Problem	*Probable Cause*	*Remedy/Comments*
Problem No. 11: Rounded Peaks		
No problem 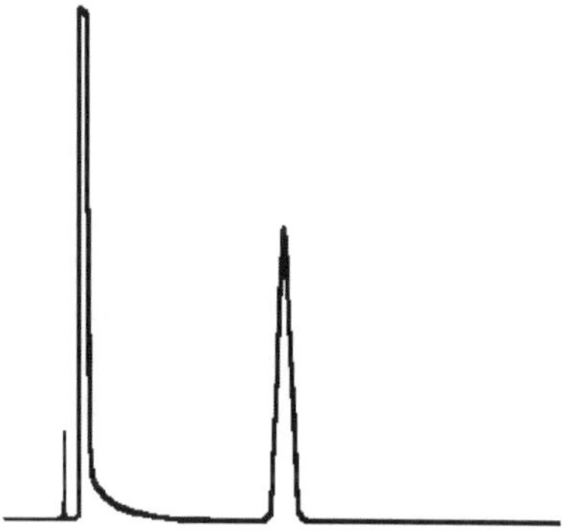	1. Detector operating outside linear dynamic range 2. Recorder gain set too low. 3. Column over-loaded. 4. Sample-column interaction 5. Detector and/or recorder time constants are set too high	1. Reduce sample volume and/or concentration. 2. Adjust gain. 3. Inject smaller volume (*e.g.*, 10 µL vs. 100 µL) or 1:10 or 1:100 dilution of sample. 4. Change buffer strength, pH, or mobile phase composition. If necessary, raise column temperature or change column type. (Analysis of solute structure may help predict interaction.) 5. Reduce settings to lowest values or values at which no further improvements are seen.
Problem 		

Problem	Probable Cause	Remedy/Comments
Problem No. 12: Baseline Drift		
Normal 	1. Column temperature fluctuation. (Even small changes cause cyclic baseline rise and fall. Most often affects refractive index and conductivity detectors, UV detectors at high sensitivity or in indirect photometric mode.) 2. Non-homogeneous mobile phase. (Drift usually to higher absorbance, rather than cyclic pattern from temperature fluctuation.) 3. Contaminant or air buildup in detector cell. 4. Plugged outlet line after detector. (High pressure cracks cell window, producing noisy baseline.) 5. Mobile phase mixing problem or change in flow rate. 6. Slow column equilibration, especially when changing mobile phase.	1. Control column and mobile phase temperature, use heat exchanger before detector. 2. Use HPLC grade solvents, high purity salts, and additives. Degas mobile phase before use, sparge with helium during use. 3. Flush cell with methanol or other strong solvent. If necessary, clean cell with 1N HNO_3 (never with HCl and never use nitric acid with PEEK tubing or fittings.) 4. Correct composition/ flow rate. To avoid problem, routinely monitor composition and flow rate. 5. Flush column with intermediate strength solvent, run 10-20 column volumes of new mobile phase through column before analysis. 6. Check make-up of mobile phase 7. Use guard column. If necessary, flush column with strong solvent between injections or periodically during analysis. 8. Change wavelength to UV absorbance maximum.

Problem	*Probable Cause*	*Remedy/Comments*
	7. Mobile phase contaminated, deteriorated, or not prepared from high quality chemicals. 8. Strongly retained materials in sample (high k') can elute as very broad peaks and appear to be a rising baseline. (Gradient analyses can aggravate problem.)9. Detector (UV) not set at absorbance maximum but at slope of curve.	
Not Normal 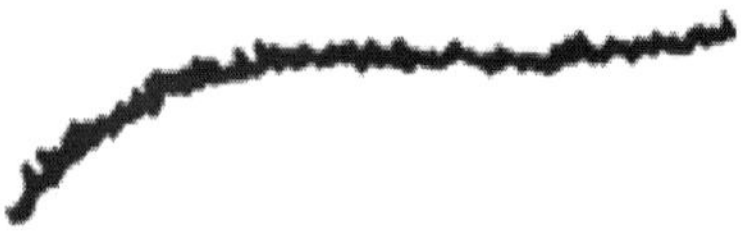		

<table>
<tr><th>Problem</th><th>Probable Cause</th><th>Remedy/Comments</th></tr>
<tr><td colspan="3">Problem No. 13: Baseline Noise (regular)</td></tr>
<tr><td>No problem
</td><td>1. Air in mobile phase, detector cell, or pump.
2. Pump pulsations.
3. Incomplete mobile phase mixing.
4. Temperature effect (column at high temperature, detector unheated).
5. Other electronic equipment on same line6. Leak.</td><td>1. Degas mobile phase. Flush system to remove air from detector cell or pump.
2. Incorporate pulse damper into system.
3. Mix mobile phase by hand or use less viscous solvent.
4. Reduce differential or add heat exchanger.
5. Isolate LC, detector, recorder to determine if source of problem is external. Correct as necessary.
6. Check system for loose fittings. Check pump for leaks, salt buildup, and unusual noises. Change pump seals if necessary.</td></tr>
<tr><td>Problem
</td><td></td><td></td></tr>
</table>

Problem	*Probable Cause*	*Remedy/Comments*
Problem No 14: Baseline Noise (irregular)		
No Problem 	1. Leak. 2. Mobile phase contaminated, deteriorated, or prepared from low quality materials. 3. Detector/ recorder electronics. 4. Air trapped in system. 5. Air bubbles in detector. 6. Detector cell contaminated. (Even small amounts of contaminants can cause noise.) 7. Weak detector lamp. 8. Column leaking silica or 9. Packing material.	1. Check system for loose fittings. Check pump for leaks, salt build-up, and unusual noises. Change pump seals if necessary. 2. Check make-up of mobile phase. 3. Isolate detector and recorder electronically. Refer to instruction manual to correct problem. 4. Flush system with strong solvent. 5. Purge detector. Install back- pressure regulator after detector. Check the instrument manual, particularly for RI detectors (excessive back pressure can cause the flow cell to crack). 6. Clean cell. 7. Replace lamp. 8. Replace column and clean system.
Problem 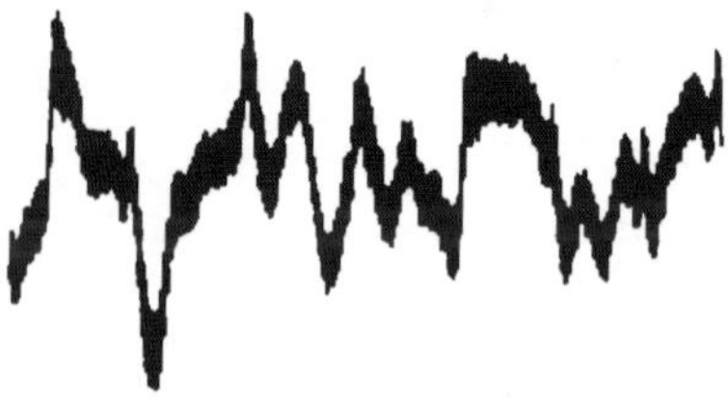		

Problem	Probable Cause	Remedy/Comments
Problem No. 15: Broad Peaks		
No Problem  Problem 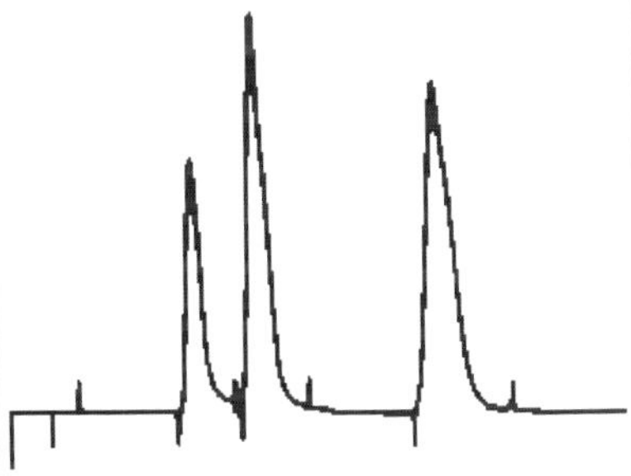	1. Mobile phase composition changed. 2. Mobile phase flow rate too low. 3. Leak (especially between column and detector). 4. Detector settings incorrect. 5. Extra-column effects: a. Column overloaded b. Detector response time or cell volume too large. c. Tubing between column and detector too long or I.D. too large. d. Recorder response time too high. 4. Buffer concentration too low. 5. Guard column contaminated/ worn out. 6. Column contaminated/ worn out. 7. Void at column inlet. 8. Peak represents two or more poorly resolved compounds. Column temperature too low	1. Prepare new mobile phase. 2. Adjust flow rate. 3. Check system for loose fittings. Check pump for leaks, salt buildup, and unusual noises. Change pump seals if necessary. 4. Adjust settings. a. Inject smaller volume (*e.g.*, 10 µL vs. 100 µL) or 1:10 and 1:100 dilutions of sample. b. Reduce response time or use smaller cell. c. Use as short a piece of 0.007-0.010" I.D. tubing as practical. d. Reduce response time. 5. Increase concentrations. 6. Replace guard column. 7. Replace column with new one of same type. If new column does not provide narrow peaks, flush old column (Table 2), then retest. 8. Replace column or open inlet end and fill void. 9. Change column type to improve separation. 10. Increase temperature. Do not exceed 75 °C unless higher temperatures are acceptable to column manufacturer.

Problem	*Probable Cause*	*Remedy/Comments*
Problem No. 16: Change in Peak Height (one or more peaks)		
No problem 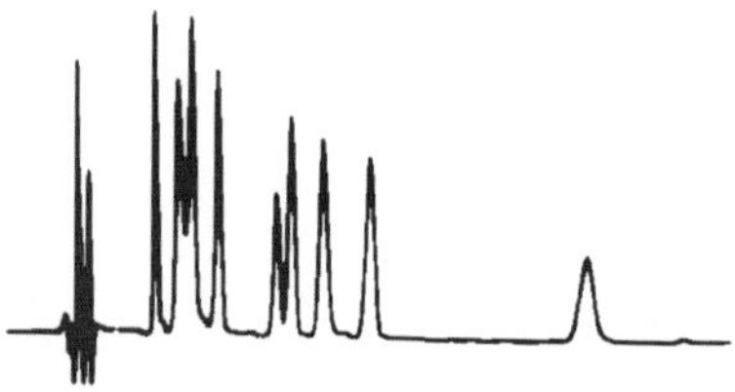 Problem 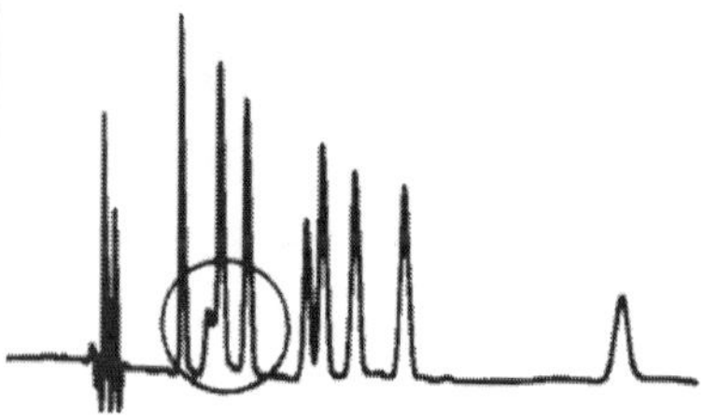	1. One or more sample components deteriorated or column activity changed. 2. Leak, especially between injection port and column inlet. (Retention also would change.) 3. Inconsistent sample volume 4. Detector or recorder setting changed. 5. Weak detector lamp. 6. Contamination in detector cell.	1. Use fresh sample or standard to confirm sample as source of problem. If some or all peaks are still smaller than expected, replace column. If new column improves analysis, try to restore the old column, following appropriate procedure (Table 2). If performance does not improve, discard old column. 2. Check system for loose fittings. Check pump for leaks, salt buildup, and unusual noises. Change pump seals if necessary. 3. Be sure samples are consistent. For fixed volume sample loop, use 2-3 times loop volume to ensure loop is completely filled. Be sure automatic sampler vials contain sufficient sample and no air bubbles. Check syringe-type injectors for air. In systems with wash or flushing step, be sure wash solution does not precipitate sample components. 4. Check settings. 5. Replace lamp. 6. Clean cell.

Problem	*Probable Cause*	*Remedy/Comments*
Problem No. 17: Change in Selectivity		
No problem 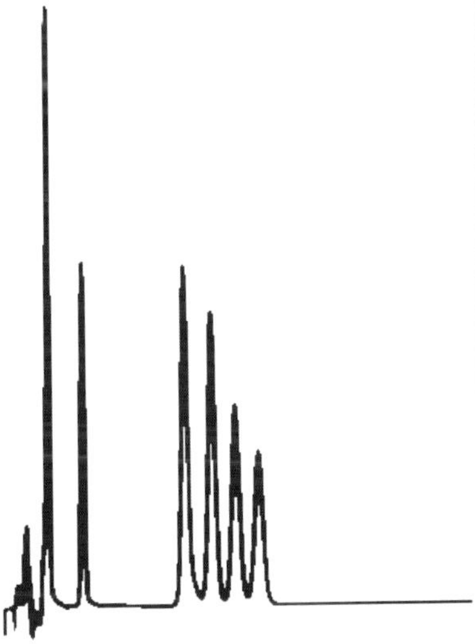 Problem 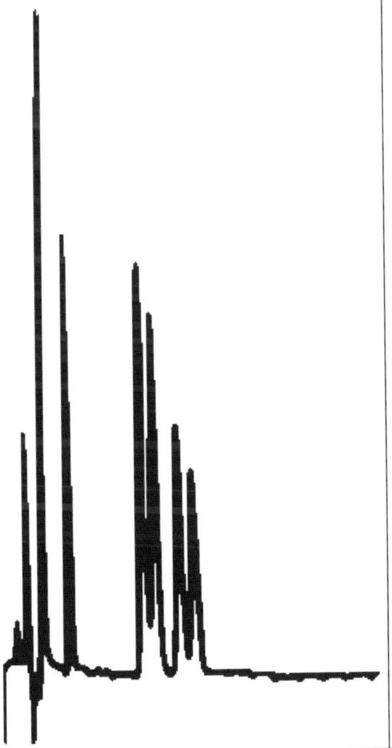	1. Increase or decrease solvent ionic strength, pH, or additive concentration (especially affects ionic solutes). 2. Column changed; new column has different selectivity from that of old column. 3. Sample injected in incorrect solvent or excessive amount (100-200 µL) of strong solvent. 4. Column temperature change.	1. Check make-up of mobile phase. 2. Confirm identity of column packing. for reproducible analyses, use same column type. Establish whether change took place gradually. If so, bonded phase may have stripped. Column activity may have changed, or column may be contaminated. 3. Adjust solvent. Whenever possible, inject sample in mobile phase. 4. Adjust temperature. If needed, use column oven to maintain constant temperature.

Problem	*Probable Cause*	*Remedy/Comments*
Problem No. 18: Negative Peak(s)		
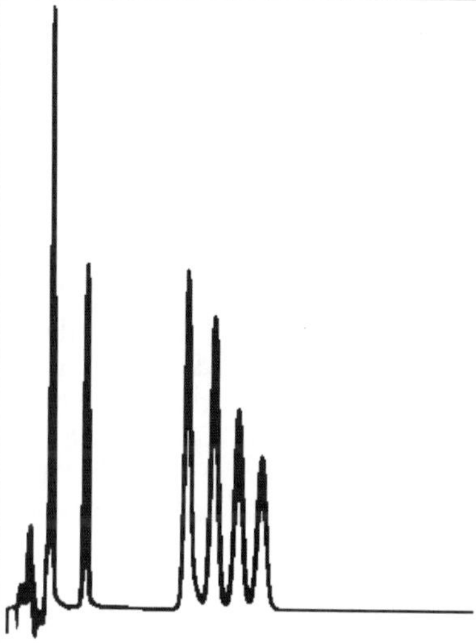 Problem 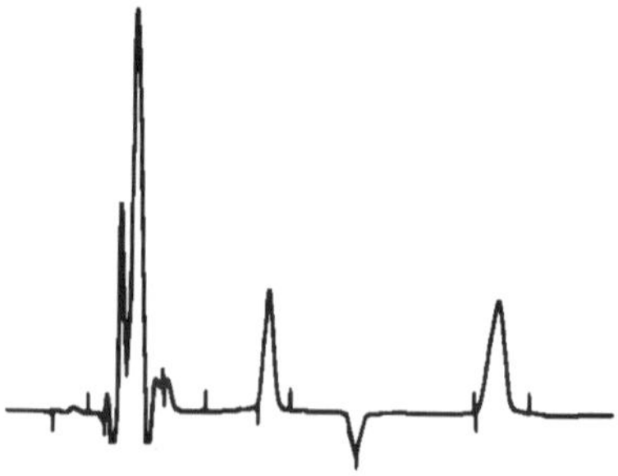	1. Recorder leads reversed. 2. Refractive index of solute less than that of mobile phase (RI detector). 3. Sample solvent and mobile phase differ greatly in composition (vacancy peaks). 4. Mobile phase more absorptive than sample components to UV wavelength.	1. Check polarity. 2. Use mobile phase with lower refractive index, or reverse recorder leads. 3. Adjust or change sample solvent. Dilute sample in mobile phase whenever possible. 4. a. Change polarity when using indirect UV detection, or b. Change UV wavelength or use mobile phase that does not adsorb chosen wavelength.

Problem	*Probable Cause*	*Remedy/Comments*
Problem No. 19: Ghost Peak		
Previous sample 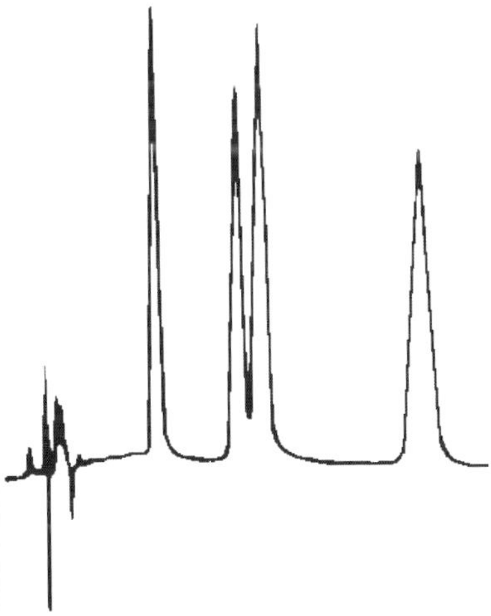 Normal 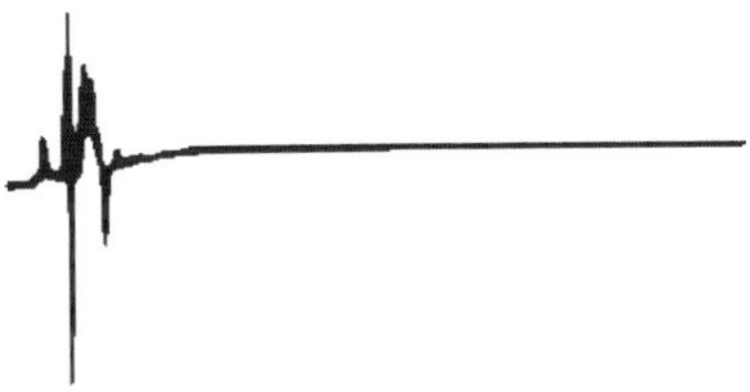 Solvent injected after sample Problem 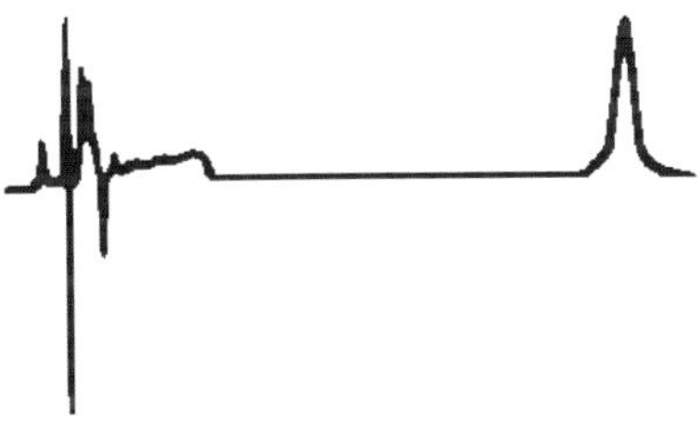 Solvent injected after sample	1. Contamination in injector or column. 2. Late cluting peak (usually broad) present in sample.	1. Flush injector between analyses (a good routine practice). If necessary, run strong solvent through column to remove late eluters. Include final wash step in gradient analyses, to remove strongly retained compounds. 2. a. Check sample preparation. b. Include (step) gradient to quickly elute component

15.3 Column Performance

The following procedures should rejuvenate a column whose performance has deteriorated due to sample contamination. Disconnect and reverse the column. Connect it to the pump, but not the detector. Follow the appropriate flushing procedure in 14.2 table, using a flow rate that results in column back pressure of 1500- 4500 psi, but never higher than the maximum recommended pressure in the manufacturer's instruction manual. If you have a SUPELCOSIL column, analyze with the test mix and the conditions listed on the data sheet. Efficiency, symmetry, and capacity should be within 10-15 per cent of the test sheet values. If not, repack the column inlet or replace the column.

Table 15.2: Column Restoration Procedure

Silica Column	**SUPELCOSIL LC-PCN Column**
Flush with the following: 1. 50 mL hexane 2. 50 mL methylene chloride 3. 50 mL 2-propanol 4. 50 mL methanol 5. 25 mL methylene chloride 6. 25 mL mobile phase Evaluate column performance according to conditions specified by the manufacturer. **Silica-Based Reversed Phase Column (alkyl%, phenyl, or diphenyl column, SUPELCOSIL LC-PAH column)** **A. Water Soluble Samples** Flush with the following: 1. Flush with warm (60 °C) distilled water 2. 50 mL methanol 3. 50 mL acetonitrile 4. 25 mL methanol 5. 25 mL mobile phase Evaluate column performance **B. Samples Not Soluble in Water** **Flush with the following:** 1. 50 mL 2-propanol 2. 50 mL methylene chloride 3. 50 mL hexane 4. 25 mL isopropanol 5. 25 mL mobile phase Evaluate column performance	**A. To Remove Protein** Flush with 10 column volumes of acetonitrile:water, 50:50, containing 0.1 per cent trifluoroacetic acid. **B. To Remove TCA** Flush with 10 column volumes of distilled water (adjust pH to 2.5 with H_3PO_4), then with 10 column volumes each of: 1. water (to remove salts) 2. methanol (to remove water) 3. methanol/methylene chloride, 50:50 (a general clean-up solution) 4. methanol If column performance still is not acceptable, prepare the mobile phase buffer at 10X the concentration used for the analysis and recycle through the column overnight. * Re-equilibrate the column with mobile phase at the normal buffer concentration and re-evaluate. *Use caution: with some buffer types and/or concentrations a 10-fold increase in concentration can cause precipitation.

Polar-Bonded Phase Column (amino, cyano, or diol column or Pirkle-type chiral columns).	Silica-Based Ion Exchange Columns(strong or weak anion or cation exchange)
For a column used in the reversed phase mode (*e.g.*, organic solvent/aqueous buffer mobile phase), follow the same cleanup procedure as for silica-based reversed phase columns. For a column used with nonaqueous mobile phases, use the following scheme:	Most analyses involving ion exchange systems use ionic mobile phases. Compounds that may affect column performance are usually insoluble or only slightly soluble in water. The following procedure should be sufficient to remove these compounds.
Flush with the following:	Flush with the following:
1. 50 mL chloroform	1. 50 mL hot (40-60 °) distilled water
2. 50 mL methanol	2. 50 mL methanol
3. 50 mL acetonitrile	3. 50 mL acetonitrile
4. 25 mL methylene chloride	4. 25 mL methylene chloride
5. 25 mL methanol	5. 25 mL methanol
6. 25 mL mobile phase	6. 25 mL mobile phase
Evaluate column performance.	Evaluate column performance
	Silica-Based Columns for RPLC of Proteins and Peptides
	Follow the protocol for silica-based reversed phase columns. Alternatively, make one or more 100 µL injections of trifluoroethanol (determine the number of injections by evaluating column performance after each injection).
	Evaluate column performance

%C8, C18, *etc.*

15.4 Non-bonded Silica Columns Exposed to Polar Solvent

Samples and mobile phases containing very strongly polar solvents, such as water or alcohols, can deactivate uncoated silica HPLC columns. This can drastically affect column performance, particularly solute retention and selectivity. (Figure 15.1). Even prolonged column flushing with a nonpolar solvent only partially restores column performance, while wasting chemicals. A silica regeneration solution quickly and inexpensively restores silica column performance by removing trapped polar material. Pump the solution through the affected column for 10 minutes at a rate of 4 mL/minute, then flush with mobile phase for 10 minutes at a rate of 2 mL/minute. Evaluate column performance by using the test mixture for evaluating silica columns. Performance should be virtually the same as before the polar solvent was introduced (Figure 15.1). Silica Column Regeneration Solution, 200 mL.

Table 15.3: Properties of Organic Solvents Commonly Used in HPLC

Solvent	*Polarity*	*Miscible with Water?*	*UV Cutoff%*	*Refractive Index at 20 °C*	*Solvent Strength, Єo (silica)*	*Viscosity at 20 °C, cP*
Hexane	Non Polar	No	200	1.3750	0.00	0.33
Isooctane	↓	No	200	1.3910	0.01	0.50
Carbon tetrachloride		No	263	1.4595	0.14	0.97
Chloroform		No	245	1.4460	0.31	0.57
Methylene chloride		No	235	1.4240	0.32	0.44
Tetrahydrofuran		Yes	215	1.4070	0.35	0.55
Diethyl ether		No	215	1.3530	0.29	0.23
Acetone		Yes	330	1.3590	0.43	0.32
Ethyl acetate		poorly	260	1.3720	0.45	0.45
Dioxane		Yes	215	1.4220	0.49	1.54
Acetonitrile		Yes	190	1.3440	0.50	0.37
2-Propanol		Yes	210	1.3770	0.63	2.30
Methanol		Yes	205	1.3290	0.73	0.60
Water	Polar	Yes	–	1.332	>0.73	1.00

%Typical values.

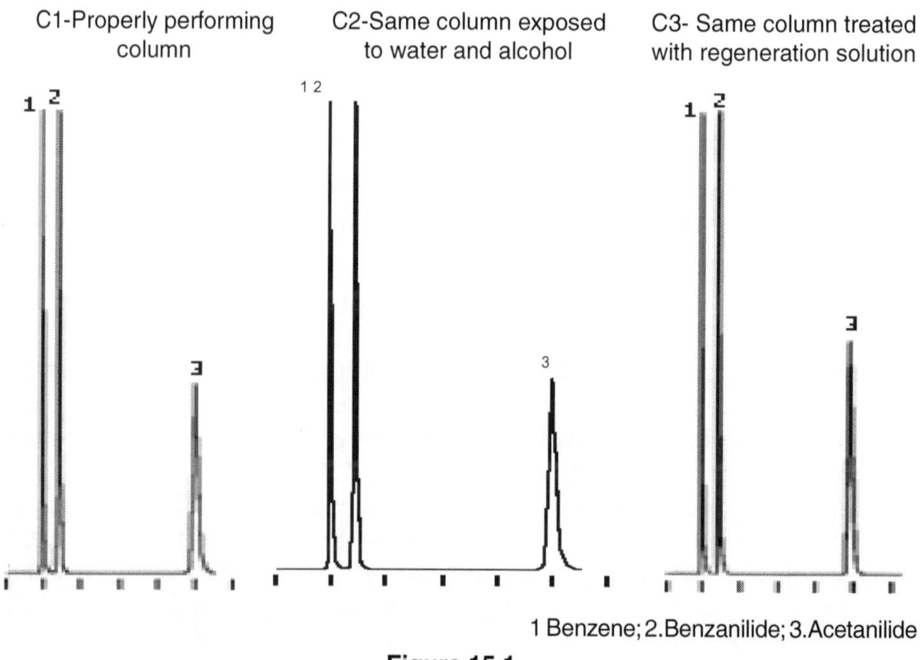

Figure 15.1

15.5 Preventing Leaks

Leaks are a common problem in HPLC analyses. To minimize leaks in your system, avoid interchanging hardware and fittings from different manufacturers. Incompatible fittings can be forced to fit initially, but the separation may show problems and repeated connections may eventually cause the fitting to leak. If interchanging is absolutely necessary, use appropriate adapters and check all connections for leaks before proceeding. Highly concentrated salts (>0.2 M) and caustic mobile phases can reduce pump seal efficiency. The lifetime of injector rotor seals also depends on mobile phase conditions, particularly operation at high pH. In some cases, prolonged use of ion pair reagents has a lubricating effect on pump pistons that may produce small leaks at the seal. Some seals do not perform well with certain solvents. Before using a pump under adverse conditions, read the instrument manufacturer's specifications. To replace seals, refer to the maintenance section of the pump manual.

15.5.1 Unclogging the Column Frit

A clogged column frit is another common HPLC problem. To minimize this problem from the start, use a precolumn filter and guard column. To clean the inlet, first disconnect and reverse the column. Connect it to the pump (but not to the detector), and pump solvent through at twice the standard flow rate. About 5-10 column volumes of solvent should be sufficient to dislodge small amounts of particulate material on the inlet frit. Evaluate the performance of the cleaned column using a standard test mixture.

15.5.2 Replacing a Frit at the Column Inlet

Sometimes neither solvent flushing (see above) nor restoration procedures (see Table 15.2) restore a column's performance. If you've isolated the column as the problem source, and other restorative procedures have failed, a void in the packing or a persistent obstruction on the inlet frit may exist. As a last resort, open the inlet end of the column. Caution: opening the inlet end, and more so opening the outlet end, can permanently damage the packing bed. Before opening columns, consult the manufacturer's literature. (Never open either end of a resin-filled column).

15.5.3 Use the following Procedure to Open a Column

1. Disconnect the column from the system. To prevent the packing from oozing out of the column, perform subsequent steps as quickly as possible.
2. Using a vise and wrench, or two wrenches, carefully remove the inlet end fitting (see Figure D). If the frit remains in the fitting, dislodge it by tapping the fitting on a hard surface. If the frit stays on the column, slide it off rather than lift it off. This will help preserve the integrity of the packing bed

3. Examine the old frit. Compression of the frit against the stainless steel tubing will leave a ring around the edge on the column side of a properly seated frit. No ring can mean the ferrule is seated too near the tubing end. The resulting loose connection can leak silica or act as a mixing chamber.
4 Examine the packing bed. If it is depressed or fractured, you need a new column.
5. Replace the frit.
6. Replace the end fitting. Screw it down fingertight, then tighten 1/4 turn with a wrench.

Column with Conventional End-fittings

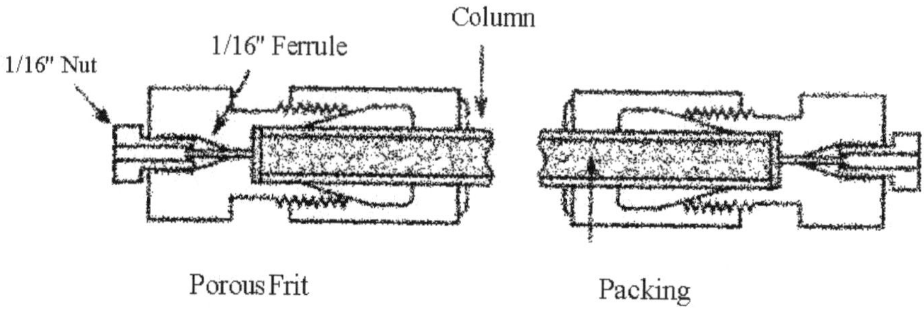

Modular Column with Reusable End-fittings

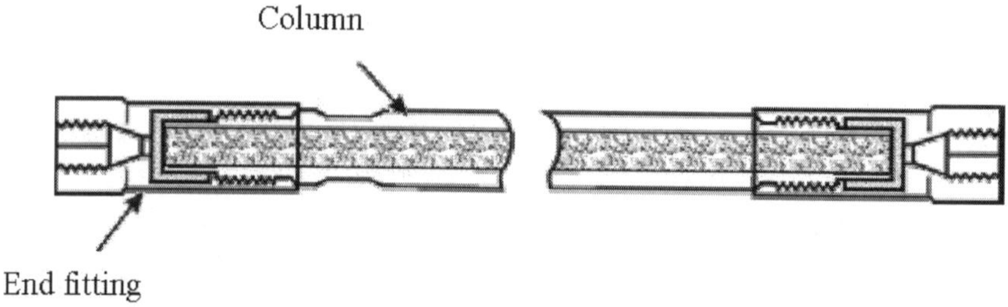

Index

R

S

T

U

V

W

Z